Understanding the Oceans

Understanding the Oceans brings together an internationally distinguished group of authors to consider the enormous advances in marine science that have been achieved since the voyage of HMS *Challenger* over a century ago. The whole book draws inspiration from the seminal contributions stemming from that voyage, and individual chapters show how succeeding generations of scientists have been influenced by its findings. We see how they have responded to the ongoing multiple challenges involved in undertaking the exploration of the ocean. The story continues up to the exciting developments in oceanography covering the late twentieth century, and glimpses prospects for tomorrow.

Covering the whole spectrum of the marine sciences, the book has been written and edited very much with the non-specialist reader in mind. Marine scientists, whether students or researchers, will welcome these authoritative accounts of significant stages in the development of their subject. Other scientists will find the book to be an accessible and informative introduction to marine science and its historical roots.

Edited by **Margaret Deacon**, **Tony Rice** and **Colin Summerhayes**, Southampton Oceanography Centre, Southampton, UK.

Understanding the Oceans

A century of ocean exploration

**Edited by Margaret Deacon,
Tony Rice and Colin Summerhayes**

LONDON AND NEW YORK

First published in 2001 by UCL Press

Reprinted 2003 by Routledge
2 Park Square, Milton Park,
Abingdon, Oxon, OX14 4RN

Transferred to Digital Printing 2005

*Routledge is an imprint of the
Taylor & Francis Group*

Typeset in Baskerville by RefineCatch Ltd, Bungay, Suffolk

British Library Cataloguing in Publication Data
A catalogue record for this book is available from the British Library.

Library of Congress Cataloguing in Publication Data
A catalogue record for this book has been requested.

ISBN: 1-85728-705-3 HB
ISBN: 1-85728-706-1 PB

Printed and bound by Antony Rowe Ltd, Eastbourne

This book is dedicated to the memory of Henry Charnock, 1920–97, who originally conceived the idea of the Southampton Oceanography Centre. He was Professor of Oceanography at Southampton University, 1966–71 and 1978–86 (and subsequently Emeritus Professor) and Director of the Institute of Oceanographic Sciences from 1971 to 1978. Colleagues worldwide have benefited not only from his many personal contributions to understanding the oceans, but also from his concern to promote oceanographic studies in the widest sense, whether they were directed to the past, present or future state of the science. The editors of this volume would further like to record their gratitude for help received during its preparation.

Contents

Figures

Tables

Notes on contributors

Harold L. Burstyn is Patent Attorney, Office of the Staff Judge Advocate, Information Directorate, US Air Force Research Laboratory, Rome, New York, USA, and Adjunct Professor of Electrical Engineering and Computer Science, Syracuse University, Syracuse, New York. During the 1950s and 1960s Dr Burstyn carried out pioneering work on the history of oceanography, concentrating on the *Challenger* Expedition.

J. Dennis Burton is a chemical oceanographer, now Emeritus Professor in the School of Ocean and Earth Sciences, Southampton Oceanography Centre, Southampton University. He was previously for many years on the staff of the University's Department of Oceanography.

Jacqueline Carpine-Lancre was for many years Librarian of the Musée océanographique at Monaco (Fondation Albert Ier). Trained originally as a historian, she has also written numerous books and articles on historical topics, in particular on the life and work of Prince Albert I.

Peter G. Challenor is on the staff of the James Rennell Division for Ocean Circulation and Climate, Southampton Oceanography Centre.

Margaret B. Deacon is an honorary visiting fellow in the School of Ocean and Earth Sciences, Southampton University, at the Southampton Oceanography Centre. She has written extensively on the history of oceanography. Her book *Scientists and the sea, 1650–1900*, originally published in 1971, was reissued in 1997.

Robert Marc Friedman is a member of the Department of History, University of Oslo, where he specializes in Scandinavian science of the early twentieth century, in particular geophysical and polar research.

Brian M. Funnell (died April 2000) was a founder Professor in the School of Environmental Sciences, University of East Anglia, Norwich, and more recently Emeritus Professor, following his early retirement in 1989.

Christopher R. German is a member of the Challenger Division for Seafloor Processes at the Southampton Oceanography Centre.

W. John Gould is head of the World Ocean Circulation Experiment (WOCE) International Project Office and Climate Variability and Predictability Study (CLIVAR), based at the James Rennell Division for Ocean Circulation and Climate, Southampton Oceanography Centre.

Trevor H. Guymer is head of the James Rennell Division for Ocean Circulation and Climate at the Southampton Oceanography Centre.

Sir Anthony Laughton FRS joined the staff of the National Institute of Oceanography (later renamed Institute of Oceanographic Sciences) in 1955, and served as director from 1978 until his retirement in 1988.

Rosalind Gunther Marsden, a retired teacher of Huddersfield, Yorkshire, is engaged on a history of the origins and work of the Discovery Committee.

Eric L. Mills is Professor of the History of Science, Dalhousie University, Halifax, Nova Scotia, Canada, where he is also Professor of Oceanography. His book *Biological Oceanography* (1989) documents how scientists during the first half of the twentieth century came to understand how nutrients in sea water sustain marine life.

A. L. (Tony) Rice is a marine biologist and author. Until his retirement in 1998 he was head of benthic biology in the Challenger Division for Seafloor Processes at the Southampton Oceanography Centre, and was previously at the Institute of Oceanographic Sciences. He has a long-term interest in the work of the *Challenger* Expedition and published *British Oceanographic Vessels* in 1986.

Meric A. Srokosz is on the staff of the James Rennell Division for Ocean Circulation and Climate, Southampton Oceanography Centre.

Dorrik A. V. Stow is a Professor and marine geologist in the School of Ocean and Earth Sciences, Southampton Oceanography Centre, University of Southampton.

Colin Summerhayes was Director of the Institute of Oceanographic Sciences from 1988 to 1995. He was subsequently deputy director of the Southampton Oceanography Centre and head of the Challenger Division for Seafloor Processes. Since 1997 he has been director of the Global Ocean Observing System (GOOS) project office, based at the Intergovernmental Oceanographic Commission, UNESCO, Paris.

Paul Tyler is a marine biologist and was in the Department of Oceanography at the University of Wales (Swansea) before moving to Southampton in 1986. He is currently Head of Academic Studies at the School of Ocean and Earth Sciences, Southampton Oceanography Centre, Southampton University.

Andrew Watson is a Professor and marine scientist at the School of Environmental Sciences, University of East Anglia, Norwich. He was formerly at the Plymouth Marine Laboratory, where he developed and first applied the techniques of ocean tracer release experiments.

T. Roger S. Wilson is a marine chemist, formerly head of the chemistry section at the Institute of Oceanographic Sciences Deacon Laboratory, Wormley.

John D. Woods CBE was director of marine and atmospheric sciences at the Natural Environment Research Council from 1986 to 1994. He is now Professor of Oceanography in the T. H. Huxley School of Environment, Earth Science and Engineering, at Imperial College, London University.

C. M. Young is on the faculty of the Harbor Branch Oceanographic Institution, Fort Pierce, Florida, USA.

Foreword

The voyage of HMS *Challenger* was a remarkable enterprise in many ways. Its historical and scientific significance, and its continuing relevance for Earth and ocean science today, are the subject of this volume, and of the conference at which these papers were first presented. The *Challenger* Expedition foreshadowed many recent developments in the field which we regard as modern, but were already present in the work of the naturalists on board over one hundred years ago. The unity of the Earth and ocean sciences was already implicit in the daily work of the ship, with soundings to establish the topography of the sea-floor, dredging for sediments and benthos, net hauls for fish and for plankton, and water sampling and hydrography all carried out in their turn. Today such multi-disciplinary science is still regarded by some as novel, but we are only breaking down the artifical barriers we have ourselves erected between the disciplines, having allowed excessive specialization to obscure the inter-relatedness of almost everything in the sea, and indeed in the whole Earth system, in the widest sense.

Our generation is the first to have been privileged to see the images of the Earth from space, which so effectively convey the special beauty and uniqueness of the Earth and convince us, as words alone cannot, that our lives depend on the functioning of a global system, which involves a complex interplay between the physics, chemistry, biology and geology of the solid Earth, the oceans, the atmosphere, and the biosphere. We find it necessary to establish new institutions such as the SOC, where scientists from all these disciplines can work closely together, in order to recreate the fundamental holism which informed the planning of the *Challenger* Expedition, and which we have under-valued for too long.

Many of the problems addressed by our forebears on the *Challenger*, including the determination of the circulation and biological productivity of the oceans, have proved to be much more difficult than anyone expected. Modern technology now offers us new techniques to solve these perennial problems, and we can hope that we may come to understand the functioning of the life support system of our planet before the consequences of other technological developments alter it beyond repair. We need to promote a truly global approach to the environmental sciences, because it is vital to discover whether the system is resilient or fragile, and in what ways, in order that we can continue to coexist within it for another million years or so.

This volume therefore promotes a broad view of the science of the Earth and its oceans, in the spirit of the *Challenger* Expedition, and I hope it will engage the interest and enthusiasm of both established experts and newcomers to the field, in helping to recreate the unified approach to Natural History which seemed so natural and appropriate to the scientists who worked so long and hard on the *Challenger*, as she inaugurated oceanography as a science on a global scale.

Professor John Shepherd FRS

Director, Southampton Oceanography Centre 1994–99

Acknowledgements

The editors would like to express their warm appreciation of the generous assistance of Brian M. Funnell towards the completion of this volume. Sadly, Brian died in April 2000.

Particular thanks are also due to Roger Jones, formerly of UCL Press, without whose encouragement and efforts in the early stages this work could not have been brought to completion.

The National Institute of Oceanography moved to this former Admiralty building in Surrey in 1953. It was later renamed the Institute of Oceanographic Sciences, and in 1987 the IOS Deacon Laboratory. In 1995, it moved to the Southampton Oceanography Centre (see below). The figurehead (now at SOC) from HMS *Challenger*, which carried out the pioneering oceanographic voyage of 1872–6, can be seen in the centre of the picture.

The present-day RRS *Discovery*, moored in the Empress Dock, Southampton, alongside the Southampton Oceanography Centre. Since 1995, SOC has housed scientists from the Natural Environment Research Council and the oceanography and geology departments (now School of Ocean and Earth Science) of Southampton University.

Introduction

Margaret Deacon and Colin Summerhayes

Understanding the oceans

This is a time of well-founded concern about the ocean, and how well we understand its processes. Specific problems such as the depletion of fisheries, the effects of marine pollution and the impact of mineral extraction, as well as broader questions such as the role of the ocean in controlling the Earth's climate, not only pose challenges to scientific understanding but also represent factors that could influence both the lives and livelihoods of many, and potentially alter the earth as we know it. In appreciating what is involved in such issues, let alone what might be done to avoid unwelcome consequences, accurate scientific knowledge of the sea is imperative. This is the province of oceanographers, but it is not only scientists who are concerned about these matters and a wider knowledge of what they are doing should form part of the public debate.

This is not as straightforward as it might seem. For one thing, late twentieth-century oceanography is a dynamic and therefore rapidly changing discipline. Developments during recent decades have transformed our scientific knowledge of the sea, with the result that present-day understanding of the ocean differs significantly from what was possible even fifty, let alone a hundred years ago. This is in many respects due to advances in technology, which have made possible new modes of exploring an environment notoriously hard to study. At the same time, wider concerns about the ocean, and growing awareness of its significance in the Earth's past, present and future, and in the origin and maintenance of life, have helped to create a more general interest in such work, beyond the immediate scientific community actively engaged in research. Too often, this wider audience lacks works able to bridge the gap between fragmented and specialized scientific literature and the needs of a broader readership, a function that textbooks only partially fulfil. Older general sources (such as Herring and Clarke 1971) continue to be valuable, but the subject has altered considerably since those days. Some books outstandingly help to change the outlook of a generation. Rachel Carson's *The sea around us*, first published in 1950, was such a one; perhaps Sylvia Earle's *Sea change* (1995) may be another. There is, however, a continuing need for works, such as Summerhayes and Thorpe (1996), which are neither popular accounts on the one hand, nor written exclusively for those working in the same field on the other, and which try to present authoritative accounts of scientific understanding of the ocean at the close of the twentieth century to a wider audience, whether scientists or non-scientists.

This book seeks to assist that process, but also to take it somewhat further. To understand the oceans, we need not only good science, but a good appreciation of science, and we can achieve a better understanding of what current scientific knowledge can tell us about the sea if we also have some information about how it was obtained. In this book recent oceanographic discoveries are presented through accounts of how, as well as why, scientists study the sea, and some of the changes that are taking place in the way they go about it. Two aspects of this process receive special attention. Many chapters deal with how technological improvements during the last fifty years have transformed possibilities in the principal fields of ocean research, leading to important advances in scientists' understanding of what goes on in the sea. Other chapters highlight a different but equally important requirement for scientific progress by showing how, over the last hundred years or so, infrastructures have been developed that are capable of sustaining the degree of scientific activity necessary to observe natural phenomena on an oceanic or even a global scale. Looking at the ways in which the organization of scientific research, as well as research methods, have changed over the years helps us to understand more about not only the objective of all oceanographic activity, the sea itself, but also why the science has developed in the way it has, for as Robert Marc Friedman points out in Chapter 9, 'Neither the present contours of science nor its relations with society are inevitable or natural.' While modern oceanography has greatly altered in the lifetime of workers still active in the field, it is valuable to identify not only the differences, but also the underlying similarities, in seeking to understand how the science operates today.

Both these themes are reflected in the discussion of past and present research, and of future possibilities, by scientists and historians in the chapters of this book. A framework to the whole is provided by a more general topic, the voyage of HMS *Challenger* between 1872 and 1876. This expedition, and still more the publication of its report, which was not completed until 1895, has played an important part in the development of ocean science. This was why, when the opening of a major research facility for the study of the ocean took place in 1995 – the Southampton Oceanography Centre, itself seen as a momentous event for the future of British marine science – the development of oceanography during the intervening century, with particular reference to the role of the *Challenger* Expedition, suggested itself as the theme for a meeting to be held as part of the opening celebrations. It was at this symposium, entitled 'The *Challenger* Legacy', held in September 1995, that the papers that form the chapters of *Understanding the oceans* were first presented.

The Southampton Oceanography Centre is intended to act as a focus for marine science (including geology) in Britain, and as a leading centre in Europe for the twenty-first century. The new institution was planned as a joint venture between the University of Southampton and the Natural Environment Research Council (NERC). It incorporates the University's departments of geology and oceanography (now the School of Ocean and Earth Science) with research divisions of the NERC's Institute of Oceanographic Sciences Deacon Laboratory, previously based at Wormley in Surrey, and its Research Vessel Services from Barry in South Wales. The SOC provides the United Kingdom with a large-scale facility for research and education in marine and earth sciences, together with technology and support services, and berths for the major UK research ships, RRSS *Discovery*, *Charles Darwin* and, until recently, the present-day *Challenger*.

The opening of a new centre on this scale is something of a landmark, and is sufficiently unusual to prompt a measure of taking stock, of looking to see where we are now and how we got there, as well as what the future might hold. So, as well as addressing the more specific themes referred to above, the presentations on this occasion also commemorated the first large-scale British oceanographic enterprise, the scientific circumnavigation of the earlier *Challenger*, and in particular the centenary of the conclusion of this epoch-making project with the publication of the final volumes of the expedition's report in 1895. The *Challenger*

Expedition was a pioneering venture that had a profound impact both on the contemporary development of marine science, and on its subsequent metamorphosis into an international scientific discipline. This is why it continues to capture the imagination of successive generations of oceanographers. But what has its true legacy been? While the scale of the scientific achievement, and of its impact on later work, are amply borne out by examples given in this book, Chapters 1 and 2 take a rather more critical look at the expedition than perhaps might have been possible at the time of the celebrations, in 1972, of the centenary of its departure. They reveal that, in spite of being a truly remarkable achievement, both in terms of its organization and in the work it carried out, the *Challenger* Expedition did not provide the sort of impetus that its subsequent reputation might lead us to expect in either of the particular aspects of marine science highlighted in this book, that is, the significance of technological innovation and of adequate institutions in scientific development.

History plainly shows that, in science, institutions as well as individuals have an important role to play. Science is not an abstract body of knowledge, but represents the human activity of observing and interpreting independently existing complex natural phenomena. To encapsulate the closest approximation possible at any one time concerning how these operate, scientists use concepts that they constantly seek to extend and refine, but do not necessarily agree over. The formulation of ideas may be the preserve of the individual, but those ideas only gain their power and influence through being promulgated and discussed in scientific societies and journals, and the detailed and interdisciplinary work needed to confirm or transform them, especially when directed towards an objective as large and complex as the ocean, needs to be done through organizations or groups. The *Challenger*'s enduring influence on marine science was due in great measure to the publication of the report of the expedition, but after this promising start there remained no public organization to carry on its work. It is only in the twentieth century that permanent institutions dedicated to oceanographic research have been established, mostly since 1945.

During the second half of the twentieth century, a combination of new ideas and opportunities has led to a period of rapid development, as part of which present-day research fields and institutions, the Southampton Oceanography Centre among them, have evolved. Whether in a hundred years' time the SOC will be deemed to have had as great an influence on marine science in the twenty-first century as the results of the *Challenger* Expedition have had up to the present, as its founders may have dreamed, can hardly be guessed at now, but hopefully it will provide, as the *Challenger* was unable to do, an infrastructure that can support the future development of oceanography in Britain. As far as the *Challenger* is concerned, the reasons for its enduring impact must be sought elsewhere, in the contribution to knowledge that pointed the way for others to follow, and provided an intellectual basis for the subsequent study of the sea.

A major part of this book deals with work being done in the present rather than the past, and consequently with much more recent discoveries about the oceans than those of *Challenger* days. The emphasis here is on oceanographers' dependence on technology, and how technological innovations, particularly during the second half of the twentieth century, have opened up important new ways of exploring the sea, and for interpreting the data thus obtained. The successful unravelling of the mysteries of the ocean has always been closely linked with the ability to devise improved methods and apparatus. Before the twentieth century's development of submarines and the aqualung, there was no way in which human beings could penetrate beyond the outer limits of this element. Even today, without the use of instrumentation, it is impossible to gain with the unaided senses more than a literally superficial impression of what is going on – and even the constantly changing surface of the sea defies the eyes' attempts to define and categorize except in the crudest terms. From the introduction of the sounding lead (which archaeologists have shown to have been in use since prehistoric times) to modern submersibles, seafarers and scientists have relied on increasingly sophisticated scientific apparatus to provide them with essential information that lay beyond the powers of their own unaided observation. In Chapter 16 John Woods develops this theme and looks at how he anticipates oceanography will change in the next twenty years when advances in technology are expected to transform possibilities still further.

But the story of oceanography is not just a question of how scientists and engineers responded to the challenge of providing apparatus that would work in an inhospitable medium. Throughout the development of marine science over the past three hundred years, scientists have had to respond to the continuing challenge, not only of how to make meaningful observations and measurements in an environment as large and apparently featureless as the sea, but also of persuading not just the scientific community in general but society at large that these activities were worthwhile, and that their findings had a wider significance. Though for individual scientists from early times the beauty and complexity of the natural world may have been sufficient to attract their curiosity, there were limits to what could be learnt unaided. The development of modern oceanography has necessarily depended on more than the intrinsic interest of scientific discoveries and achievements and these have increasingly been achieved only through society's growing realization of the importance of the ocean in both the Earth's history and in human affairs. This has been largely a twentieth-century development, and should be reinforced as oceanography moves from its exploratory research phase to one in which the knowledge being gained can increasingly be applied in ways that will benefit a world civilization that is dependent on the ocean and its resources.

The reason why this should be so, together with the other topics raised above, is explored more fully in the remainder of this introduction, which reviews how marine science has changed between the time of the *Challenger* and the late twentieth century. Clearly, in the most obvious sense, the differences between what was possible then and modern achievements and understanding are enormous, but underlying similarities also continue to influence the way in which oceanographers work. Perhaps the most unexpected of these, to someone unfamiliar with the subject, would be the continuing sense of excitement and discovery. Marine scientists at the time of the *Challenger* Expedition had a sense of being on the threshold of a new world, which has been strikingly reinforced during the past century, and particularly during the past two decades. Until shortly before the *Challenger* set sail, very little was known about the depths of the ocean, and there was little expectation that they held anything of direct interest to mankind in general. The work of the *Challenger* scientists and their contemporaries showed how false this assumption was; their researches established, broadly speaking, the main subject areas within which modern oceanographers still work. Yet in spite of the important discoveries made then, and by their successors in the early and middle years of the twentieth century, the sea still had many undiscovered features, which have only recently begun to come to light. It is not yet fifty years, for example, since the first observations of internal eddies led to a revolution in ideas about ocean circulation (see Chapter 10), and barely twenty years since the discovery of hydrothermal vents and the novel life forms associated with them, described in Chapter 8. This continuing sense of wonder and surprise is only one of the ways in which parallels still exist between modern science and the work of the *Challenger* Expedition over a century ago.

The *Challenger* legacy

It should be emphasized that scientific interest in the sea existed long before the time of the *Challenger* Expedition. Though not as old, probably, as the science of navigation, it can be seen in the earliest scientific writing by philosophers and geographers of classical antiquity and, centuries later, inspired researches undertaken during the first flowering of the scientific revolution of the seventeenth century (Deacon 1997). However, the origins of modern oceanography date from the middle years of the nineteenth century, when a number of technological innovations, particularly those linked to the development of steam-powered vessels, transformed what had hitherto been chancy, difficult and therefore rarely attempted forms of deep-sea observation into an activity that could be undertaken systematically and with a reasonable degree of success (McConnell 1982). Development of these new methods was further boosted by the technical challenge of laying and retrieving cables in deep water when submarine telegraphy was introduced in the 1850s,[1] but their first large-scale deployment for scientific purposes was not until the 1870s, during the voyage of HMS *Challenger*.

Present-day oceanography is a far more comprehensive and wide-ranging activity, and is furthermore an international undertaking, with research centres spread around the globe, a diversity of heroes and links to sciences unguessed at when the *Challenger* set sail. Nevertheless, in looking at what oceanography is today, the legacy of the past can also help, not only in understanding our current view of the sea, but also how science functions in presenting this picture to us. Because the *Challenger* Expedition was the first large-scale expedition to have as its primary objective the scientific study of the sea, its results have left an enduring mark upon most of the broad divisions of modern oceanographic science. While this book does not attempt a catalogue of individual cases, which would be impressive but lacking in meaning for the general reader, numerous examples occur in the text. In this introduction we look in a more general way at how the expedition was novel in its time, and how it has influenced the pattern of future research.

In spite of the enormous improvements in our understanding of the sea since its time, the *Challenger* Expedition is rightly seen as a turning point. The scientific legacy of this historic voyage to oceanography has been rich and varied, and still has a significant place in present-day knowledge, a theme explored in different ways throughout this book. The most obvious connection lies in continuing interest in specific areas and problems first highlighted by the expedition, covering subjects as diverse as the origin of manganese nodules (see Chapter 14) and the distribution of life in the depths of the sea (see Chapter 15). But the expedition's legacy is also more pervasive. Both through its contribution towards reshaping contemporary ideas about the sea, and in laying down the broad outline that more recent work is filling in, it has played a major role in creating the twentieth century's picture of the ocean. Yet its precise relationship with the subsequent development of oceanography is somewhat paradoxical, and has been widely misunderstood. The *Challenger* has to some extent been idealized by subsequent writers, and enjoys an almost mythical status. Not surprisingly, it succeeded better in some areas than in others, as will subsequently become clear.

But what led to the expedition being sent out at that particular moment? Was it the availability of new methods or were other factors involved? It is understandable that there was a time-lag between the early cable voyages and the expedition; it took time for the new technology to become established. But changes in scientific thought ultimately led scientists to make use of these new methods for their own purposes. One of the paradoxes of the *Challenger* Expedition was that it actually represented the culmination of a process that was serving to turn the attention of scientists back towards the study of the ocean after a period of neglect. This had been partly due to the technical difficulties involved in the study of the deep sea, but was reinforced by the belief that the prevailing cold and darkness and intense pressure of the ocean depths would make life there impossible, and that there would consequently be nothing to observe. The person whose name is particularly associated with this view was Edward Forbes (Rehbock 1979), an influential biologist of the middle years of the nineteenth century. He pioneered valuable new studies of marine organisms in shelf seas, and also drew attention to evidence for climate change to be found by comparing the geographical distribution of modern forms with that of similar species from fossil deposits. During a rare opportunity to dredge in deeper water, in a naval survey ship working in the Mediterranean, Forbes found that the number of species declined with depth and suggested that perhaps no life existed below 300 fathoms (about 1000m). There were other observations contradicting this view, but the idea of an azoic zone, once established, proved hard to unseat, and it was not until the first of the group of shorter voyages that preceded the *Challenger*, in HMS *Lightning* in 1868, that it was generally accepted that life flourishes at far greater depths. This discovery, by C. Wyville Thomson and W. B. Carpenter, together with the latter's growing interest in how their findings also suggested the existence of deep circulation within the ocean (see introduction to Part 3), and the controversy that these ideas aroused, led to the campaign for a voyage of circumnavigation to broaden the scope of these new investigations to take in all the major oceans of the world (Deacon 1997). At a time when scientists were still absorbing the implications of Charles Darwin's ideas on evolution, published little over a decade earlier, there were intriguing possibilities. What kind of creatures would be found in the deep ocean? Did ancient forms, known only from fossils, still linger there, or would they even discover the origins of life itself, as suggested by the recent discovery of *Bathybius*?[2]

If the scientific possibilities suggested by the voyage were exciting, as an undertaking it was revolutionary. Harold Burstyn shows in Chapter 2 how the '*Challenger* Expedition, unique in its century and scarcely duplicated in magnitude in ours, stands as a beacon on the route to what science has become'.[3] The voyage and the subsequent compilation of its report extended over twenty-three years and was a precursor of the kind of large-scale research project that now plays such an important part in many areas of modern science. This is a sufficient reason for historians of science and of science policy to study the expedition, but modern oceanographers also feel moved to write: 'We still live with the legacy of the *Challenger* expedition' (Summerhayes and Hamilton 1996). What leads scientists today to feel that the expedition's work is still so relevant to their endeavours?

As has already been stated, the *Challenger*'s circumnavigation was the first major expedition to have the scientific study of the sea as its primary objective. Previous scientific voyages of exploration from the time of Captain Cook onwards had yielded important information about the marine environment, but never before had those interested in the science of the sea had the opportunity to plan an undertaking on such a scale. Small wonder that the rise of modern oceanography is often dated from this event. Yet are there cogent reasons why an expedition that took place over a century ago should be remembered today, not just as a 'first', or as a triumph of Victorian technology? The latter it undoubtedly was, but even here, as shown in Chapter 1, the expedition was not particularly innovative in terms of its technology. Because, rather than experiment with new and untried apparatus, the *Challenger* relied on more familiar methods, the expedition has since been charged with conservatism. However, Tony Rice argues that there were sound scientific reasons for this choice, as it enabled the most effective use to be made of the expedition's unique resources and opportunities. The ultimate justification must lie in the fact that the resulting discoveries and observations, made both by expedition members and by the scientists of different nationalities who later contributed to the report, were on such a scale and formed such an influential body of knowledge that the work of the expedition played a fundamental role in the early development of oceanographic ideas and of how we look at the sea in general. Even over a century later, scientists still appreciate that the expedition 'established a paradigm for the study of the ocean' (Summerhayes and Hamilton 1996), in spite of the radical changes that have since taken place.

Though the dispatch of the expedition was a milestone in itself, its lasting impact on understanding the ocean would have been far less effective if these results had not been adequately published. At the time, it was not a foregone conclusion that money would be made available to complete the project, which turned out to be far larger than anyone had anticipated (see Chapter 2). Wyville Thomson, the expedition's scientific leader, died before this situation could be resolved. His assistant John Murray, another member of the *Challenger*'s scientific team, took over the task of editing the report and overcame the difficulties to bring it to a successful completion. During the voyage, Murray himself had specialized in the study of sea-bed deposits and he also co-authored a volume of the report containing the first comprehensive classification of marine sediments and worldwide chart of their distribution. This is only one of the areas in which the expedition results established a framework for subsequent studies and interpretation that is still reflected in modern researches, despite diversification into research areas not identified until later. In other aspects, too, such as observations of the physical and chemical properties of both the surface and the interior of the ocean, the *Challenger* scientists, if not necessarily the first in the field, also produced observations extensive enough to form a standard for future work (see for example Chapter 14). This is not to say that their efforts were comprehensive, or that they were equally successful in all areas they attempted. Ocean circulation was one topic in which results were particularly disappointing, and where more could have been achieved with existing technology. Yet even here, as in the other areas that were open to them, data and observations from the *Challenger* are still embedded in modern knowledge of the deep sea, and can still form the basis for studying long-term changes occurring there.

This explains why the *Challenger* not only had a great influence on the science in its own time, but has bequeathed a legacy of information that scientists still find valuable today. Another less obvious aspect of its influence had important consequences for the future development of marine science, as an example useful for coaxing funding from governments. Though its influence on later developments in the United Kingdom was not always wholly beneficial (for example, see Chapter 4), it was used by scientists in other countries to press for support from their own governments. In scientific terms it remained an inspiration even as late as the 1940s when Hans Pettersson, organizer of the Swedish Deep-Sea Expedition that revolutionized the study of deep-sea sediments (see Chapters 6 and 7), acknowledged his debt to Murray's example. Even today, the name of the *Challenger* still conveys to the international oceanographic community the sense of a heritage shared. This can still be helpful to individual scientists, working in widely differing geographical locations, as well as in the ever more fragmented areas within wider specializations that go to make up present-day oceanography. As Robert Marc Friedman tells us (Chapter 9), 'Professional and social identities are created in part through historical awareness', and the experience of a sense of belonging to a more unified tradition is an important part of this. But does what the *Challenger* might represent for a modern scientist equate with the historical reality of the expedition, or has it become an icon, as Eric Mills (1993) has already suggested?[4] If this is so, it may not necessarily affect its effectiveness as a symbol, but historically we ought to be aware of where the differences lie.

The *Challenger* was a British venture but the wider impact of its work was assisted by the determination of Wyville Thomson, followed by Murray, to make the report an international project, in spite of the chauvinistic objections of some British zoologists. It is not inappropriate therefore to regard the *Challenger* Expedition as part of the common international heritage of oceanography, though the science has since become, to a far greater extent than was true even then, one in which national boundaries have little meaning. Even the scientific team taking part in the voyage had had an international flavour.[5] Few people were working on the deep sea at that time, and links between them were important intellectually and correspondingly valued. This situation is one that would be not at all unfamiliar to modern oceanographers where, though the overall numbers involved have grown enormously since the 1950s, the greater degree of specialization within the broader branches of the science means that the number of people working in any particular area may still be quite small.

As we have already seen, Chapter 1 shows that the expedition's contribution to technological developments for exploring the ocean was actually less than might be expected, given its prestige. The other area in which the *Challenger* Expedition most significantly failed was to create long-term opportunities for British scientists to continue the work it had begun. For the science of the sea to develop, new opportunities for research were essential. This need was widely recognized, but suggestions for further more modest researches to augment the work of the expedition were almost all rejected by successive governments.[6] Murray ensured that the Challenger Office in Edinburgh, where the expedition's report was edited, also acted as a research institution, briefly in conjunction with the Scottish Marine Station for Scientific Research, founded in 1883 (Deacon 1993). However, by the time of the report's completion in 1895 public funding had already dried up, so Murray continued research at his own expense. In 1898 he received a knighthood for his work, but it is probable that being known more informally as John Murray of the *Challenger* actually meant even more to him. Through his work on the results, his forceful but endearing personality, and his numerous other scientific projects, undertaken both concurrently with the editorship and subsequently, until his death in 1914, Murray kept the legacy of the *Challenger*'s work alive to an expanding network of researchers in both Europe and North America. This was notwithstanding the fact that until, late in life, he became a wealthy man (Burstyn 1975), there were long spells when he had no opportunity to work at sea. When, shortly after 1900, some British marine scientists established a national society, they naturally called it the Challenger Society.[7]

What Murray, and others at the time, were largely unable to do as individuals was to create enduring institutions. After Murray's death, his personal network of contemporaries and younger researchers fell apart. This minor but keenly felt dissolution was reinforced and became more general a few months later following the outbreak of World War I. Murray had drawn many younger recruits to the field, even after the formal completion of the *Challenger* work, and his personal leadership of ocean science continued into the twentieth century, but this was not enough to ensure continuance of the new discipline. The reason why he was unable to found a viable institution for marine research was lack of funding, and similar problems were experienced by others. The first permanent arrangements for oceanographic research, as opposed to marine biological stations,[8] which had a rather different emphasis, only began to appear from 1900 onwards (see later in this introduction), and whether as laboratories or university departments or other research organizations, remained small and few in number until after World War II. Not until the more general development of government funding for science came about in the second half of the twentieth century were the institutional and other developments on a scale to make possible the scientific advances described in the remainder of this book. During the intervening period a number of different solutions were tried. Chapters 3, 4 and 9 describe some of the attempts that were made in the early twentieth century to organize marine research on a more continuous basis.

In Chapter 3 we see two examples of the rare occasions when individual enthusiasm was combined with access to resources on a national scale. This was achieved most notably by another of the leading figures in marine research in the post-*Challenger* age, Prince Albert I of Monaco, who between the 1880s and the outbreak of war in 1914, carried out an impressive series of researches in the Atlantic and the Mediterranean in yachts that were all, save the first, designed to act as research vessels and equipped with the latest technology. Like John Murray, the Prince built up an informal network of fellow researchers throughout Europe and America; unlike him, as Jacqueline Carpine-Lancre describes, his great wealth enabled him to ensure the continuation of his work by founding research institutions in both Paris and Monaco. However, even a head of state could find difficulty in achieving more limited objectives, as Prince Albert's friend and colleague Dom Carlos of Portugal discovered. As the science and its technological possibilities developed still further, even deployment of resources on the scale of Prince Albert's work would no longer be adequate to carry the subject forward.

A further stage in development can be seen in Chapter 4, where Rosalind Marsden shows how in the 1920s a voyage in the *Discovery*, originally built for Captain Scott but re-equipped as an oceanographic research ship, was intended like the *Challenger* as a single expedition but led to another large-scale project, which continued up to and beyond World War II, with a series of voyages by two purpose-built research vessels, the RRS *Discovery II* and the smaller *William Scoresby*. The original purpose of the work had been to study the life history of Antarctic whales, and to establish guidelines for a sustainable fishery, but it developed into a wide-ranging survey of the physics, chemistry and biology of the Southern Ocean, which formed the basis for significant advances in understanding the waters of this influential circumpolar ocean, providing as it does a link between the other major oceans of the world. The author shows how this research programme gradually evolved through the sometimes conflicting aims of the various government and other bodies who sponsored the project, their representatives and the scientists themselves. It illustrates the kind of trade-off between scientific objectives and the not necessarily identical aims of funding bodies that may equally underlie more recent advances, such as those discussed in the remainder of the book. Through the creation of an organization that employed scientists beyond the immediate duration of expeditions, to write up their results, the Discovery Committee formed an intermediate stage between the *Challenger* and modern scientific institutions. Its experience shows, as does Chapter 9 in a different context, how, irrespective of prevailing policies, national agencies were already becoming drawn in to support research at sea during the interwar period. The reasons why this should be so change over time, but are crucial to understanding not only the history of marine science but also how it operates today.

Oceanography and its hinterland: science and the community

Though the *Challenger* laid down a foundation for the scientific study of the sea, for oceanography to be able to develop as a science in the full sense of the term, it was necessary to obtain long-term funding both to support work at sea and to permit the development of permanent research institutions, whether in universities or in independent or government-financed laboratories. Only in this way could an ongoing tradition of scientific work be provided, offering both career prospects and education. But why should society in general be prepared to make resources available to allow scientists to work in an area that might not appear to impinge greatly on most people's lives? Why should we study the sea? The answer is of course that it does have a major influence, not only on those groups who rely on the sea for their livelihood, but on human existence in general, and indeed on life itself. This is something that has only become evident over time and the implications are still sinking in. Thus reasons why the study of the sea is perceived to be important have changed over time. Even the earliest scientists attracted to this field, in the seventeenth century (Deacon 1997), stressed that their studies had a more serious motive than mere satisfaction of curiosity. Support for modern research still depends on finding the common ground between scientists and society at large, thus enabling divergent interests to be reconciled.

In getting the *Challenger* project under way, W. B. Carpenter showed an excellent under-standing of how to do this, but for his immediate successors it was not so easy. During the first half of the twentieth century sea-going was interrupted by World War I and the sub-sequent economic depression further restricted new possibilities. This was how a scientist of great potential, like H. U. Sverdrup, could become in the 1920s, like John Murray in the 1890s, an oceanographer without a ship, as Robert Marc Friedman shows in Chapter 9. Space does not permit a look at the more general reasons for the growth in state funding for science in the second half of the twentieth century, which has been essential for the rapid development of oceanography since World War II, but there are also more specific reasons why at different times society has perceived the need for research at sea and its applications. As in earlier times when the problem of longitude occupied eighteenth-century astronomers, or with nineteenth-century initiatives to construct lighthouses and harbours, and to provide reliable tide-tables, improvements in navigation have always had a high priority in maritime nations. Both commerce and national security reaped benefits from such developments, and from the improvement in communications provided by submarine telegraph cables, which formed a worldwide network by 1900.

However, scientific research does not take place in a vacuum; as it develops it invariably requires greater resources, of both manpower and equipment. This is oceanography's dilemma. As Harold Burstyn points out in Chapter 2, it has always been an expensive science. But if the *Challenger* Expedition investment was large in its day, modern research ships are far more sophisticated, and therefore costly. It is possible to chart the development of oceanography by looking at how ideas about the sea have changed over time, as new information became available – slowly in the past, more rapidly in the century between 1850 and 1950, and extremely rapidly during the last half century. It can also be argued that technological breakthroughs were the driving force, providing opportunities for new dis-coveries and changing theories. However, availability of resources is also a crucial factor; as John Woods says, in Chapter 16, it is 'all a matter of money'. Effective ideas and methods for exploring the ocean were in existence well before 1872, but funding continued to be prob-lematic long after that date. To pursue the scientific study of the sea, as its earliest protagon-ists were well aware (Deacon 1997), sufficient funds had to be found, and to do this it has usually been necessary to convince others that such expenditure is worthwhile.

In the mid-nineteenth century, even before the telegraph engineers turned their attention to the deep ocean, there was a growing interest in maritime meteorology, stimulated by the publication of Matthew Fontaine Maury's wind and current charts in the United States. As the century drew towards a close, many countries began to experience problems with fisheries, which demanded specialized knowledge that led to the setting up of dedicated research organizations. This was soon followed by recognition of a need for more concerted action, resulting in the foundation of the International Council for the Exploration of the Sea in 1902. However, before long, after almost a century without major international conflict, the sea again became a battlefield. In both world wars, and during the Cold War era following, submarines added a new dimension to naval warfare, posing a major threat to national security and involving scientists as well as engineers in the search for detection and avoidance mechanisms. For the first time navies found themselves having to think seriously about physical conditions in the interior of the sea, and not just about underwater topography. In many instances oceanography's debt to military research is huge (see for example Chapter 5), and goes largely unrecognized. During the second half of the twentieth century such concerns have been joined by those arising out of diversified economic opportunities linked to the sea. Rapidly developing industries, such as those extracting minerals, especially oil and gas, from the sea bed, have brought many benefits, but also parallel anxieties about the dangerous consequences of over-fishing and pollution. As late as 1900 some scientists still believed that the ocean was so vast that the activities of man would have little effect on it. However, such a point of view soon ceased to be tenable, even before present-day concerns about the greenhouse effect, global warming and possible sea-level rise came to the fore.

In the late twentieth century it has not just been a question of how to increase utilization of marine resources, or even of the related and parallel growth of environmental concerns about the sea, important though these are. Gradually it has become apparent that knowledge of the oceans is also of far greater significance than formerly appreciated in understanding Earth processes as a whole. Oceanographic research has shown that the sea, the oceans, and the crust beneath it, are active rather than inert, inherently dynamic rather than stable, on different scales of magnitude, and that the impact of changes can extend worldwide, beyond the ocean boundaries. What seemed to be a system broadly speaking in equilibrium, now with increased understanding of the multiplicity of factors involved, appears rather less so, both in the short and in the long term. While theories of continental drift and plate tectonics have revolutionized understanding of the whole of the earth's geology, improved knowledge of ocean circulation has shown that the sea itself has a much more far-reaching effect on the world's weather and long-term climate than meteorology traditionally acknowledged, and in ways that vary over time depending on the relative positions of land and sea. As our knowledge of the wider universe expands, the ocean, once taken for granted, is now appreciated as a unique resource, essential for the creation and maintenance of life on earth. This shift in attitude, in some measure leading to improvement in access to resources, and allied to new ideas and more advanced technology, has turned oceanography into a high-profile science at the end of the twentieth century.

The growth of oceanography in the twentieth century

For the sake of convenience, in this introduction the term 'oceanography' has been treated as synonymous with marine science. The *Challenger* Expedition was, as we have seen, a significant turning point in the study of the sea, helping to lay the foundation of major fields of modern ocean research. This has sometimes been equated with founding oceanography itself, but paradoxically the *Challenger* did not bequeath a fully fledged science to the twentieth century, despite the fact that the use of 'oceanography' as the collective term for marine sciences was well established by the 1890s. One reason is that the largely descriptive nature of the work then being carried out in physical oceanography, or hydrography as it was then known, led contemporaries to view it more as a branch of physical geography than as an independent science. However, one of the foremost historians of oceanography, Eric Mills (1993), argues that a science in the full sense should not just be seen as the existence of a broad set of shared ideas and interests, but as a professional activity, involving scientists in undertaking work on problems of common concern through organizations and higher education institutions, which can provide a range of resources together with opportunities for exchange of ideas and information, and the training of new recruits to the field (see Chapter 9). According to this view it was from about 1890 to even as late as 1960 that the '"pioneer period" of the development of oceanography' took place, as it was during this time that the institutional framework that has sustained the growth of modern research was being put in place.

This is because only since 1900 has oceanography in general, as opposed to marine biology, gradually acquired the institutional bases, together with the core texts, specific research problems and agreed methodology, necessary for it to function fully as a science. The twentieth century opened promisingly with the foundation of the Institut für Meereskunde in Berlin (Lenz and Streicher 1997) and Prince Albert I of Monaco's Institut océanographique, comprising a research institute in Paris and a museum in Monaco, as described by Jacqueline Carpine-Lancre in Chapter 3. The first truly international collaborative projects, such as the International Council for the Exploration of the Sea (ICES) and the bathymetric chart of the oceans (see Chapters 3 and 5), also date from this period.

For a time ICES ran its own laboratory (Schlee 1973) but it also provided an important impetus for the establishment of new national centres for marine research, such as the Internationale Meereslaboratorium at Kiel in Germany (see Chapter 13). However, this period of expansion in marine science was interrupted, first by World War I and then by subsequent economic crises that restricted possibilities for further development during the interwar years. Local factors too could influence events, as Eric Mills (1989) demonstrates in his book *Biological oceanography*, where he examines the reasons why the promising work on marine productivity at Kiel was abandoned, only to be revived later at Plymouth. Yet, while activity was patchy, this was nevertheless an important time in the development of oceanography, during which new studies, joined to existing work, resulted in a period of consolidation when the field matured and foundations were laid that, when the opportunity arose, would underpin the rapid advances of the 1940s and 1950s.

During the *Challenger* era, the branches of marine science brought together by the expedition had still functioned largely independently of each other as a set of distinct fields, linked to their parent disciplines, though applied to a common objective. Connections were certainly made, and many more modern discoveries can be seen prefigured in the writings of, for example, Murray's *Challenger* colleague, the chemist J. Y. Buchanan, but the subject was too new, the possibilities too vast, with too few people involved and too little funding, and the data too sparse for their ideas to be followed up in any coherent fashion. By the 1920s and 1930s, with more focused enquiries, and new methods yielding more and improved data, a generation of scientists was emerging, of whom Sverdrup was one (see Chapter 9), which saw that the physics, chemistry and biology of the ocean could not be studied in isolation. Their work laid down much of the intellectual foundations for postwar development and they may be regarded as the true founders of modern oceanography. More recently, new understanding of processes in marine geology has brought about an appreciation that what is going on at, or beneath, the sea bed has more to tell us about other aspects of ocean science than was earlier realized (see Chapters 6–8). Today marine scientists can study not only the oceans as they are now, but also their past history, and possible future development.

We can therefore see that though the development of oceanography accelerated after World War II, through increased resources for institutions, manpower and fieldwork, key intellectual developments that made possible the emergence of oceanography as an integrated discipline had taken place somewhat earlier. On the one hand there were the chemical and biological studies of how life in the sea is maintained, as chronicled by Eric Mills (1989) (see Chapters 13 and 14). On the other hand, modern studies of ocean circulation, an area in which the *Challenger* had been least successful, date from pioneering studies made in the early twentieth century by Scandinavian scientists. This work was followed during the interwar period by the scientists of the German *Meteor* expedition (Schlee 1973), and in Britain through the operations of the Discovery Committee (see Chapter 4). This field now began to take over as the central discipline of ocean science, with scientists employing dynamical methods to show how the various forces acting on sea water operate, and also improved methods of data analysis to chart the distribution of water masses in the ocean. It was from this area, rather than as hitherto from biology, that the future leaders of mid-twentieth-century oceanography typically came (see for example Chapter 9).

Marine science during the latter half of the twentieth century owes much to recent technological progress, as will be seen in the next section, but the reason that oceanography did not develop further in the first half of the century was not simply the lack of such advantages. More could have been done using methods that would not have been unfamiliar to the scientists of the *Challenger*, but improvements in instrumentation (McConnell 1982), as well as in theory, continued to be made. Oceanography did not progress further at this time because of the difficulty of obtaining funding, rather than lack of problems to work on and methods to tackle them (see Chapter 9). The situation was made worse because it is such an expensive science and pressure to reduce sea time is one reason why there is now such interest in using unmanned remotely operated vehicles to carry instrumentation packages (see Chapter 16).

Oceanographers, like other scientists, are primarily interested in ideas, and in explaining the features they observe in the sea. To do this they need data, and this could be hard to get hold of (as the career of H. U. Sverdrup detailed in Chapter 9 illustrates) when, during the interwar period, economic depression meant that public spending budgets were slashed. To governments, civil servants and the public at large clearly the collection of such data appeared to be of low priority, unless linked to some aspect of national interest, whether political, economic or military. Only when such specific and more general interests combine has marine science undergone periods of rapid development, both in more recent times and in pre-*Challenger* days.

During the early twentieth century such links between scientific research and public utility were most often made in the field of fisheries science, which has continued to develop as an important though somewhat separate branch of marine research. As early investigators soon realized, understanding the behaviour of fish populations involved looking not only at human activities but at environmental factors too, so that what began in the nineteenth century as enquiry into fishing methods led by degrees to encompass more broadly based enquiries into fish biology and life histories that could, as with the early ICES work, have a significant impact on work in other areas of marine research. Thus the Discovery Investigation voyages began (see Chapter 4) with the limited objective of supplying figures on whale populations, but grew by degrees to cover the oceanography of the entire Southern Ocean. Similarly, Sverdrup's investigations of the declining Californian sardine fishery led to fundamental discoveries about Pacific Ocean circulation (see Chapter 9). Scientists have not been unaware of these links, and have indeed made use of them in pursuit of scientific objectives, sometimes successfully as did W. B. Carpenter when lobbying for the *Challenger* Expedition (see Chapter 2), sometimes with a less favourable outcome (see for example Chapter 9, p. 166). As the experiences of a research programme and an individual scientist show, in Chapters 4 and 9 respectively, oceanographic leaders in those days, doubtless as now, had to fight stiff battles not only with the elements but with organizations and committees. Attempts to combine scientific research with projects more likely to win widespread support have not always proved of unmixed benefit to the science. This is clearly demonstrated by the uneasy relationship that existed between ocean science and polar exploration in the early twentieth century, of which examples can also be found in Chapters 4 and 9.

World War II proved a turning point for marine science, but many of the previous observations continue to apply, in spite of the great changes that have since taken place. Oceanographers were then still making observations using methods closer to those employed in the *Challenger*, with one or two exceptions, such as the recently developed bathythermograph, than to those available today, and their picture of the deep ocean, though far more detailed, remained in many respects unchanged. The introduction of new technology, as outlined later in this introduction, often as a result of developments in other sciences such as electronics, then began the transformation of observations and ideas that has enabled oceanography to become qualitatively as well as quantitatively different today from its state barely fifty years ago. But such developments only became possible because of the general increase in levels of funding for science, from which oceanography also benefited. Clearly, money on its own will not produce good or meaningful science, but without the opportunities it creates, as the history of oceanography has shown many times, little could be achieved.

A growing perception of the value of marine science in the postwar period made possible both the expansion of existing institutions, such as the Scripps Institution of Oceanography where Harald Sverdrup was still director (Shor 1978), and the establishment of new ones, in both the New World and the Old. At this time national defence considerations were often paramount. In the United Kingdom the impetus for the foundation of the National Institute of Oceanography (NIO) came initially in 1944 from the Hydrographer of the Navy, Vice-Admiral Sir John Edgell, who was concerned that the valuable contribution made by oceanographers to the war effort should not be lost when peace returned (Rice 1994). It was not until 1949 that NIO was formally established, but in the interim, a group of scientists established at the Admiralty Research Laboratory in Teddington had made a fundamental breakthrough in the study of ocean waves, undertaken with the aim of achieving forecasts for military landings. This group, led by George Deacon, one of the physical oceanographers employed by the Discovery Committee (see Chapter 4), together with the rest of the remaining staff from the prewar investigations, formed the nucleus of the new institute, which in 1953 moved to a wartime Admiralty building in Surrey. In the 1960s, as the motivations for research began to change, control of NIO passed from the Admiralty to the newly created Natural Environmental Research Council, and in 1973 the laboratory became a part of the Institute of Oceanographic Sciences. In 1995 the Institute of Oceanographic Sciences Deacon Laboratory moved from the Surrey site to become part of the Southampton Oceanography Centre.

During the forty years of its independent existence, NIO played an important part in new oceanographic developments (Rice 1994), many of which are referred to in the later chapters of this book. One notable innovation from the early days, which led to fundamental changes in understanding the circulation of the ocean, was John Swallow's neutrally buoyant float for measuring subsurface currents (see Chapter 10). This illustrates the point that since that time it is not only in its technological capabilities that oceanographic research has altered. Whereas much of the early Institute's success came through individual initiative, over the years, in common with similar institutions elsewhere, it became subject to 'increasing emphasis on strategic as opposed to curiosity-driven research, coupled with growing financial stringency' (Rice 1994: 142), which has led to a growing dependence on commissioned work. Related to this has been the practice of working in larger interdisciplinary groups, which have the advantage of a more varied approach to problems, and probably also a more effective use of resources. The need to do this has also encouraged external cooperation. In later chapters, many references will be found to international organizations and joint programmes that, since the International Indian Ocean Expedition of the early 1960s, have become a significant feature of ocean science, permitting oceanographers to share finite resources. However, if the need to maximize potential, as well as new research possibilities, have led to new methods of operation, in some respects oceanography does not change all that much. It still depends on the 'encounter aboard ocean-going vessels, of scientists, engineers and sailors', like that which produced the effective deep-sounding gear on which the *Challenger* Expedition relied (Rozwadowski 1998).

Oceanography today: a science on the threshhold

In the chapters dealing with modern research on the ocean it will be apparent that the middle decades of the twentieth century were a crucial period for oceanography. Before this time, wider support for scientific study of the sea had been generally lukewarm, even among scientists working in other fields. As a result, although exceptions did occur (as seen in Chapters 3 and 4), resources were hard to come by and the subject remained scattered and small-scale. By the late 1930s, lack of investment in marine science was causing concern in some quarters, especially among the scientific community in the United States (see Chapter 9). However, the events of World War II and the subsequent Cold War period accelerated its development though other factors have since contributed. During the 1940s and 1950s the significance of what oceanographers had to offer began to be more widely appreciated. In common with other disciplines in the postwar period, marine science began to benefit from increased public spending on science throughout the developed world, a process that has continued, more or less, to the present day. This meant that institutions could be founded or expanded, new posts created, and ships built. As concerns of national security have become less pressing with the ending of the Cold War, others, broadly linked first to economic and then to environmental themes, have arisen to take their place.

But it is not through the resulting expansion of resources and manpower alone that oceanography has been able to achieve the important breakthroughs of the past fifty years. If resources had permitted, more could have been done with the techniques available in the first half of the twentieth century, which were more sophisticated than those employed on the *Challenger*, though not on the whole greatly differing from them. However, the often spectacular developments in methods for the collection and interpretation of data made possible by modern technological developments have been responsible for advances that have radically transformed both the range and scale of observations from what was possible forty or fifty years ago. In this process, oceanography has greatly benefited from developments being made elsewhere in both science and technology. This of course is not entirely new, and we have already seen how *Challenger* scientists employed methods developed in other fields. Eric Mills (Chapter 13) shows how, in a less well known episode, chemists adapted methods used in agricultural and industrial processes to measure the trace elements in sea water that act as nutrients and are essential to marine life, an important case study of how oceanography has been indebted to its parent sciences.

The nature of the medium in which oceanographers work meant that many types of observation were difficult, if not impossible for the early pioneers. Progress in oceanography is heavily dependent on access to a wide range of advanced technologies, since the ocean (or inner space) is hostile and inaccessible. As in outer space, most observations must be made remotely, using costly instruments. Sound replaces light as the medium for communication, navigation and observation in the sea, requiring the deployment of sophisticated acoustic and sonar systems of various kinds. Intellectual leadership in oceanography increasingly lies with those organizations possessing state-of-the-art technological capability.

Oceanographic techniques are constantly evolving as we realize just how complex the ocean system is. In the past, observational methods were relatively simple and relied on stopping a ship to make a measurement at a point. The demand for improved spatial and temporal resolution, and the associated requirement for higher rates of data capture, have been increasingly met by the deployment of towed instruments. These record observations automatically while the vessel is under way, necessitating real-time processing of the data by ship-board computer. This is commonly done only for physical variables such as temperature and salinity, but similar methods are being developed for measuring biological and chemical properties (e.g. fluorescence proxying for chlorophyll-a, and nitrate sensors for nutrients). There is a growing requirement for chemical sensors and acoustic and optical methods of detecting plankton while under way. The high data-capture requirements necessary to underpin the growing Global Ocean Observing System (GOOS), sponsored by the Inter-governmental Oceanographic Commission, will make it impossible to employ expensive research vessels to satisfy all needs for subsurface data; instead, automatic underwater vehicles equipped with multi-sensor packages, along with the deployment of large numbers of subsurface floats capable of profiling the properties of the water column and telemetering information back to shore via satellite, are likely to provide cost-effective technological solutions for surveying entire ocean basins (see Chapter 16).

The most significant development in oceanography over the past twenty years has been the advent of satellite technology, with observational satellites enabling a broad spectrum of environmental data to be collected with an efficiency never before possible (see Chapter 11). Satellites are equipped with a wide range of instruments to provide global-scale synoptic observations of the sea surface that are needed to map and model ocean processes. Navigational satellites have revolutionized position fixing, again improving spatial analysis of ocean properties at fine scales. Communications satellites provide a now essential means of passing information from ship or buoy or float to shore and vice versa.

The other major revolution has been the arrival of the computer and its miniaturization, enabling vast amounts of data to be captured, stored and processed in real time as well as encouraging the numerical modelling of hitherto intractable complex problems. As a result, entirely new research horizons have opened up for oceanography – for example the possibility of keeping multi-sensor recording instruments *in situ* for several years, enabling extremely accurate data to be collected on variability in ocean processes over time. In this last instance the challenge lies in keeping sensors clean by preventing biofouling.

The enormous advances in late twentieth-century oceanographic exploration made possible by these new techniques have not only fundamentally altered our perceptions of the ocean but have also had an influence extending beyond the confines of marine science itself. One example can be seen in the impact of marine research on the evolution of ideas of continental drift and plate tectonics, which now underpin the whole of modern thinking on the earth's geology. The *Challenger* had pointed to the existence of a median ridge running the entire length of the Atlantic Ocean (Deacon 1997), and similar features were later discovered in other oceans, but, hidden beneath great depths of water, their true significance as spreading centres where new oceanic crust is created was not appreciated until the 1960s. Marine geophysicists, carrying out mapping of magnetic anomalies on the deep-sea floor, provided the first direct evidence that this was occurring, an example of how the exploitation of new technologies may play a role in setting the scene for paradigm shifts – the name given by Thomas Kuhn about the same time to major upheavals in scientific theory. The widespread use of magnetometry for military purposes after World War II led to the technique's being imported into oceanography, where it revealed magnetization of the sea bed in symmetrical stripes on either side of a mid-ocean ridge. This showed where newly formed crust had taken on the prevailing direction of the earth's magnetic field, which undergoes periodic reversals over time. Technological advances have played a fundamental part in the huge advances made in all branches of oceanography during the past fifty years and further examples of this are described in more detail in Parts 2–4 of the book.

Though there is much to say about the legacy of the *Challenger* in terms of its improvement in our understanding of the way in which the oceans work, we cannot forget that now, as in the late nineteenth century, the implication of enhanced understanding is improved application. Application was also the thrust of the interest of navies in ocean processes and properties during and after World War II. Enough is now known about how the oceans work for government and industry alike to invest in making systematic, routine and sustained measurements as the basis for monitoring the present state of the sea and its contents, including its pollutants and living resources, and for providing forecasts of the future condition of the sea and its contents for as far ahead as possible. Such information is now widely demanded by managers of all kinds in industry and government, especially in coastal waters and seas, as the basis for the wide, safe and sustainable use of seas and oceans that was called for in Agenda 21, the Agenda for the 21st Century produced by the United Nations Conference on Environment and Development in Rio de Janeiro in 1992. As yet there are not enough of these kinds of observations to provide adequate and accurate warnings of storms, high waves and surges; to alert managers of ports and harbours to changes in water level; to help optimize the design and operation of offshore installations; to forecast sea-ice and other hazards on shipping routes; to detect poor water quality; to manage fisheries and aquaculture efficiently; and to facilitate safe and healthy marine recreation. We have made progress: some sea areas are well monitored – but many more are not, and there is much to do to make the dream a reality.

Because the ocean has such a tremendous capacity to store heat and move it around the world, thereby influencing climate, operational observations sustained over periods of years are essential to underpin our understanding of climate change and our ability to forecast it. Television screens and newspapers in late 1997 and early 1998 were full of stories of El Niño, the periodic phenemenon which then, as every four years or so, made the eastern equatorial Pacific Ocean unusually warm, shutting down fisheries off Peru and bringing heavy rains to the South American coast. El Niño is a global phenomenon, in which the intimate connection between ocean and atmosphere leads to droughts in places as far away as Australia, Indonesia, northeast Brazil and southern Africa, while other places, like Argentina, become unseasonally wet.

Using data from the array of buoys placed along the equator in the Pacific Ocean in the TOGA project between 1984 and 1994, supplemented by data from tide-gauges, satellites and commercial ships, numerical modellers are now able to use coupled ocean–atmosphere models run on advanced supercomputers to predict the extent and magnitude of El Niño events up to nine months in advance. Given that the El Niño of 1982–3 did US$13 billion worth of damage through droughts, floods, storms and effects on fisheries, such advanced warnings can help planners to minimize its effects considerably. It is because of the success of TOGA as a research programme in demonstrating this potential application that the array of buoys is now supported largely with operational rather than research funds by NOAA.

The Tropical Atmosphere Ocean (or TOA) array of buoys, as this operational system is now called, is an integral part of the Initial Observing System of GOOS, the global system for operational oceanographic services for managers of all kinds. In a very real sense, GOOS is the equivalent for the ocean of the World Weather Watch (WWW), through which the World Meteorological Organisation works with individual nations to collect and exchange the data essential for weather forecasting worldwide. GOOS will provide data that contributes to the success of the WWW. Unlike the WWW, however, GOOS will also provide a wide variety of information to meet the needs of the broad community at sea and along the coasts that depends on operational oceanographic observations.

In the climate arena, the GOOS-IOS will provide information essential for governments wishing to monitor how successful they have been in implementing the requirements of the Framework Convention on Climate Change. Because of the power of the ocean to influence climate everywhere, GOOS will also underpin climate forecasts useful to the suppliers of food, energy, water and medical supplies on land.

Operational oceanography is an idea whose time has come, and whole communities of oceanographers have embraced its tenets in signing up to develop GOOS in their regional seas, notably through the EuroGOOS Association, which covers European seas, and through NEAR-GOOS, which covers the seas of northeast Asia. In these bodies and in the development of national GOOS committees in nations as far apart as the USA, Brazil and Australia, among others, we see yet another of the *Challenger*'s legacies, the turning of ocean research into widespread applications for the general good.

Conclusion

Scientific knowledge and ideas, as well as technical capacity, have developed enormously during the past half century, let alone over the longer period since the *Challenger* sailed. Much of this has been made possible by technological advances during the intervening years, not only by those designed specifically for oceanographic projects, but also by utilization of advances in other fields of science and technology, which have been successfully applied to marine exploration in many instances. However, in spite of the differences, many underlying similarities remain. Scientific research still has to be planned, organized and paid for. Oceanographers need laboratories and institutions to work in, ships and apparatus, or other methods of collecting data, and societies and journals to act as media for discussing and publishing results, not to mention access to funding to make the whole process possible in the first place. They need to work together to share finite resources and to maximize efforts, leading to the move towards joint expeditions and international research programmes, many of which are referred to in this book. As oceanographers look at the ocean to find out how it works, so historians study scientific processes to find out how they operate. Looking at episodes in the earlier development of science allows us to explore aspects of science as an activity that are either omitted or seen as peripheral in standard accounts of research and results, but that crucially affect their progress. Some are intrinsic to the scientific process (for example research methods, techniques, expeditions and institutions) while others illumine relationships with the wider scientific community or society at large, such as national and international organizations, joint programmes, patronage and funding.

Developments during the period between the mid-nineteenth and the mid-twentieth centuries led to the foundation of the modern science of oceanography, but discoveries and research carried on during the second half of the twentieth century have transformed knowledge and ideas about the oceans in ways unforeseen barely a generation ago. Even so, present-day oceanography still has close links with contributions made fifty or a hundred years ago, although then marine science was little known or appreciated beyond the small group of scientists involved in such work. Though convinced of its significance, few if any could then have predicted how the subject would develop by the year 2000. Lack of resources and technical difficulties held back effective exploration of the deep sea until the middle of the nineteenth century (Rozwadowski 1998), but there was also an ideological barrier to be overcome. When the scientists on board HMS *Challenger*, and similar ventures in the 1860s and 1870s, began to make serious investigations of this virtually unknown environment their observations changed the general perception of the ocean depths, previously envisaged as totally inhospitable to life. Because the sea conceals, both at the surface by its continual changeability, with an apparent lack of permanent features, and below it by great depths of water, the process of filling in these blanks was slow, and often had to wait for new methods of quantifying surface features and of penetrating the depths. Researches during the final decades of the twentieth century have yielded vastly more knowledge, and perhaps even greater excitement, than was possible during that first flush of enthusiasm for the scientific exploration of the deep sea a century earlier.

It was nevertheless during the late nineteenth century that the foundations were laid for a coordinated science of the sea, and developments in the first half of the twentieth century built on these to produce the modern science of oceanography. The decades immediately before World War II were critical, but it was only after this event, and partly because of it, that the science reached critical mass. Concepts developed during the previous half century then provided the bases for new investigations, which in turn stimulated a golden age of great inventiveness in both theory and techniques during the 1950s and 1960s. The importance of the work done during this period is amply affirmed in this book, and should not be lost sight of amid, pardonably, sometimes euphoric reactions to more recent discoveries and possibilities, which could scarcely have taken place without it. When funding again began to falter in the 1970s and early 1980s, it seemed as though marine science might again fall victim to inbuilt expense and limited popular appeal and suffer a further period of marginalization, but a combination of pressing scientific problems and spectacular results made possible through the application of new and improved technology have reversed the trend and kept oceanography in the public eye. But will this state of affairs continue, with the drawing to a close by the year 2000 of the shipborne, data-collecting, phases of a number of international programmes, including two of the largest ever undertaken, the World Ocean Circulation Experiment (WOCE) (see Chapter 10) and the ODP (Ocean Drilling Project; see Chapter 7)? Oceanographers (see Chapter 16) have plenty of ideas about what to do next, but do the rest of us need them to do it? The answer can only be yes; as already seen, applied oceanography has many uses that can benefit mankind, but these are early days. We are still far from understanding the ocean and are only now just beginning to appreciate its nature and how it can change over time, partly in response to natural processes and cycles, and increasingly also in response to human activities. Furthermore, as a result of the work that has taken place over the past half century, our perception of the ocean has undergone yet another transformation. As Tony Rice (1994: 146) writes:

> If you looked for a single word to encapsulate the changed view of the oceans that we have now, compared with that of ... 1953, it would have to be variability. For in all disciplines we now believe that the oceans and the deep sea floor are much more patchy, much more changeable and, indeed much more complex than was thought to be the case then. If anything, however, in terms of the rational use and protection of our environment, understanding the oceans seems even more important.

It has been suggested that a hidden cultural bias exists that causes widespread indifference towards the state of the ocean (Earle 1995), even though our very lives, and those of our descendants, depend on it. Furthermore, as Robert Marc Friedman points out in Chapter 9, 'Neither the present contours of science nor its relations with society are inevitable or natural. Changing political, economic, and cultural circumstances [provide] opportunities and constraints for scientists to define and pursue oceanographic research.' So what effect will these circumstances have on the development of marine science in the future? Oceanographers can see their way forward into the twenty-first century (Chapter 16) but they cannot foresee how far these aims will win support, nor can they dictate to society at large. They can, however, do their best to see that the general public as far as possible has an appreciation of the issues involved, and facts rather than positions on which to base its judgement. As part of this process it helps to understand how science works, what it can do and what it cannot. The Discovery Investigations voyages, described in Chapter 4, did not prevent wider political and economic considerations leading to scientific warnings being disregarded. The cry for 'more research' by administrators can sometimes look suspiciously like unwillingness to take a particular course of action, but scientists who have to respect their data may equally come under fire when they decline to extrapolate from their findings to support a particular position, as the Brent Spar episode illustrated (Rice and Owen 1999). They in their turn are human, and therefore not infallible, but also, as science grows, increasingly involved in social interaction with other scientists and its corresponding pressures, in a way that many early marine researchers were not, at least as regards this aspect of their scientific work.[9]

In the past, interest in the science of the sea was very much an individual activity, constrained by individual resources that even on the grandest scale (see Chapter 3) were never quite sufficient to carry out major investigations. During the past century and a half, for both scientific and practical reasons, oceanographers have increasingly moved from the private to the public domain, and towards national and international research progammes. At the same time their work has increasingly taken on a dimension of more general relevance, as the science of the ocean becomes viewed not just as a curiosity lacking wider significance, but as a body of knowledge with import for many aspects of human activity and understanding – something that enthusiasts for the subject could have told them, and often did, any time from the seventeenth century onwards. Yet perhaps the transfer has not all been one way. Just as acceptance of the concept of the *Challenger* Expedition, it has been suggested (Rozwadowski 1996), may have been assisted by the growth of a wider maritime culture, so in the second half of the twentieth century, the growth of mass global air and sea travel for leisure purposes, together with growing familiarity with the marine environment, either directly through popular sports such as yachting and scuba diving, or indirectly through pictures from underwater cameras and submersibles, may be helping to familiarize a much larger group of people with an environment hitherto remote, and often terrifying. It is likely that these trends may have played a part in the growth of popular awareness of the sea, expressed both in interest about how knowledge is developing and as a part of more general concern about the environment. It can hardly be a coincidence, that as public awareness of and interest in the sea develops, yet another popular feature of late nineteenth-century cities, large marine aquaria, have made a strong comeback in the 1990s. Nineteenth-century works related the wonders of the sea, and often cited them as evidence of the care of the Creator. Today we have the benefit of seeing the wonders in full colour, from our living rooms, and the help of experts in understanding what we see. Whatever the nature of individual views on ultimate causes, we have no excuse for failing to appreciate both the remarkable nature of the earth we live on and our responsibility not to trash it, but to use its resources in a sensible and sustainable manner.

In spite of the great changes that have taken place in oceanography over the past century, close links still exist between the thoughts and aspirations of scientists in the past and those working at the present day, and the underlying requirements of successful ocean science remain essentially the same: enthusiastic and persistent individuals, challenged by ideas, able to gain access to financial and other resources, and in possession of technology capable of working in a difficult environment to yield the necessary data. These factors were combined in the voyage of HMS *Challenger*, 1872–6, which perhaps more than any other single event paved the way for the growth of oceanography as an independent science in the twentieth century, and have continued to characterize its development ever since, as may be seen in the following pages.

Notes

1 As Helen Rozwadowski (1998) and others have pointed out, much of this technology was adapted from items already in use, by fishermen and navigators among others, rather than developed specifically for the purpose of scientific research.

2 *Bathybius haeckelii*, which appeared to be a primitive organism, had been identified by T. H. Huxley in samples of deep-sea sediment collected by cable survey ships in the 1850s. Since Huxley was perhaps the most prominent British zoologist of his day excepting Darwin, as well as a keen proponent of the latter's ideas, it was potentially embarrassing when, after much fruitless examination of freshly obtained samples, the *Challenger* Expedition's chemist, J. Y. Buchanan, found *Bathybius* in a stored specimen and showed that it was not organic after all, but a chemical precipitate (Rehbock 1975). However, a century or so further on, scientists are again seriously considering the idea that life on earth may have originated in the sea, perhaps in association with hydrothermal vents (see Chapter 8). The *Challenger* found that, broadly speaking, deep-sea species were different from but mostly related to present-day forms already known from shallower water. However, events such as the more recent discovery of the coelacanth have shown that some species of fish as well as marine invertebrates can survive little changed in the sea for very long periods of time.

3 The building of the Southampton Oceanography Centre, at a cost of £48 million to construct and equip, has been described as 'Britain's single largest investment in marine science since the *Challenger* expedition' (Summerhayes and Hamilton 1996). However, it is not just the actual sums involved but the boldness of the concepts that are remarkable in both instances. Both projects were large-scale ambitious scientific enterprises, undertaken and completed in spite of rather than because of the prevailing political ethos of the day.

4 In doing so, he is not necessarily conveying approval. Indeed, he gives a salutary warning against paying too much attention to expeditions in general, and the *Challenger* Expedition in particular, when studying the history of oceanography, and readers of the present book should bear in mind that, in spite of its special status, the subsequent development of marine science, even in those early years, was achieved through a wide range of individual, national and international initiatives, some of which are also featured here.

5 Half the small scientific team were Scottish, reflecting the position of Wyville Thomson, himself a Scot, as a professor at Edinburgh University, but one of these, John Murray, had been born in Canada. Of the other naturalists, H. N. Moseley was English, while R. W. von Willemoes-Suhm was German. The artist, J. J. Wild, was Swiss and had worked with Thomson in Ireland. Suhm's appointment was made late in the day after another (Scottish) scientist had dropped out rather

than as a result of deliberate planning, but Thomson's insistence on involving overseas researchers in the writing of the report was the result of a deliberate decision to adhere to a practice already established in work on collections from the voyages that preceded the expedition.

6   Parliament felt that a sufficient, and indeed generous contribution had been made to the scientific study of the sea through the *Challenger* Expedition. Its reluctance to commit further funding to this area was reinforced by the policy of the Treasury (Alter 1987), which, as Roy MacLeod (1976) shows, exercised an increasingly tight control on expenditure during this period. For the impact on marine science see also Deacon (1994).

7   The Challenger Society for the promotion of the study of oceanography, founded in 1903 (Deacon 1990), still exists and publishes a journal, *Ocean Challenge*, which contains articles on recent developments in oceanography written with the more general reader in mind. Membership is open to anyone interested in the science of the sea, and further details can be obtained from the Membership Secretary, Challenger Society for Marine Science, c/o Southampton Oceanography Centre, Southampton SO14 3ZH.

8   Many marine biological stations were founded around the world during the second half of the nineteenth century and in the early 1900s. Many, in Europe especially, like the Stazione Zoologica at Naples, founded by the German zoologist Anton Dohrn in 1872 (the year of the *Challenger*'s departure) and perhaps the best known, were established as a result of the excitement caused by the publication of Darwin's theory of evolution, to provide laboratory accommodation for zoologists who were interested in developing his ideas through studies of the physiology and embryology of marine organisms. Some stations, such as the Plymouth Laboratory of the Marine Biological Association of the United Kingdom (opened 1888), also carried out fisheries research that was more likely to attract government funding. Many of these institutions survived to play an expanded role in marine science in the second half of the twentieth century, but few developed into major centres for ocean research, the prime example being the Scripps Institution of Oceanography in the United States.

9   Early students of the science of the sea generally operated in isolation, as regards that part of their work. There were exceptions in the subject's early history when aspects of marine science briefly enjoyed a more central position, and an institutional focus, such as that provided by the Royal Society in the seventeenth century (Deacon 1997) and by the British Association in the nineteenth (Rehbock 1979). However, this isolation ended as soon as the scientist went to sea, and engaged in social interaction with a quite different group of people, the ship's crew, not to mention other seafarers. In historical times the personal effects of these encounters range from comic to a good deal more serious; the collection of data was as much likely to be affected by conflicts of interest on board ship as by any innate defects in the scientific equipment employed.

References

Alter, P. (1987) *The reluctant patron: science and the state in Britain, 1850–1920*, trans. A. Davie. Oxford: Berg.

Burstyn, H. L. (1975) 'Science pays off, Sir John Murray and the Christmas Island phosphate industry, 1886–1914', *Social Studies of Science* **5**: 5–34.

Carson, R. (1950) *The sea around us*. London: Staples.

Deacon, M. B. (1990) 'British oceanographers and the Challenger Society, 1903–1922', in W. Lenz and M. Deacon (eds) *Ocean sciences: their history and relation to man*, Deutsche Hydrographische Zeitschrift, Ergänzungsheft, **B22**: 34–40.

Deacon, M. B. (1993) 'Crisis and compromise: the foundation of marine stations in Britain during the late nineteenth century', *Earth Sciences History* **12**: 19–47.

Deacon, M. B. (1994) 'British governmental attitudes to marine science', in S. Fisher (ed.) *Man and the maritime environment*. Exeter: University of Exeter Press, pp. 11–35.

Deacon, M. B. (1997) *Scientists and the sea, 1650–1900: a study of marine science*. Aldershot: Ashgate (first published 1971. London: Academic Press).

Earle, S. A. (1995) *Sea change: a message of the oceans*. New York: Putnam.

Herring, P. J. and Clarke M. J. (eds) (1971) *Deep oceans*. London: Arthur Barker.

Lenz, W. and Streicher, S. (1997) 'Das Institut und Museum für Meereskunde in Berlin, 1900–1946', *Historisch-Meereskundliches Jahrbuch* **4**.

McConnell, A. (1982) *No sea too deep: the history of oceanographic instruments*. Bristol: Adam Hilger.

MacLeod, R. M. (1976) 'Science and the Treasury: principles, personalities and policies, 1870–85', in G. L'E. Turner (ed.) *The patronage of science in the nineteenth century*. Leiden: Nordhoff, pp. 115–72.

Mills, E. L. (1989) *Biological oceanography: an early history, 1870–1960*. Ithaca, New York: Cornell University Press.

Mills, E. L. (1993) 'The historian of science and oceanography after twenty years', *Earth Sciences History* **12**: 5–18.

Rehbock, P. F. (1975) 'Huxley, Haeckel, and the oceanographers: the case of *Bathybius haeckelii*', *Isis* **66**: 504–33.

Rehbock, P. F. (1979) 'The early dredgers: "naturalizing" in British seas, 1830–1850', *Journal of the History of Biology* **12**: 292–368.

Rice, A. L. (1994) 'Forty years of land-locked oceanography; the Institute of Oceanographic Sciences at Wormley, 1953–1995', *Endeavour* **18**: 137–46.

Rice, A. L. and Owen, P. (1999) *Decommissioning the Brent Spar*, London: E. and F. P. Spar.

Rozwadowski, H. M. (1996) 'Small world: forging a scientific maritime culture for oceanography', *Isis* **87**: 409–29.

Rozwadowski, H. M. (1998) '"Simple enough to be carried out on board": the maritime environment and technological innovation in the nineteenth century', in M. Hedin and U. Larssen (eds) *Teknikens landskap*. Stockholm: Atlantis, pp. 83–97.

Schlee, S. (1973) *The edge of an unfamiliar world. A history of oceanography*. New York: Dutton.

Shor, E. N. (1978) *Scripps Institution of Oceanography: probing the oceans 1936 to 1976*. San Diego: Tofua Press.

Summerhayes, C. P. and Hamilton, N. (1996) 'The Southampton Oceanography Centre and the legacy of the *Challenger* expedition', *Geoscientist* **6**(2).

Summerhayes, C. P. and Thorpe S. A. (eds) (1996) *Oceanography: an illustrated guide*. London: Manson.

Part 1
The historical context

1 The *Challenger* Expedition
The end of an era or a new beginning?

A. L. Rice

Introduction

The scientific circumnavigation of HMS *Challenger* from 1872 to 1876 is acknowledged as representing an important stage in the development of ocean science, if not its very beginning. This recognition is certainly justified. Prior to the *Challenger* Expedition, no circumnavigation had been devoted solely to scientific research; no government had invested so much effort and funding in a single scientific undertaking; no marine expedition had collected so much data and so many samples; no marine collection had been so well curated; and no expedition had produced such a mass of published scientific information or had such an influence on future research.[1]

Yet the expedition has not been without its critics, either at the time or since. The contemporary criticisms concentrated on what might be called the sociopolitical issues, such as the cost of the venture and the controversial involvement of non-British scientists in working up the results (M. B. Deacon 1971: 371). Later, the quality of some of the *Challenger* results, and particularly the physical data, was questioned. George Deacon (1968), for example, suggested that uncertainty about the effect of pressure on the thermometers used, coupled with limited confidence in their accuracy, may have prevented the *Challenger* scientists and the authors of the reports from making the most of the results collected. More generally, several authors have criticized the expedition, overtly or tacitly, for being somewhat old-fashioned and conservative, either for failing to try new technologies or for failing to persevere with them. It is this technological aspect of the expedition that I will concentrate on, rather than the specific scientific achievements, in the hope that we may still learn something from it even after more than a century.

So much has been written about the origins, progress and results of the voyage, and the personalities involved (see Rice 1986 and references therein, Jones 1990, Rehbock 1992), that a detailed account here would be superfluous. However, to place the technological topic in context, particularly for readers unfamiliar with the voyage, a short summary of the expedition and its background is justified.

The expedition

Credit for the instigation of the expedition must go mainly to W. B. Carpenter, who became convinced as early as 1869 that the British Government should finance a major scientific circumnavigation. Although primarily a physiologist by profession, Carpenter had a great interest, and expertise, in deep-sea biology. Along with Charles Wyville Thomson (Fig. 1.1), who was to become the scientific director of the *Challenger* Expedition, he led a series of short cruises in the steam paddle vessels HMS *Lightning* and *Porcupine* during 1868–70 to the north and west of the British Isles and into the Mediterranean. Following the success of these cruises, during which animals had been collected from deeper than 4000m and intriguing subsurface temperature measurements had been made, Carpenter wrote to the Royal Society (of which he was himself a Vice-President) suggesting that a major proposal for oceanic research should be submitted to Parliament.

The proposal fell on fertile ground in the Royal Society, in the Admiralty and even in the Treasury, and the application received formal approval in April 1872. A number of factors contributed to the success of the application (see Chapter 2, Burstyn 1968), including the support of the Hydrographer, G. H. Richards, whose department was being encouraged to extend its interests to the deep ocean by the needs of submarine telegraphy, the fear that Britain's lead in deep-sea science would be eclipsed by other nations, particularly the USA, and even the fact that the government's financial structure recently had been reformed.

A ship was quickly selected, Captain G. S. Nares was given command of the expedition, Thomson was appointed Scientific Director, and the necessary refitting of the vessel put in hand at Sheerness. The *Challenger* (Fig. 1.2) was a 14-year-old steam-assisted screw corvette, 226 feet long, 2306 tons displacement and with a 400 nominal horsepower steam engine driving a twin-bladed screw that could be lifted clear of the water when the vessel was under sail. Her sail area, and therefore her manpower requirements,[2] were reduced, and all but two of her 22 guns were removed. The space thus freed below decks[3] was used for a zoological laboratory, a chemical laboratory and a photographic darkroom, and for living accommodation for the six civilian scientists.[4]

The main modifications were to the upper deck where an 18 horsepower donkey engine was fitted forward of the main mast to drive a shaft mounted across the deck with drums on

Figure 1.1 Charles Wyville Thomson (1830–82), scientific director of the *Challenger* Expedition.

Figure 1.2 HMS *Challenger* under sail. (Tizard et al. 1885: 1.)

either end to haul the trawls and dredges (Fig. 1.3). A dredging platform, a sort of extra bridge, was installed over the donkey engine and shaft for the deployment and recovery of the gear, both to separate these activities from those of the seamen and to keep the mud and other refuse off the navy's precious deck! Small sounding platforms protruded from either side of the deck near the main mast, while three storage racks, each holding 2000-fathom coils of ready-use dredging rope, were built on the foredeck. Finally, when the ship reached warm latitudes some activities, and particularly bird skinning, became so antisocial below decks that a deck house, a sort of 7 foot by 8 foot shed, was built at the after end for this purpose and also as a wet laboratory in which to deal with the plankton net catches.

All these changes, as well as the accumulation and loading of the vast quantity of equipment and stores and the appointment of the naval and civilian staffs, were completed in less than six months. The *Challenger* sailed from Sheerness to Portsmouth in early December 1872, to leave finally on 21 December only nine months after the formal approval had been received. By the time she returned to Spithead, on Queen Victoria's 57th birthday on 24 May 1876, she had in the meantime spent 713 days at sea out of a total of 1606, and had covered 68,890 nautical miles.

During the many port calls the scientists made extensive shallow-water collections and observations of the terrestrial flora and fauna, including the human inhabitants. But the main work was conducted during the long ocean passages. A total of 362 official stations were occupied, on average one every two or three days at sea. At each station a sounding and a sample of the bottom sediment were taken, while the water temperature was determined at the surface and the bottom, and often at several intermediate depths. Water samples were also obtained, usually from near the bottom and from mid-water, and finally biological samples were collected. These always consisted of a trawl or dredge sample from the sea floor, while plankton nets were generally fished in the upper 50 fathoms (90m) and occasionally as deep as 800 fathoms (1460m). The speed and direction of the surface currents were

Figure 1.3 The port side of the upper deck of the *Challenger* looking forward. This frequently
reproduced view (from Tizard et al. 1885: fig. 12) shows the dredging and sounding
arrangements. The line from the gear was led to the winding drums on one end of the
shaft driven by the donkey engine, then via blocks on the deck (one shown in the
foreground) to the other end of the shaft. By the time the line came off the second set
of drums, all tension on it would have been removed, so that it could be coiled by hand.

recorded fairly routinely, whereas attempts to measure the shallow subsurface currents were
made rather less regularly.

The biological specimens were dispatched to Edinburgh periodically during the voyage to
await the ship's return.[5] Shortly after the end of the expedition the 'land and other incidental collections' were transferred to the British Museum, the Natural History section of
which was about to move to new premises in South Kensington (Rice 1992). The marine
collections were retained in Edinburgh where a Challenger Office was established to organize their dispatch to a multinational group of experts. The resulting monographs appeared
in a series of Official Reports published by the Stationery Office, initially under the super-

intendence of Wyville Thomson and then, after Thomson's death in 1882 at the age of only 52, under John Murray, who had sailed as a junior naturalist on the expedition but later was to become the most celebrated of the participants (see also Postscript). The original expectation was that the reports would consist of fourteen or fifteen volumes to be published within about ten years of the end of the expedition. In the event, they filled fifty volumes,[6] of which the last did not appear until 1895. The vast majority, though not all, of the marine collections eventually returned to the Natural History Museum, where they remain a crucial reference collection used by taxonomic specialists from throughout the world.

The *Challenger* technology

Despite the contemporary criticisms, at least some undoubtedly based on sour grapes, and the parsimony of the Treasury in fulfilling its obligations both with regard to the publication of the official reports and funding further oceanographic research (see Burstyn 1968, M. B. Deacon 1971), the expedition generally was considered a great success.[7] Nevertheless, it is clear that by the time the Narrative volumes of the report were published in 1885, the authors (Tizard, Moseley, Buchanan and Murray 1885) were conscious that several of the techniques and scientific instruments used on the *Challenger* had been superseded, and that in some cases better alternatives had been available when the ship sailed. Indeed, as McConnell (1982) has pointed out, the Narrative includes accounts of several instruments that could have been carried but were not, and of others that were not developed until after the *Challenger* had returned. To examine this theme a little further, I will consider four distinct aspects of the expedition's technology: the use of rope as opposed to wire; the gears used to obtain biological samples; the sounding technique; and the measurement of subsurface temperatures.

Rope versus wire

Nothing epitomizes the nineteenth-century technology of the *Challenger* Expedition more than her use of rope. She was supplied with a staggering total of 220,000 fathoms (more than 400km), ranging from relatively thin (but unspecified) current drogue line to 3 inch (*c.* 75mm) circumference dredging rope; during the voyage she 'expended', that is lost or discarded, no less than 125,000 fathoms!

Responsibility for relying upon rope rather than wire lay firmly with Thomson, though he would have had a stalwart supporter in Nares, who was himself definitely a sail and rope man.[8] By the time Thomson wrote the general introduction to the zoological series of the reports, published in 1880 only two years before his death and five years before the first Narrative volume appeared, it is clear that he realized that the expedition was open to criticism for its dependence on rope. For he writes (p. 8): 'There can be no doubt that in any future expedition, on whatever scale, it would be an unjustifiable waste of time and space to neglect the use of wire for sounding, and wire rope for dredging and trawling.' He had already explained why he had chosen rope, for after introducing the earlier *Lightning* and *Porcupine* cruises he writes (pp. 2–3):

> When it afterwards became my duty to superintend the more important arrangements for the scientific work on board the *Challenger*, although many changes were suggested, I made it a principle to deviate as little as possible from a plan of working which had been shown to be at all events practicable. During the progress of our voyage many

alterations – some of them, such as the use of wire for sounding introduced by Sir William Thomson [see below], and employed with such signal success by Captain Belknap in the US ship *Tuscarora* and of wire-rope for dredging adopted by Professor Agassiz, evidently decided improvements – were introduced elsewhere; but as we had become accustomed to our own plans, and could depend almost with certainty upon the amount of work we could do within a certain time, I thought it better to continue steadily throughout as we had begun, and to secure the largest possible series of similar and comparable observations, rather than run the risk of losing time through possible failures.

In other words, in this context Thomson was using the 'devil we know' argument, recognizable also in several other areas of *Challenger* technology. For apart from the successful use of wire for sounding from the USS *Tuscarora* in 1874 while the *Challenger* was still at sea, Alexander Agassiz, Director of the Museum of Comparative Zoology at Harvard and an experienced mining engineer, had introduced the use of wire for deep trawling during the cruises of the US Steamer *Blake* during 1877–80 (Sigsbee 1880, Agassiz 1888). Rope would never again be used for over-the-side work on any serious deep-sea expedition, but should the *Challenger* have been the first of these rather than the last to rely on rope?

Biological sampling

The basic *Challenger* instrument for sampling the sea floor was the dredge, as developed during the cruises of HMS *Porcupine* (see Thomson 1873 and Figs 1.4 and 1.5D). This, in turn, was a development of Ball's dredge, a modification of the commercial oyster dredge in which the mouth was provided with an angled scraper on each of the long sides so that it would fish effectively either way up. The 34 iron dredge frames supplied to the *Challenger* had mouths ranging from 5 feet by 15 inches (150 by 37cm) to 3 feet by 12 inches (90 by 30cm), the smallest generally being used at the greatest depths. The twenty nets provided consisted of a bag about 4 feet 6 inches (135cm) long made of 1 inch (2.5cm) mesh, the lower third lined with finer meshed 'bread bag stuff' to catch the smaller animals. When fitted to a dredge, the net was protected by longitudinal iron rods and ropes attached to the frame and to a cross-bar carrying a number of 'hempen tangles', bundles of teased-out hemp rope.

The addition of the tangles had been suggested by Captain Calver of the *Porcupine* and had greatly improved the efficiency of the dredge by entangling all manner of organisms, whereas the net often came up containing nothing of interest. The obvious reason was that the scrapers on the mouth of the dredge were far too divergent so that they dug into the soft bottom sediment and either flipped the dredge over or quickly clogged the mouth with mud, thereby making it ineffective. The problem was largely solved by Captain Charles Sigsbee during the cruises of the *Blake* simply by making the dredge mouth less divergent. But the *Challenger* staff failed to find an answer and, apart from during the early part of the expedition, most of the deep sea floor samples were taken with a trawl rather than a dredge.

The *Challenger* was supplied with 22 beam trawls, similar to those used by commercial fishermen in shallow waters (see Fig. 1.5A). A heavy hardwood beam, usually 10 feet (3m) long for the deepest trawlings, was attached to the top of a D-shaped iron runner at either end and dragged a 20-foot-long conical net. Although the trawls successfully collected the *Challenger*'s remarkable samples, it was always difficult, if not impossible, to know whether they had landed right way up or upside down on the seabed. Again, the problem was solved on the *Blake* by the simple expedient of attaching the beam and the net to the back, rather

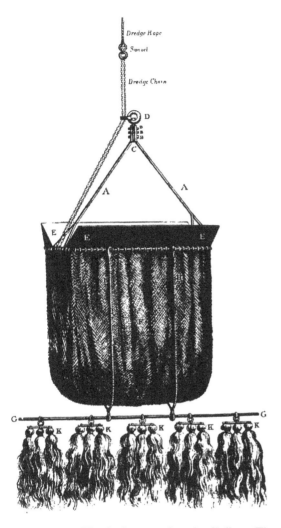

Figure 1.4 The dredge as used on the *Challenger* (Tizard et al. 1885: fig. 19) showing the weak link
    arrangement (D) and the 'hempen tangles' (K).

than the top of the runners so that the trawl worked equally well whichever way it landed.
The resulting 'Agassiz' trawl was rapidly adopted as a standard deep-sea gear, still widely
used today. In contrast, the deep-sea use of conventional beam trawls more or less ended
with the *Challenger.*[9]

Finally, the nets used to collect midwater animals from the *Challenger* were very small (no
more than 16 inches, about 40cm, in diameter) and exceedingly simple tow-nets, likened by
Murray and Hjort (1912) to long muslin or calico night caps. Within twenty years of the end
of the *Challenger* Expedition, plankton sampling had been raised to a totally new level with
the development of opening and closing nets to investigate the vertical distribution of mid-
water organisms and particularly the introduction of quantitative sampling by the German
planktologist Victor Hensen (Mills 1989). Again, the *Challenger* Expedition represented the
end of a technological era rather than a new beginning.

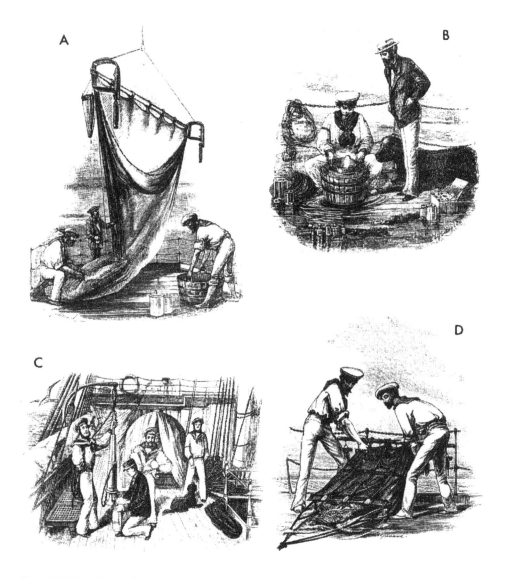

Figure 1.5 Scientific activities on board the *Challenger*. All from illustrations by Elizabeth Gulland used as tailpieces in Tizard et al. 1885: (A) beam trawl catch being landed on the dredging platform; (B) net samples being sieved; (C) water sample being taken from a 'slip' water bottle, designed to collect water from close to the seabed; (D) dredge catch being emptied onto the deck.

Sounding

Stimulated by the needs of the new technology of submarine telegraphy, deep-sea sounding went through something of a vogue during the middle decades of the nineteenth century, particularly with the British and US navies (see McConnell 1982). Any acceptable sounding technique had to fulfil several criteria: accuracy/reliability, speed, cheapness and the ability to retrieve a sample of the bottom sediment. More than half a century before the successful development of echo-sounding, a number of significant problems had to be overcome in satisfying these criteria.

The general technique was to lower a weighted line from the ship and to measure the amount of rope out when the weight reached the bottom. Position-keeping for the several hours that it took to obtain a deep sounding had been almost impossible in sailing ships, but had been largely solved in powered vessels such as the *Challenger*. Steam power for hauling in the sounding line also made the process less arduous, but the other problems remained.

To get a sounding line to the bottom as quickly and vertically as possible demanded a heavy weight and a light line. But such a line would be incapable of retrieving the weight and a bottom sample. This problem had been partly overcome as early as 1852 when a young US naval officer, John Brooke, devised a technique for detaching the weight when the sounder reached the seabed. Brooke's apparatus consisted of a metal rod, armed with tallow to collect a sediment sample, which passed through the centre of a spherical weight suspended from the top of the rod on a wire sling. When the sounder reached the bottom the sling was released and the weight slipped off the rod. The rod and the sample could now be retrieved on a relatively light line.

Although the general principle was retained, many variants were developed over the next twenty years, including tubes to collect the sediment, valves to retain the sample, different forms for the weight and the release mechanism, and the use of thinner lines, including wire, to increase the speed of the operation. Trials with both copper and iron wire had been made much earlier (see McConnell 1982) but had not been successful because of the weaknesses resulting from the splicing necessary to join the maximum lengths that could be manu-factured. But just before the *Challenger* sailed, Sir William Thomson reported the first success-ful deep-sea sounding using wire, for in June 1872 he had obtained a sounding in 2700 fathoms (almost 5000m) in the Bay of Biscay. In doing so, and aided by the compactness of the wire, he opened a new and fruitful avenue in sounding technology by developing a sounding machine in which the various elements, power drive, storage drum and accumula-tor system, were combined in a single unit. By putting a variable resistance or brake on the drum to compensate for the weight of wire out, the system stopped suddenly when the weight struck the bottom, thus indicating the depth. A modification of Thomson's technique was used successfully from the USS *Tuscarora* in 1873–4 for a cable route survey across the northern Pacific and gave rise to a plethora of increasingly sophisticated sounding machines used on American survey vessels and cable ships (McConnell 1982).

A Thomson sounding machine had been placed on board the *Challenger* at the outset, but the drum collapsed the first time it was tried and it was never used again. Instead, the expedition reverted to the more traditional technique using hemp line. When the ship sailed, the currently fashionable version of Brooke's apparatus was the *Hydra* sounder (Fig. 1.6) in which the sling retaining the disposable weights was released by a spring mechanism. But this mechanism was unreliable and tended to pre-trip near the surface. Early in 1873 an improvement, the Baillie rod and sinkers (Fig. 1.6), not depending upon a spring, was developed and was sent out to the *Challenger* later that year to replace the *Hydra* sounder.[10] From this point on, the Baillie rod, weighing 35lbs and carrying 3 or 4cwt of detachable weights for the deepest soundings, was used routinely and with almost total success (Tizard et al. 1885).

The Baillie rods were used with 1-inch circumference hemp line, an alternative $\frac{3}{4}$-inch line proving to be too weak. The 125-fathom lengths of the line, as manufactured, were spliced together into 3000-fathom lengths and stored on reels near the sounding platforms. During deployment, except for the stops necessary to attach thermometers to the line, the rope was

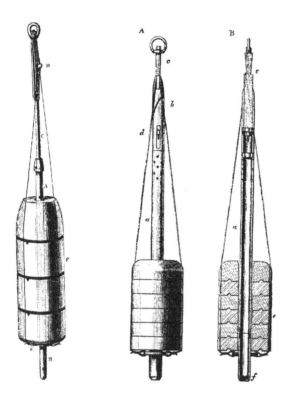

Figure 1.6 The *Hydra* sounder (left), with which the *Challenger* was originally supplied, and the Baillie sounder (right) which replaced it. (Tizard et al. 1885: figs 13 and 14.)

allowed to run freely from the drum once a few hundred fathoms had been paid out via the donkey engine. The time taken for each 25 fathoms to run out was recorded, a sudden increase in this interval indicating the arrival of the weight on the bottom. The sounding tube, hopefully with a sample of the seabed sediment, was now retrieved using the donkey engine with stops to remove the thermometers. The process was very time-consuming; the deepest *Challenger* sounding, at a depth of 4475 fathoms (almost 8200m), took two-and-a-half hours, compared with less than half this time with a wire sounding machine.

When Tizard, Moseley, Buchanan and Murray (1885) wrote the first Narrative volume of the reports they were clearly conscious, like Thomson before them, of the wire versus rope controversy. Accordingly, though they accepted that the future of oceanography lay with wire,[11] they quite forcefully defended the use of rope on the *Challenger* on the grounds that time was not at a premium during the voyage and that the risk of wire parting, with the resultant loss of valuable instruments such as thermometers, piezometers and water bottles, made rope much the safer option.

Subsurface temperatures

The normal technique for obtaining subsurface temperatures at the time of the *Challenger* Expedition was the use of developments of the self-registering thermometer invented by James Six in the late eighteenth century (McConnell 1982). In these thermometers, the central part of a U-shaped or three-limbed narrow-bore tube was filled with mercury while either end was filled with alcohol. One limb ended in an alcohol-filled bulb, while the bulb at the other end was partly air-filled. As the temperature changed, the alcohol in the filled end expanded and contracted, moving the mercury column backwards and forwards. The extent of this movement was marked by small glass or metal 'indexes' designed to retain their position within the tube when the mercury moved away from them, thus recording the maximum and minimum temperatures to which the thermometer had been exposed.

The version supplied to the *Challenger*, the Miller–Casella pressure-protected thermometer (Fig. 1.7), had been developed following the *Lightning* cruise, during which Wyville Thomson had found that the standard Hydrographic Office pattern instruments were very unreliable in deep water. The new instruments, designed by W. A. Miller and made by the firm of Casella, consisted of a simple U-tube with a large bulb at one end and smaller one at the other. The larger or primary bulb and part of the tube were filled with the thermometric fluid, a solution of creosote in alcohol. The secondary bulb and the adjacent tube were partially filled with the same fluid, the space being filled with its vapour and slightly compressed air. The central part of the U was filled with mercury, with an index at either end. Each index consisted of a small piece of steel wire enclosed in glass and held in place within the thermometer tube by the spring in a small length of hair tied to one end of the index. Close to the bend, each limb had a small bulge or aneurism in its bore to prevent the indices being carried into the bend. To protect the thermometer from the effects of pressure, the primary bulb was enclosed in a second one, with the space between partly filled with alcohol. The thermometer was mounted on an ebonite sheet to which were attached the scales (in degrees Fahrenheit) engraved on slips of white glass. The thermometer was enclosed in a protective cylindrical copper case with a door in the side to allow the thermometer to be read without removal. In use, the indices were first drawn along each limb with a strong magnet until they rested against the end of the mercury thread. After retrieval, the lower end of each index marked the extreme (high or low) of the temperature range to which the thermometer had been exposed.

The *Challenger* was supplied with 104 'protected' thermometers (presumably the Miller–Casella instruments), of which 48 were lost or broken in use. But although they were used for hundreds of apparently successful subsurface temperature determinations, the instruments had a number of inherent problems. First, the indices were prone to being displaced if the thermometer was jerked suddenly, though this problem could be reduced by attaching the instrument to the sounding line via an elastic 'stop' to absorb any shocks. Second, to reduce the resistance to their passage through the water, the instruments were purposely kept very small, only 9 inches (22cm) in overall length. Consequently, the Fahrenheit degree divisions on the scales of some of the instruments were so small that it would have been impossible to read them to an accuracy of more than a quarter of a degree even if they had been engraved directly onto the thermometer stem.[12] Engraved on separate slips of glass, as the scales were, it was quite impossible to read the temperature even to this accuracy, whereas it soon became apparent that much greater accuracy was required to understand the subtleties of oceanic temperature distributions. Moreover, with the thermometer and scale not rigidly attached to one another there was a strong possibility of relative movement in use, producing even greater errors. 'Errors from this source' noted Tizard et al. (1885: 97) 'are very liable to occur, and are due solely to defective instrument-making. No instrument of this kind should be sent out of the workshop, to be used on such important work as deep-sea investigation, which has not a scale etched on the stem.'

But by far the most serious potential systematic source of error, and the most difficult to deal with, was the effect of pressure. Although the primary bulbs of the Miller–Casella thermometers were pressure-protected by being enclosed in a second one, the instruments were still thought to be affected by pressure on the unprotected stems. Since the stems were not uniform in thickness, and varied between instruments, the precise pressure effect would be unique to each thermometer. Accordingly, the *Challenger* staff were provided with pressure corrections for each instrument showing the amount to be deducted from each reading according to the depth to which the thermometer had been sent. In the early stages of the expedition these corrections were applied, without question, to all the subsurface temperature measurements and the corrected figures used in the reports and temperature sections sent home to the Admiralty. However, the *Challenger* staff soon began to doubt the validity of the corrections, which seemed to be much too large. They eventually concluded that the corrections had been obtained by compressing each thermometer in a hydraulic system and noting the elevation of the recorded temperature at each pressure step, but failing to take account of the actual temperature rise within the equipment as a result of the compression itself (see Tizard et al. 1885: 99). Despite these doubts, the provided corrections continued to be applied to the *Challenger* data 'in order that they might be comparable with those that had gone before'. But this must have affected the shipboard confidence in the temperature measurements very seriously. For it was not until Professor P. G. Tait had examined very carefully some thirty of the surviving thermometers after the expedition (Tait 1882, McConnell 1982) that the true, and relatively small, magnitude of the pressure effect was appreciated.[13] In fact, the calculated correction was so small that the temperature sections published in the *Challenger Reports* were based on the temperatures 'as read', that is, without correction, and the same figures were used by Buchan (1895) in his report on oceanic circulation published along with the Summary volumes.[14] From the temperature sections, Buchan inferred, correctly, a major northward flow of cold deep water from the Antarctic into the other oceans, and particularly into the Atlantic. He was also able to infer a southward flow of deep water from the North to the South Atlantic, but it is not clear whether he realized that these water masses were part of the complex layered thermohaline deep circulation system or thought that they would simply collide somewhere in mid-ocean (see Warren 1981). Writing some thirty years later and with the benefit of hindsight, Merz and Wüst (1922) suggested that the complete picture of the meridional circulation, at least of the Atlantic, might have been deduced from the *Challenger* physical observations despite the limitations of the thermometers. But neither Buchan (1895) nor Buchanan (1884), who wrote up the specific gravity results, went much beyond a general description of the results with very little interpretation or discussion. G. E. R. Deacon (1968) suggested that the *Challenger* scientists could well have made more of their results and speculated about the details of the general oceanic circulation if they had been more confident in their temperature data. The absence of such confidence must be attributed mainly to the Miller–Casella thermometers; but were there any alternatives?

In January 1874, a year after the *Challenger* sailed, Casella's rival instrument-making firm of Negretti & Zambra patented a quite new deep-sea thermometer containing only mercury as the thermometric fluid and abandoning the use of indices to record the temperature. Instead, the new thermometers employed a principle already used in medical and industrial applications (and familiar today in clinical thermometers) in which a narrow constriction in the bore of the tube causes the mercury thread to break, and the temperature at this instant is read from the amount of mercury retained beyond the constriction. In the early versions of the new instrument, the thermometer was in the form of a U and had to be rotated vertically through 360° to break the mercury thread and register the temperature. As in the Miller–Casella thermometers, the bulb was pressure-protected, but since the new instruments were considerably larger, with the scales engraved directly on the stem, they immediately overcame some of the most serious deficiencies of the Miller–Casella ones; they were easier to read, they avoided the problem of moving indices, and they recorded the temperature at depth irrespective of the temperature in the more superficial layers.

Several of these reversing thermometers and a frame or frames for holding them (Fig. 1.7) were sent out to the *Challenger* and, from January 1875 onwards, were employed in a series of comparative tests with the Miller–Casella instruments by Staff Commander T. H. Tizard. After initial problems with the reversing apparatus, which were rectified by one of the ship's engineers (see Tizard et al. 1885, McConnell 1982), four thermometers were used to take a total of 48 temperature readings at depths down to 2530 fathoms (*c.* 4600m). Thirty of these readings were higher than those obtained at the same localities with the Miller–Casella instruments, three lower, and fifteen similar. Tizard assumed, possibly correctly, that these discrepancies were attributable to faults in the reversing thermometers, and he does not seem to have been overly enthused by the improvements to the system introduced by Negretti & Zambra after the end of the *Challenger* Expedition, including reducing the

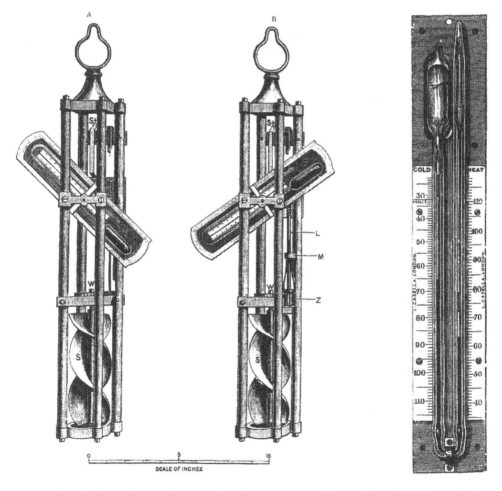

Figure 1.7 The Negretti & Zambra reversing thermometer and frame (left) and the Miller–Casella protected thermometer (right). (Tizard et al. 1885: figs 27 and 29.)

reversing requirement to only 180° (Tizard et al. 1885: 93–5). In this he was, of course, wrong, for the principle of the reversing thermometer was to dominate physical ocean-ography for at least the next six decades, totally replacing the maximum–minimum principle (McConnell 1982). But on the *Challenger*, the reversing thermometers were used only for these trials, and the results were not incorporated into the scientific results.

In preparation for the expedition, the *Challenger* chemist, Buchanan (Fig. 1.8), had made some piezometers with the objective of determining the compressibility of fresh water and of a range of saline solutions. But the realization that temperature inversions (in which the water temperature does not decrease consistently with increasing depth) could not be detected with the other instruments available led him to develop piezometers that might be able to do this. Piezometers are basically fluid-filled tubes, one end of which is open to the ambient environment. They are used usually to measure pressure or the effects of pressure, such as the change in volume of a fluid. Buchanan reasoned that, lowered into the sea, any change in volume of a piezometer fluid would result not only from the increasing pressure, but also from the changing (usually decreasing) temperature. He therefore constructed two types of piezometer (Fig. 1.9), one in which the main fluid was mercury, the volume of which is greatly affected by temperature but not by pressure, and the second in which the main fluid was water, on which the relative effects of pressure and temperature are reversed, at least at the low temperatures encountered in the deep sea. By sending a pair of such instruments to the same depth, two independent assessments of the combined effects of pressure and temperature could be obtained. Using the length of sounding line out as a first approxima-tion of the depth, the reading of the mercury instrument could be corrected for pressure to give a first approximation of the *in situ* temperature. This could then be applied to the reading of the water piezometer to give a second approximation of the depth. This, in turn, could be applied to the mercury piezometer reading to obtain a second approximation of the temperature. The process could, of course, be continued, but Buchanan thought that the

Figure 1.8 John Young Buchanan, the *Challenger* chemist.

Figure 1.9 Mercury piezometer (left) and water piezometer (right) produced by Buchanan during the voyage. (Tizard et al. 1885: figs 33 and 35.)

second approximations would be so close to the 'truth' that further iterations were not necessary. Some sets of observations in the Pacific apparently gave quite good results, though Tizard et al. warn that water piezometers should not be used at great depths because quite small errors in the temperature determination could result in large errors in the depth estimate.

After the expedition, Buchanan patented a pressure sounder based on the same principle (see McConnell 1982), but reversing thermometers were by that time so dominating sub-surface temperature measurements that the piezometer idea was not pursued.

The *Challenger* was also supplied with an electrical resistance thermometer designed by C. W. Siemens. The use of such thermometers for deep-sea work had been first suggested by Siemens and his brother Werner in 1866. An instrument with a depth capability of 1000 fathoms (*c.* 1800m) had been tried on the *Porcupine* in 1869, but with little success because the galvanometer included in the system could not be read adequately on a moving ship (McConnell 1982). The *Challenger* resistance thermometer was provided with a Thomson (Lord Kelvin) marine galvanometer to overcome this problem.

Resistance thermometers depend upon the effect of temperature on the resistivity of a conductor, in this case a small coil of iron wire. This 'resistance' coil is lowered into the sea on two long insulated wires and forms one arm of a Wheatstone bridge, the other arm consisting of an exactly similar 'comparison' coil immersed in a water-filled vessel on the ship. A galvanometer indicates the potential difference between the two sides of the bridge, which is brought into balance by changing the temperature of the water surrounding the comparison coil by the addition of ice or hot water. When the bridge is balanced, the temperature of this water is the same as that surrounding the resistance coil in the sea.

After stating that 'Several more or less successful observations were made with this instrument during the cruise', Tizard et al. (1885) make the amazing statement that 'No permanent place was fitted for the galvanometer or apparatus, and in consequence continuous and careful observations were not made'. By the time this was written, Tizard et al. knew that a resistance thermometer had been used with considerable success from the US Steamer *Blake* in 1881 and they forecast that 'it will in all probability be extensively used in future deep-sea investigations'. So although they were convinced that this was an important technological advance, the *Challenger* officers and scientists did not persevere with it and no details of their results were published.

In fact, although resistance thermometers were tried several times for deep-sea work during the 1880s and 1890s, the problems of using long insulated wires in the ocean, of operating a galvanometer on a moving ship and of controlling the temperature of the comparison coil were so great that the technique was more or less abandoned by the end of the century. In any case, several reversing thermometers on a single sounding line would provide the same or better results much faster. It was to be a further fifty years before the general principle behind the resistance thermometer was to be used again in oceanography, leading to the extremely important development of the STD and the modern CTD[15] (Warren 1981). In the meantime, the reversing thermometer that the *Challenger* staff had also more or less shunned, held sway.

Conclusion

From a purely technological point of view it is difficult to escape the conclusion that the *Challenger* Expedition was conservative rather than innovative. In all the areas examined, mid-water and benthic biological sampling, sounding, subsurface temperature measurement and the widespread reliance on rope, the expedition tended to be the last occasion on which a particular technique or instrument was used rather than the first. Even where the ship was supplied with the latest technology, such as wire for sounding, the new reversing thermometers or electrical resistance thermometers, the *Challenger* personnel seem to have been discouraged by early problems and rarely persevered in attempting to correct them. Although the *Challenger* experience stimulated interest in the deep sea in general, and the development of new technology, the *Challenger* officers and scientists were not usually directly involved in these developments. The single exception seems to have been Buchanan, the chemist. He developed a number of pieces of equipment, both during the *Challenger* Expedition and subsequently, including the piezometers mentioned earlier and an ingenious stopcock water bottle that successfully collected mid-water samples during the expedition and was used, at least by Buchanan, for some years afterwards (McConnell 1982).[16]

If the *Challenger* Expedition was technologically conservative, is this a serious criticism and might things have been different if the personnel had been more innovative? Clearly, it is possible that some of the results might have been 'better'. An obvious example is the subsurface temperature determinations which, as we have seen, suffered from the limitations of the Miller–Casella thermometers and the *Challenger* personnel's lack of confidence in the instruments. But any shortcomings in the published physics reports are due at least as much to the prevailing philosophy as to the limitations of the technology, though the two are obviously related. The fact is that biology, sounding and sea-floor sediment sampling were the main priorities during the voyage. Despite Carpenter's personal interest in oceanic circulation, physics (and chemistry) were much lower down on the agenda.

It is also possible that had wire been used routinely for over-the-side work, these operations might have been faster and rather more stations could have been worked. But although the expedition has been referred to several times as a rather leisurely affair, a station was worked on average for every two days at sea; despite the quite protracted port calls, it is doubtful whether a significantly higher work rate could have been sustained for the duration of a cruise of this length. In any case, the significance of the expedition lay in its total devotion to science over such a wide geographical range; data and samples from a few extra stations would have made little difference to this.

On the other hand, what might have been the disadvantages of a more innovative or forward-looking approach? The *Challenger* Expedition was emphatically not an instrument-testing cruise. It had the specific objective 'to investigate the physical and biological conditions of the great Ocean Basins', particularly to determine in broad outline the contours of the deep-sea floor and the nature of the seabed biological communities. It would have been unthinkable to have risked failing in these general aims by using techniques that were not known to be reliable.

To obtain a good series of soundings was at the top of the *Challenger* wish list. Indeed, the official instructions (see Tizard et al. 1885: 23) state that obtaining soundings 'is an object of such primary importance that it should be carried out whenever possible, even when circumstances may not admit of dredging, or of anything beyond sounding'. Wire soundings in deep water were achieved successfully on the USS *Tuscarora* within months of the *Challenger's* departure and the supplied Thomson wire sounding machine might have been made to work by the ship's engineers. But were they supplied with enough wire to replace the inevitable losses and, if not, could they have obtained replacements at their port calls at a time when the manufacture of long lengths of wire was still a difficult and specialized activity? Probably not. Under these circumstances, the decision to rely on the established rope technology seems not unreasonable.

Similarly, if they had anticipated Agassiz's use of wire rope for dredging and trawling, inevitably there would have been significant losses, as there were of rope. Could they have carried enough wire from the start, or could they have obtained replacement wire during the voyage? Again, probably not. On the other hand, abundant supplies of hemp rope were available at any port or shipyard throughout the world. Moreover, it was much easier to handle than wire, requiring only a relatively simple capstan to haul it aboard, after which it could be coiled and stored almost anywhere. In contrast, the use of wire in deep-sea work demanded the development of massive winches and storage drums (Fig. 1.10).

Thomson's (1880) arguments for the techniques and instruments used, essentially supported by Tizard et al. five years later, were that, despite their shortcomings, they were familiar and were known to work, and that having started the cruise with one technology, for example the Miller–Casella thermometers, any change might have resulted in non-comparable data, a point specifically made in the instructions. Setting out, as they were, for a cruise expected to last for more than three years, into a nineteenth-century world with limited communications, and often into the least frequently visited parts of that world, it is difficult to fault Thomson's caution. If he had gambled, and lost, neither he nor the *Challenger* Expedition would be remembered with such pride, respect and affection today. In

Figure 1.10 The hauling winch (upper) and the storage drum (lower) employed for trawling with wire
on the *Talisman* in 1883 (Filhol 1885: figs 16 and 17). A similar arrangement had been used
by Agassiz on the *Blake*, and a combined winch and storage drum was used later on the
Michael Sars. (Murray and Hjort 1912.)

other words, any technological conservatism on the part of the *Challenger* scientists seems largely justifiable.

So are there any lessons to be learnt? The modern oceanographic world is very different from that of the *Challenger*. Cruises are now, thankfully, rarely longer than a few weeks. News and data, good and bad, can be flashed across the globe in seconds instead of weeks or months. Equipment and expertise can be transported, at a price, almost anywhere in the world in a matter of hours. Clearly, many of the constraints on the *Challenger* personnel do not apply today. But we still have the same concerns for technical reliability and intercomparability. In the exploding technology, particularly electronic, of the late twentieth century it is inevitable, and desirable, that oceanography should exploit the new techniques, including autonomous underwater vehicles and world-ranging satellites transmitting synoptic data to land-based laboratories. But we abandon all of our tried and tested simpler technologies at our peril. There will, I suspect, always be the need for ground truthing with the equivalent of the *Challenger*'s trawls, dredges, thermometers and water bottles. But above all, I hope that scientists will always from time to time go to sea. For one crucial area has seen no improvement since the *Challenger*. The minds, eyes, ears and hands of Thomson and his shipmates were just as good as anything we have today.

Acknowledgements

I am grateful to the late Professor Henry Charnock FRS for his valuable comments on an early manuscript and for convincing me that my initial assessment of the technology of the *Challenger* Expedition was far too critical. I am also grateful to Mike Conquer for his excellent work, as ever, in producing the illustrations.

Notes

1 Consequently, the ship's name has been used in a number of contexts ranging from the intrepid Professor George E. Challenger in Conan Doyle's *The lost world*, through Royal Naval hydrographic and specialist vessels, NERC and US research/survey ships to the ill-fated US space shuttle. But it might have been quite different. Two more or less equally suitable vessels were apparently available at the time, HMS *Challenger* and HMS *Clio* (see Burstyn 1972). Had the *Clio* been chosen, it is questionable whether this rather less macho name would have been quite so attractive for at least some of these applications.

2 The ship's normal complement, as built, was 290 (see Rice 1986). The vessel's ledgers for the expedition (Public Record Office, ADM 117) are somewhat confusing, but they suggest that she had a complement of only 21 officers, 136 petty officers and seamen, 21 marines of all ranks (that is a total of 178, though she seems to have sailed with 175). In addition, she carried a supernumerary list of 50 boys first class, six scientists, a laboratory assistant (see note 4) and a domestic servant for Thomson, making a total of 233. Since there were no midshipmen, there was no requirement for a separate gun room to accommodate them, the boys presumably messing along with the seamen.

3 The hold remained largely unaltered except that the magazines, no longer needed for powder, were used instead to store rope and alcohol for sample preservation. Then, as now, the presence of a large quantity of alcohol aboard was considered a potential temptation. Thomson (1880: 22) explains how they avoided any problem. 'The ammunition was removed from the fore-magazine of the *Challenger*, and about a hundred cylindrical iron vessels, each containing four gallons of 85 per cent. alcohol, were stowed in the racks. A cistern, holding about thirty gallons, was fitted into the nettings immediately above the work-room; a pipe led the spirit down to a tap in the work-room, and the spirit, which was under the charge of the gunner's mate, was handed up in the cylinders when required and emptied into the tank, which was, of course, kept carefully locked. The key of the spirit tap in the work-room I kept on my own bunch, and I may add, that under this

simple arrangement we never had the least difficulty about spirit, although several thousand gallons passed in this way through the work-room during the voyage.'

4 Thomson was supported by three naturalists (John Murray, Henry Nottidge Moseley and Rudolf von Willemoes Suhm), a chemist (John Young Buchanan) and an artist and secretary to Thomson (J. J. Wild). Finally, a 17-year-old, Frederick Pearcey, was employed as 'A ward room officers' servant to be borne as supernumerary for duty in the chemical laboratory and the naturalists' workroom'. Compared with the annual salaries of £1000 to Thomson, £400 to Wild and £200 to the more junior scientists (see Rice 1989), Pearcey's pay was not generous. He was initially paid 7d (*c.* 3p) a day, the same as the boys, but was re-rated as a 'Domestic 3rd class' before the ship sailed, earning 1/1d a day (compared with 1/3d a day for ordinary seamen and marine privates). Finally, from July 1874 he was classed as a 'writer 3rd class' on 2/- a day.

5 Material was sent back from Bermuda, Halifax, the Cape, Sydney, Hong Kong and Japan. Thomson (1880: 23) writes: 'after the contents of the ship had been finally cleared out at Sheerness, we found, on mustering our stores, that they consisted of 563 cases, containing 2270 large glass jars with specimens in spirit of wine, 1749 smaller stoppered bottles, 1860 glass tubes, and 176 tin cases, all with specimens in spirit; 180 tin cases with dried specimens; and 22 casks with specimens in brine'. He adds that 'of upwards of 5000 bottles and jars of different sizes sent from all parts of the world to be stored in Edinburgh, only about four were broken, and no specimens were lost from the spirit giving way'. However, I know from personal experience as a Curator at the British Museum (Natural History) that at least some of the samples received erroneous locality labels at some point in their history.

6 Of the total of some 30,000 pages of the reports, no fewer than 18,668 were devoted to the accounts of the marine biological collections. More than 4500 metazoan species previously unknown to science were described, an average of about 15 species from each of the official stations.

7 The biological results received particular praise, though H. N. Moseley (1892: 508) claimed that 'the zoological results of the deep-sea dredgings were rather disappointing', since the hopes of many naturalists 'of finding almost all important fossil forms existing in life and vigour at great depths' were unfulfilled.

8 Although George Strong Nares (1831–1915) had served in steamships as early as 1856, his basic naval training had been in sail and he had spent seven years training young men destined to become naval officers. During this time he wrote *The naval cadet's guide* (1860) (known simply as *Seamanship* from the second edition), which became the standard naval text on all aspects of rope and sail work until well into the present century (see Deacon and Savours 1976).

9 Although these improvements were significant, they did not change the nature of the sampling gears fundamentally. They collected mainly the larger animals living on or just within the sea-floor sediments and, as on the *Challenger* Expedition, for a further eighty years gave the impression that the deep-sea floor in any locality is populated by rather small numbers of individuals belonging to just a few species. It was not until the development by American scientists of gears to take large samples of the bottom sediments in the 1960s and 1970s (see Gage and Tyler 1991) that the abundance and enormous variety of the smaller deep-sea organisms began to become apparent.

10 Amazingly, the Baillie rod and sinkers were still in use in the Royal Navy in the 1950s, Admiral Steve Ritchie using this gear to obtain a sediment sample from the bottom of the Marianas Trench in HMS *Challenger* in 1951 (Ritchie 1992).

11 Tizard et al. (1882: 72) point out that a tapered wire may be (and indeed is) necessary for dredging because of the weight of the wire itself. 'It is conceivable that a sea might be so deep that it would be impossible to reach its bottom with a line of any known material.'

12 Tait (1882: 7) was extremely scathing about the *Challenger* thermometers which, he wrote, 'cannot be said to do more than furnish rough and ready means of approximating to temperatures within about a quarter of a degree', though 'Had they been more nearly what would be called "scientific" instruments, they might have altogether failed on account of the rough treatment to which they were necessarily subjected during use'.

13 Tait concluded that the ebonite or vulcanite slabs on which the thermometers were mounted were unexpectedly susceptible to heating up under compression and were probably responsible for most of the errors in the original pressure corrections.

14 The report on the temperature observations (Murray 1884) consists of 263 temperature sections prepared by T. H. Tizard. The first 132 sections are based on 'raw data' as recorded on the ship

without corrections of any kind. From these data, curves of temperature against depth were drawn by hand, correcting for obvious gross errors of reading or recording. Finally, a new set of temperatures against depth were read from the curves, and it was these 'temperatures from curves' that were used by Buchan. From section 123 (10 February 1875) onwards, the original temperature readings were adjusted using Tait's generally very small corrections before the curves were drawn. Tait's corrections made no difference to the 'curve temperatures' that were again produced.

15 The STD (salinity, temperature, depth) and later CTD (conductivity, temperature, depth) became the main workhorses of physical oceanography. Lowered vertically on conducting cables, they continuously measure these physical parameters with great accuracy. The data are transmitted to the ship along the conducting cable to provide very detailed measurements.

16 The *Challenger* had been supplied with 'a waterbottle of peculiar and ingenious construction, used by Jacobsen in the German North Sea Expedition in the *Pomerania* in 1872 ... but was unfortunately mislaid at the fitting out, and notwithstanding repeated searches was not found till the ship returned' (Tizard et al. 1885: 117, McConnell 1982). Buchanan's bottle replaced the lost one.

References

Agassiz, A. (1888) *A contribution to American thalassography. Three cruises of the USCGS Steamer* Blake *in the Gulf of Mexico, in the Caribbean Sea, and along the Atlantic Coast of the United States from 1877–1880*, 2 vols. Boston and New York: Houghton Mifflin.

Buchan, A. (1895) 'Report on oceanic circulation', *Report on the scientific results of the voyage of HMS* Challenger *during the years 1872–1876*, Summary **2**, appendix. London: HMSO.

Buchanan, J. Y. (1884) 'Report on the specific gravity of samples of ocean-water, observed on board HMS *Challenger*, during the years 1873–1876', *Report on the scientific results of the voyage of HMS* Challenger *during the years 1873–76*, Physics and Chemistry **1**, section 2. London: HMSO.

Burstyn, H. L. (1968) 'Science and government in the nineteenth century: the *Challenger* Expedition and its Report', *Bulletin de l'Institut Océanographique, Monaco*, Special number **2**(2): 603–13.

Burstyn, H. L. (1972) 'Pioneering in large-scale scientific organisation: the *Challenger* Expedition and its Report. I. Launching the Expedition', *Proceedings of the Royal Society of Edinburgh* **B 72**: 47–61.

Deacon, G. E. R. (1968) 'Early scientific studies of the Antarctic Ocean', *Bulletin de l'Institut Océanographique, Monaco*, Special number **2**(1): 269–79.

Deacon, M. B. (1971) *Scientists and the sea, 1650–1900; a study of marine science*. London: Academic Press.

Deacon, M. B. and Savours, A. (1976) 'Sir George Strong Nares (1831–1915)', *Polar Record* **18**(113): 127–41.

Filhol, H. (1885) *La vie au fond des mers*. Paris: Masson.

Gage, J. D. and Tyler, P. A. (1991) *Deep-sea biology; a natural history of organisms at the deep-sea floor.* Cambridge: Cambridge University Press.

Jones, R. W. (1990) 'The *Challenger* Expedition (1872–1876), Henry Bowman Brady (1835–1891) and the *Challenger* Foraminifera', *Bulletin of the British Museum Natural History* (hist. ser.) **18**(2): 115–43.

McConnell, A. (1982) *No sea too deep: the history of oceanographic instruments*. Bristol: Adam Hilger.

Merz, A. and Wüst, G. (1922) 'Die Atlantische Vertikalzirkulation', *Zeitschrift der Gesellschaft für Erdkunde zu Berlin* **1/2**: 1–35.

Mills, E. L. (1989) *Biological oceanography, an early history, 1870–1960*. Ithaca, New York: Cornell University Press.

Moseley, H. N. (1892) *Notes by a naturalist: an account of observations made during the voyage of HMS* Challenger *round the world in the years 1872–1876*. London: John Murray.

Murray, J. (1884) 'Report on the deep-sea temperature observations of ocean-water, taken by the officers of the Expedition, during the years 1873–1876', *Report on the scientific results of the voyage of HMS* Challenger *during the years 1873–76*, Physics and Chemistry **1**, section 3. London: HMSO.

Murray, J. and Hjort J. (1912) *The depths of the ocean*. London: Macmillan.

Rehbock, P. F. (1992) *At sea with the scientifics. The* Challenger *letters of Joseph Matkin*. Honolulu: University of Hawaii Press.

Rice, A. L. (1986) *British oceanographic vessels, 1800–1950*. London: Ray Society.

Rice, A. L. (1989) 'Oceanographic fame and fortune; the salaries of the sailors and scientists on HMS *Challenger*', *Archives of Natural History* **16**(2): 213–20.

Rice, A. L. (1992) 'The politics of scientific moves; the transfer of the British Museum scientific collections from Bloomsbury to South Kensington', *Ocean Challenge* **3**(2/3): 48–52.

Ritchie, G. S. (1992) *No day too long – an hydrographer's tale*. Edinburgh: Pentland Press.

Sigsbee, C. D. (1880) *Deep-sea sounding and dredging*. Washington, D.C.: US Government Printing Office.

Tait, P. G. (1882) 'The pressure errors of the *Challenger* thermometers', *Report on the scientific results of the voyage of HMS* Challenger *during the years 1873–76*, Narrative **2**, Appendix A.

Thomson, C. W. (1873) *The depths of the sea*. London: Macmillan.

Thomson, C. W. (1880) 'General introduction to the zoological series of reports', *Report on the scientific results of the voyage of HMS* Challenger *during the years 1873–76*, Zoology **1**: 1–59. London: HMSO.

Tizard, T. H., Moseley, H. N., Buchanan, J. Y. and Murray, J. (1885) 'Narrative of the cruise of HMS *Challenger* with a general account of the scientific results of the Expedition', *Report on the scientific results of the voyage of HMS* Challenger *during the years 1873–76*, Narrative **1**(1). London: HMSO.

Warren, B. A. (1981) 'Deep circulation of the world ocean', in B. A. Warren and C. Wunsch (eds) *Evolution of physical oceanography*. Cambridge, Mass.: MIT Press, pp. 6–41.

2 'Big science' in Victorian Britain

The *Challenger* Expedition (1872–6) and its *Report* (1881–95)

Harold L. Burstyn

'Big science' and the *Challenger* Expedition

In 1964, not long after I began teaching at university, I met the late Jerome Wiesner. Dr Wiesner had just left the White House, where he had been Science Adviser to Presidents Kennedy and Johnson, to return to the Massachusetts Institute of Technology.[1] 'Tell me', Dr Wiesner said to me when he found out that I was a historian of oceanography, 'why, when they have already collected so much data, do oceanographers keep needing to go to sea to collect more and more?'

What bothered Wiesner, from his vantage point of advising the President on the federal budget for science, was the enormous expense of seagoing oceanography. Why couldn't oceanographers spend less time at sea and more time working up the data they and their predecessors had already collected? Oceanography, in other words, had become another big science, demanding ever larger appropriations for the newest and best ships with which to study the seas and for the people to sail in them.

The oceanographers' answer, I am sure, was the same as that of the high-energy physicists or the space scientists: the only investigations that can yield the ultimate truths of science are the kind done at sea. Theoreticians might predict. They might verify their predictions as well as they could from data gathered long before. But the ultimate test of their results comes only from the ocean itself, as explored by those who sail on its ever-changing surface or investigate its hidden depths.

Wiesner should not have been surprised to find oceanography among the so-called big sciences. As I have argued elsewhere, oceanography, virtually alone among the sciences, has always been either big science or no science at all (Burstyn 1968a, 1968b). What are the characteristics of big science, as we now understand them? Am I correct that oceanography is the first of the big sciences? And does the historiography of science recognize my claim that big science originates in the nineteenth century with the *Challenger* Expedition and its fifty-volume *Report*?

In 1961, not long before Wiesner and I met, Alvin Weinberg, director of the Oak Ridge National Laboratory, coined the term 'big science'. Defining it as 'a supreme outward expression of our culture's aspirations', comparable to Egypt's pyramids, the Middle Ages' cathedrals, or Louis XIV's Versailles, Weinberg felt that big science was inevitable. But he had three concerns. The demands of public accountability in a democracy were turning science into journalism, where 'the spectacular rather than the perceptive becomes the scientific standard'. The abundance of resources (in 1961!) was pushing scientists into 'spending money instead of thought'. And 'the huge growth of Big Science ... greatly increased the number of scientific administrators'. 'The Indians with bellies to the bench are hard to discern for all the chiefs with bellies to the mahogany desks' (Weinberg 1961).

Though historians of science are probing the history of big science, their accounts so far begin in the mid-twentieth century. They concentrate on particle accelerators and the other instruments of high-energy physics (Galison and Hevly 1992), with some attention to such astronomical projects as the Hubble space telescope (Smith et al. 1989). They seem to believe that only with World War II did scientists learn to work in groups (ibid.). Yet these studies, for all their foreshortening of the historical record, have value. They suggest a set of characteristics of big science, characteristics against which we can measure the *Challenger* Expedition, thereby broadening our understanding of big science. For, just as extrapolating from a single point is unlikely to give us a curve that represents reality, so our perception of historical reality is sharpened the longer the historical record we study.

The *Challenger* Expedition and its *Report*

Before I discuss what the historians of big science have to tell us, let me support my claim that the *Challenger* Expedition and its *Report* are indeed worthy of the name of big science. By launching in 1872 the voyage round the world of HMS *Challenger* in the cause of pure science, Great Britain, the world's leading maritime power, made a major commitment of government support for basic research. The cruise of 41 months, almost three and a half years, covered 69,000 nautical miles in its circumnavigation of the globe. In 710 days at sea the ship burned 4600 tons of coal, despite the use of as much sail as possible. The civilian scientific staff under C. Wyville Thomson, Regius Professor of Natural History at the University of Edinburgh, assisted by the ship's company, sounded the ocean bottom 370 times and took 255 serial temperatures. From the hauls of trawl and dredge at 240 stations came 600 cases of specimens animal, vegetable and mineral (see also Chapter 1).

Describing the collected specimens, working up the data and publishing the results took twenty more years. Wyville Thomson and John Murray became the entire scientific staff after the return of the expedition; their shipmates dispersed. Together with Alexander Agassiz, who came from America to Edinburgh especially for the purpose, Thomson and Murray sorted the specimens into groups and arranged their distribution to the experts who were to describe and illustrate them. Organizing the *Challenger* Expedition collections took place first in houses belonging to the university and later at 32 Queen Street, a private house rented as the Challenger Office. Thomson was not a good administrator, and the Treasury with its rigid rules did not make his task easy. When, early in 1882, Thomson took to his bed, the five-year government grant had nearly run out with almost nothing of the *Report* in print. Thomson died just before the financial year ended with the Treasury still dithering about renewing the grant. His death forced favourable action. John Murray was appointed to finish the project along the lines laid down by Thomson.

Murray began with great vigour to distribute the specimens still in Edinburgh and to prod the always procrastinating scientists to complete their tasks. As the preparation of reports by others came firmly under his control, Murray turned to his own part in writing up the results of the expedition. On joining the expedition staff in 1872, John Murray had quickly taken over the geological research, since none of the other natural historians claimed any expert knowledge of the sea floor. The Abbé Alphonse Renard, in a series of visits to Edinburgh, had begun the sedimentary studies that he and Murray were to share. Murray himself worked with rock specimens, collected from islands lying far off the continents, that might help fill out the scientific picture of the globe. Unfortunately, Murray had only a few isolated hand specimens. Their lithology and mineralogy could be described, but not their stratigraphic relations. By combining his laboratory identification of these rocks with the general descriptions of the islands, Murray hoped to be able to characterize them geologically.

The problem that preoccupied him was the origin of coral reefs. The first substantial contribution to the theory of their origin came from Charles Darwin's observations aboard HMS *Beagle*. James Dwight Dana had made further observations on the Wilkes Expedition. Sceptical of some of Darwin's ideas, John Murray began his own efforts to explain coral reefs by systematically setting out all the information he could find about them. Murray's study of coral reefs led him to found the Christmas Island phosphate industry, whose revenue more than repaid the British Government for the *Challenger* Expedition, a clear example of science paying off (Burstyn 1975).

No doubt the prospects for phosphate mining, clear to Murray by the beginning of 1888, had a great deal to do with the last stage of his long and only narrowly successful fight with a parsimonious Treasury over the subsidies for the *Challenger Report*. Murray took a firm stand against the government in 1889 over the still-incomplete *Report*, forcing £1600 for its completion out of a reluctant Treasury by the threat of paying it out of his own pocket. So, unlike the woefully incomplete record of the Wilkes Expedition, the *Challenger* Expedition has left us as complete an account of its researches as could be put together. This vast addition to the literature established oceanography as an independent scientific discipline.

Moreover, the resources of the British Government made possible not only the publication of the expedition's results in ample form, but also their wide dissemination. Whole sets were distributed without charge to scientific institutions all over the world. From 1872, when the cruise began, until 1899, when the last of the free copies of the *Report* were distributed, the British Government spent approximately £171,000 ($855,000) at current prices for this single research project, one of the largest in history prior to World War II. The fifty royal quarto volumes published over sixteen years may make the *Report on the scientific results of the voyage of HMS* Challenger *during the years 1872–76* the major single research project of all time, with their nearly 30,000 pages of letterpress and more than 3000 lithographic plates, 200 maps, and copious woodcuts. For more than two decades, from its announcement at the British Association meeting of 1871 to the banquet held to commemorate the publication of the final volume in 1895, the *Challenger* Expedition and its *Report* occupied the entire lives of some dozen men and substantial portions of the lives of many hundreds more.

The *Challenger* Expedition, unique in its own century and scarcely duplicated in magnitude in ours, stands as a beacon on the route to what science has become. My claim, therefore, is that the *Challenger* Expedition points the way to the new age of big science.

The *Challenger* as a precursor of big science: the argument in favour

Let me turn now to the characteristics of big science as the students of our massive recent projects delineate it. 'Big science is characterized by large multidisciplinary teams, a division of labour, team commitments, agreements and negotiations on common purposes, and hierarchical organization' (Smith et al. 1989). By the time the *Challenger Report* was finished in 1895, a large team had certainly participated. Though each member of this team had made a commitment, only constant nagging by John Murray brought the commitments to fulfilment. And the scientific staff on the vessel itself was quite small: only five, of whom one died at sea. Modern sociological research suggests that the scientific party was an optimal size (Payne 1990).

Big science is 'the type of science usually associated with large multidisciplinary teams and big machines' (Smith 1992, in Galison and Hevly 1992). Though the *Challenger* Expedition certainly required such a team, it lacked any connection to what we mean by a big machine. Nor was there any of the 'fascination with hardware . . . to spread from accelerator laboratories to other parts of physics and from the United States to the rest of the world, and to make science and technology one' (Seidel 1992, in Galison and Hevly 1992). The *Challenger* Expedition did not require building a piece of machinery unlike any ever built before. Unlike a particle accelerator or a space telescope, HMS *Challenger* was an ordinary surveying vessel of the British Royal Navy. Her basic equipment was not very different from what such a ship would use for a hydrographic survey of the ocean depths. In addition to the standard surveying equipment, *Challenger* carried the dredges and thermometers that scientists had found useful for the summer-long so-called dredging cruises that preceded her circumnavigation. Therefore the technology deployed in the *Challenger* Expedition did not require any substantial inventiveness. Indeed, as Tony Rice points out in Chapter 1, in many ways *Challenger*'s technology was backward, or at least not cutting-edge.

Another characteristic of big science is big money. Without big money, in fact, there can be no big science. The handful of scientific men who reawakened interest in marine exploration about 1870 did not understand how much money they would spend over the next quarter century. They regarded the deep sea as the last unexplored zoological province, to be conquered for science by the brief voyages they promoted in the United States, Germany and Scandinavia, as well as Great Britain. These scientists, as we now call them, anticipated a series of short cruises whose results could be published in the regular scientific literature. (This literature of course included occasional sets of a few volumes from a major expedition as well as periodicals.) Only in Britain, the world's leading nation-state, could these ambitions be channelled into a circumnavigation. British power was rooted in maritime commerce and the naval force that served it, and Britain was willing to put its navy at the disposal of scientific research. From the point of view of the men of science, it was purely fortuitous that all the forces working for scientific discovery could be orchestrated in Britain into a single oceanic expedition. To these men the *Challenger* Expedition represented merely the chance to continue doing – on a larger scale, to be sure – more of what they had been doing, a chance to take a bigger step in the same direction. The duration of the *Challenger*'s voyage was set by the length of a surveying vessel's typical commission in the Royal Navy, rather than by the scientists' estimate of the time needed to explore the world's oceans. And no-one in his wildest dreams envisaged a report of fifty volumes. Yet the richest country in the world was prepared to finance the expedition in an avowed attempt to dominate the science of the sea as she already dominated the sea itself. Though those who served in Her Majesty's Treasury might grumble and groan, in the end they paid the bills for this bold scientific venture.

Students of big science suggest further that it demands vertical integration. A hierarchical team runs each facility (Kargon et al. 1992, in Galison and Hevly 1992). But the *Challenger* Expedition had no central facility once the scientific men left the ship, and the Challenger Office in Edinburgh was more like the editorial office of a journal than a modern observatory or high-energy physics laboratory. Hierarchy there certainly was on board the *Challenger*, as on any naval vessel. But once the rivalries were settled of who among the widely dispersed scientific men would do what, once the collections and data came into the hands of those who worked them up, there could no longer be any hierarchy.

Another feature of big science is the choice of project leader. Such a choice is crucial, because only a leader who commands respect in the community from which the project's scientists will come can get the best and the brightest to join up. Science is a 'reputational system of coordinating and controlling research', a 'subset of the more general craft mode of work administration'. Though it 'shares some key features of professional systems of work control', the 'high level of task uncertainty', which results from the continual production of novelty by scientific research, means that science cannot be truly bureaucratic (Whitley 1984). Because the results of research are often unexpected, scientists have to be much more open to new ideas than bureaucrats do. And among scientists, one's reputation is all.

Thus, for the *Challenger* Expedition as for the Hubble space telescope, the choice of a particular leader was crucial. And a study of five large-scale projects funded by the International Decade of Ocean Exploration came to a similar conclusion: once leaders of a project were chosen, whether or not the project succeeded depended on how persistent those leaders were in staying involved (Mazur and Boyko 1981). Wyville Thomson was as ideal a leader for the task at sea as was John Murray for the task ashore.

Thomson was an ideal academic leader. Though ambitious enough to learn about the creatures of the sea and to communicate his knowledge to his students and the general public, he had the tact and charm necessary to lead the scientific party of a seagoing scientific expedition that lasted three-and-a-half years. Yet this very amiability, this willingness to defer sufficiently to be congenial, was his downfall when he came up against the Treasury's inflexible bureaucracy. Here the qualities needed were John Murray's grit and determination, which enabled him to bring to successful completion the project that Thomson had begun.

Because their coming into being requires legislative appropriations, the high cost of big-science projects can bring them down. One parallel between, for example, the Hubble space telescope and the *Challenger* Expedition was that each saw the shifting of costs into the future to keep down the size of this year's appropriation. That is, by shifting costs from the current year to a few years ahead, the project leaders effectively hide the size of the real commitment. Had the Treasury known in 1872, or even 1876,[2] how much the total cost would be, they would never have sanctioned publication of the *Challenger Report* on such a grand scale. So shifting the costs of the expedition into the publication period, though not planned, succeeded in defeating bureaucratic penny-pinching.

Differences from comparable twentieth-century projects

The essence of big science is the building of coalitions that make a large-scale project possible. These coalitions bring together scientists, politicians, bureaucrats and industrialists. The industrialists are especially important in the building of the new machines of our time. Yet, though their interests in submarine cables lay behind the *Challenger* Expedition, British industrialists were virtually invisible in launching it. The key to launching the *Challenger*'s circumnavigation was the coalition among scientists and politicians.

The dredging cruises began in 1868 at the instigation of the Royal Society. They received strong support in the highest quarters of the Admiralty. This support encouraged the boldest among the dredging scientists, William B. Carpenter, to plan on a larger scale. Carpenter and the Hydrographer, Captain George H. Richards, were both members of the Royal Society's Council. Carpenter learned from Richards that the cruise of a Royal Navy vessel lasted about three years from commissioning to paying off. In that time the whole world could be circumnavigated and the depths of all its oceans explored. For Captain Richards a circumnavigation could provide the soundings necessary for a network of submarine cables to link the far-flung British colonies to the seat of empire in London. For Carpenter it could provide general knowledge of the physics and zoology of the deep sea. A broad survey of all the ocean regions would show which ones were of greatest interest for further detailed exploration.

Carpenter, formulating plans with Richards' encouragement, was well placed to press them upon the reform-minded Liberal Government. Through their common membership in the Metaphysical Society, that 'high-water mark of the century's intellectual life' (Webb 1970), in whose meetings the ultimate issues of faith and truth were debated, Carpenter was acquainted with the Prime Minister, William Ewart Gladstone, and his Chancellor of the Exchequer, Robert Lowe. Carpenter found both Lowe and his cabinet colleague, the First Lord of the Admiralty, H. C. E. Childers, entirely receptive, as he wrote General Edward Sabine, President of the Royal Society (Burstyn 1972). Before the *Challenger* set sail, both Lowe and George J. Goschen, Childers' successor as First Lord of the Admiralty, had been elected Fellows of the Royal Society, an honour that came to a few distinguished politicians from time to time.

A democratic government requires public consent before it can promote the progress of science. In a twentieth-century democracy, the general public has a substantial role, which is why Weinberg worried that big science forces scientists to behave like journalists to promote their pet projects. In the late nineteenth century, however – and especially in Britain with its strong class divisions – scientific and political élites could make their decisions with almost no public attention. Dependence on public opinion was only a polite fiction in that 'Golden Age of ministerial administration' (Willson 1955). So, though even the editors of *Punch*, with their eyes on their many readers, lost interest once the *Challenger* set sail (Burstyn 1972), the lack of public interest was no barrier to completing the expedition and its *Report*.

Thus one vital facet of big science as we know it today was missing from the *Challenger* Expedition. The fact that the money must be publicly accounted for affects the kinds of projects that we undertake and their scale. Machines may be modified or their construction stretched out, as with the Hubble space telescope. Or they may even be abandoned, as with the Superconducting Super Collider. The *Challenger* Expedition had no such problems. The kind of science that the *Challenger* scientists did was controlled entirely by the scientists themselves. Her Majesty's Government was under neither political nor financial pressure to influence what the scientists did. Hence the fact that the money was public rather than private had no perceptible effect on the kind of science done.

I don't mean to say that accounting procedures weren't niggardly. Or that the Treasury couldn't have treated Sir Wyville Thomson better and perhaps prolonged his life. Rather, I suggest that the niggardly way in which public expenditure was managed then, and perhaps now, had no serious effect on the *Challenger*'s science, because that science was wholly in the hands of scientists to whose words the government ultimately listened. The scientists' responsibility for their product was complete. The bureaucratic straitjacket in which they had to operate did not limit their scientific endeavours. What the scientists got from the British Government was every opportunity to do what they felt should be done in the name of science. Surely no-one could have asked for more. Would that scientists were so free in our own time!

I believe I have now answered the questions I raised at the beginning. I have sketched out what scholars see as the characteristics of big science: big machines, big money, large multi-disciplinary teams, a division of labour, team commitments, the building of coalitions, strong leadership, hierarchical organization, and the making of budgetary compromises with an eye on public opinion. Except for the big machines, the *Challenger* Expedition had each of these characteristics to some degree. The smaller role of public opinion in Victorian Britain compared to today should not blind us to its place in the rhetoric of politicians. As Carpenter wrote to Sabine, the politicians would do whatever the scientists wanted 'if they think that the *Country* will support them' (Burstyn 1972). And, if you accept my claim that the

Challenger Report is the foundation stone of oceanographic science, the volumes to which each of us turns, as we turn to Newton's *Principia*, to find the first statement of the problem whose solution we seek, then it follows that oceanography is the first of the big sciences.

My final question remains: does the historiography of science recognize my claim that big science originates in the nineteenth century with the *Challenger* Expedition and its fifty-volume *Report?* Here I can only repeat the suggestions I made to historians of science in 1966, suggestions whose fruits Eric Mills has recently charted in authoritative detail (Mills 1993). If oceanography is indeed the first of the big sciences, it demands more attention from historians of science than they have so far given it.

Notes

1 Where he was successively Dean of Science (1964–6), Provost (1966–71), and President 1971–80).
2 Editorial note: Indeed, the various budgets submitted by John Murray during the 1880s always underestimated the ultimate time and cost of the work, though there is no firm evidence that this was done deliberately.

References

Allen, D. E. (1976) *The naturalist in Britain. A social history*. London: Allen Lane.

Burstyn, H. L. (1968a) 'The historian of science and oceanography', *Bulletin de l'Institut Océanographique, Monaco*, Special number **2**(2): 665–72.

Burstyn, H. L. (1968b) 'Science and government in the nineteenth century: the *Challenger* Expedition and its Report', *Bulletin de l'Institut Océanographique, Monaco*, Special number **2**(2): 603–11.

Burstyn, H. L. (1972) 'Pioneering in large-scale scientific organisation: the *Challenger* expedition and its Report. I. Launching the expedition, *Proceedings of the Royal Society of Edinburgh* (B) **72**: 47–61.

Burstyn, H. L. (1975) 'Science pays off: Sir John Murray and the Christmas Island phosphate industry, 1886–1914', *Social Studies of Science* **5**: 5–34.

Galison, P. and Hevly, B. (eds) (1992) *Big science. The growth of large-scale research*. Stanford: Stanford University Press.

Kargon, R. H., Leslie, S. W. and Schoenberger, E. (1992) 'Far beyond big science: science regions and the organization of research and development', in P. Galison and B. Hevly (eds) *Big science*. Stanford: Stanford University Press, pp. 334–54.

Mazur, A. and Boyko, E. (1981) 'Large-scale ocean research projects: what makes them succeed or fail?', *Social Studies of Science* **11**: 425–9.

Mills, E. L. (1993) 'The historian of science and oceanography after twenty years', *Earth Sciences History* **12**: 5–18.

Payne, R. (1990) 'The effectiveness of research teams: a review', in M. A. West and J. L. Farr (eds) *Innovation and creativity at work. Psychological and organizational strategies*. Chichester: Wiley, pp. 101–22.

Price, D. J. DeS. (1986) *Little science, big science . . . and beyond*. New York: Columbia University Press.

Seidel, R. (1992) 'The origins of the Lawrence Berkeley Laboratory', in P. Galison and B. Hevly (eds) *Big science*. Stanford: Stanford University Press, pp. 21–45.

Smith, R. W. (1992) 'The biggest kind of big science: astronomers and the space telescope', in P. Galison and B. Hevly (eds) *Big science*. Stanford: Stanford University Press, pp. 184–211.

Smith, R. W. et al. (1989) *The space telescope. A study of NASA, science, technology, and politics*. Cambridge: Cambridge University Press.

Webb, R. K. (1970) *Modern England*. New York: Dodd, Mead.

Weinberg, A. M. (1961) 'Impact of large-scale science on the United States', *Science* **134**: 161–4.

Whitley, R. (1984) *The intellectual and social organization of the sciences*. Oxford: Clarendon Press.

Willson, F. M. G. (1955) 'Ministries and boards: some aspects of administrative development since 1832', *Public Administration* **33**: 43–58.

3 Oceanographic sovereigns
Prince Albert I of Monaco and King Carlos I of Portugal

Jacqueline Carpine-Lancre
(trans. Margaret Deacon)

Rulers have often played the part of a Maecenas, either from a desire to establish their own prestige, both present and future, or for their own enjoyment. Their education, wealth and eminent social position have enabled them to become patrons to artists, musicians and writers. Sometimes they have been attracted to geographical discovery or the sciences. Is it necessary to recall the enterprises of the Portuguese prince, Henry the Navigator, the Empress Josephine's passion for botany, or the contribution of Emperor Hirohito of Japan to marine biology?

At the end of the nineteenth century, after the voyage of HMS *Challenger* and the publication of its results had given a powerful impetus to research into the marine sciences, two monarchs established themselves as true oceanographic sovereigns: Prince Albert I of Monaco, and King Carlos of Portugal (Carpine-Lancre and Saldanha 1992).

Figure 3.1 HMS *Challenger*, at the time of her oceanographic voyage, 1872–6. (From a drawing by Christian Carpine.)

PRINCE ALBERT OF MONACO

The heir of the Grimaldi dynasty, Prince Albert was born in Paris in 1848. He received a classical education at leading schools in Paris and Orleans. At the age of 17, he embarked on the career he had chosen many years earlier, that of a naval officer. His initial training took place in France but continued in the Spanish Navy. The cruises which he later undertook on board his own ships made him an experienced navigator. His travels throughout Europe and North Africa reveal a spirit of enquiry about the world and its inhabitants. He followed the development of science and technology with an enthusiasm that was all the greater because – like some of his contemporaries – he was convinced that their progress would bring mankind not only material benefits but also social justice, intellectual freedom and peace. (Novella 1992, Albert of Monaco 1998, Carpine-Lancre 1998)

It is true that the Prince had no formal university education. However, thanks to his childhood friend, the physiologist Paul Regnard, he met the professors at the Sorbonne, at the Faculty of Medicine and at the Paris Museum of Natural History. The discovery of several prehistoric skeletons in caves near Monaco led to the development of his interest in human palaeontology. Well informed by reading and by his contacts with men of learning, he followed the discussion of topical ideas, in particular Darwin's theories. The origin of life, the evolution of organized beings, natural selection, and the struggle for existence were all themes that occupied his mind.

A decisive event for the future activities of Prince Albert occurred at the beginning of 1884. This was an exhibition at the Paris Museum of the results obtained during the

Figure 3.2 His Serene Highness Prince Albert I of Monaco.

expeditions, made during the previous four years, by the *Travailleur* and *Talisman*, vessels utilized by French scientists with government support, to study the deep waters of the northeastern Atlantic and of the Mediterranean during the 1880s. His study of the exhibition and the encouragement he received from Professor Alphonse Milne-Edwards, who had directed the expeditions, led to the Prince's decision: he would henceforth devote all the time and means at his disposal to the study of oceanography. The following summer he carried out a trial voyage, a prelude to twenty-eight scientific expeditions that he organized and directed almost annually until the outbreak of World War I.

Prince Albert's expeditions

During this time four ships in succession were employed in Prince Albert's oceanographic researches. Until 1888 his expeditions were made on board the *Hirondelle*, a schooner of 200 tons, built at Gosport by Camper and Nicholson. The ship was not particularly well adapted to her new role. As she had no engine, operations carried out at depth were prolonged and difficult.

In 1889 Prince Albert succeeded his father. He placed an order with the shipbuilders Green of Blackwall, near London, for a ship specially designed and equipped for research. This three-masted schooner, equipped with an auxiliary engine, was launched at the beginning of 1891. She was 53m long and had a displacement of 650 tons. Prince Albert called her *Princesse Alice* in honour of his second wife, born Alice Heine.

The ship soon turned out to be inadequate, in both size and power, to carry out the work he had planned. The shipbuilders Laird of Birkenhead, near Liverpool, were therefore charged with constructing a ship incorporating all the improvements suggested by previous experience and by recent technological progress. Launched in the autumn of 1897, she was also named *Princesse Alice*. With a length of 73m and a displacement of 1394 tons, she was an outstanding success. Twelve seasons of work were carried out on board.

Finally, in February 1911, a fourth ship was launched, constructed by Forges et chantiers de la Méditerranée, a shipyard at La Seyne, near Toulon. In her general design and

Figure 3.3 Hirondelle, Prince Albert's first oceanographic vessel. (From a drawing by Christian Carpine.)

Figure 3.4 Prince Albert's third research ship, *Princesse Alice II*. (From a drawing by Christian Carpine.)

arrangement of the laboratories, the *Hirondelle II* closely resembled the second *Princesse Alice*, but exceeded her in size, power and speed.

Prince Albert commanded his ships himself. Apart from a second officer and, from 1902, an officer of the French Navy acting as aide-de-camp to the Prince, the staff consisted of scientists (whose number varied between one and five over the years), a doctor and an artist (Carpine-Lancre 1993). For almost ten years the Prince's principal collaborator was Baron Jules de Guerne. Dr Jules Richard was recruited in 1887, and several years later took charge of the laboratories on board, and of the scientific collections. But Prince Albert himself decided the location and programme of research of his expeditions.

On several occasions these explored the Mediterranean Sea, but more often they had as their destination the temperate or tropical regions of the North Atlantic. Each had its own character, linked to the presence of a specialist on board, inspired by a topical problem or associated with a new research field.

The first three voyages of the *Hirondelle*, between 1885 and 1887, were principally devoted to the study of the currents of the North Atlantic, between the Azores, Newfoundland and Europe. The last mission of the little schooner, in 1888, concentrated on the zoological exploration of the waters around the Azorean archipelago, one of the sectors that would be explored most frequently and with the greatest attention during subsequent voyages (Carpine-Lancre 1992).

The work of the *Princesse Alice* began in 1891. Five years later she discovered, to the south of the Azores, a bank rising to within 50m of the surface, which Prince Albert christened the Princesse Alice Bank (Carpine-Lancre 1996). Its exploration was renewed and completed the following year. Prince Albert was interested in investigating the special features of the Arctic regions and chose Spitsbergen as the objective of the first two seasons' work in the *Princesse Alice II*, in 1898 and 1899.

The 1901 season, spent in the neighbourhood of the Canary Islands, Madeira and the Cape Verde Archipelago, was particularly fruitful. It was there that the southernmost and deepest operations carried out on the Prince's cruises took place. On 6 August, a trawl brought up from a depth of 6035m a sea anemone, an annelid, three ophiurans, a starfish and a fish that received the name of *Grimaldichthys profundissimus*.

Also on the 1901 cruise, two French physiologists, Charles Richet and Paul Portier, studied the nature and effects of the venom of the jellyfish *Physalia* (the Portuguese man-o'-war), a problem suggested to them by the Prince. The initial results obtained on board led them to continue their experiments back in their Paris laboratories. They went on to discover the phenomenon of anaphylaxis, which led to an explanation for many allergic reactions. Richet received the Nobel prize for medicine and physiology for this work in 1913 (Schadewaldt 1965, May 1985).

In 1903 Prince Albert confined his operations to the Bay of Biscay and the coasts of Brittany. He was attempting to discover the causes that had brought about the almost complete disappearance of the sardine off the French coast, an event entailing a disastrous crisis for fishermen and the canning industry.

From 1904 to 1907 marine meteorology dominated the expedition programmes. The Prince experimented with kites and several types of meteorological balloons, some of which reached an altitude of over 16,000m (Carpine 1999).

In 1911 *Hirondelle II* made the first of five cruises in the Atlantic. These expeditions (Richard 1934) were halted by the outbreak of war and the Prince was unable to resume them when hostilities came to an end. He died on 26 June 1922.

The programme of researches carried out by Prince Albert was extremely varied, but two themes were always represented, and often dominated the programme of individual expeditions. One was the collection of organisms from the surface down to depths below 6000m; the other was the study of the environmental conditions in which they lived: temperature, salinity, circulation of water masses and the nature of the sediment.

All categories of fauna interested the Prince, but he attached particular importance to studying the existence and abundance of organisms inhabiting the intermediate layers. This aspect of biological oceanography was giving rise to many hypotheses and much discussion at that time (Mills 1980, 1983).

The creatures obtained were subjected to a preliminary survey at the time of collection. At the end of the cruise they were sorted into their zoological groups and sent to leading specialists, both French and foreign, for a definitive examination.

Apparatus

The success of these expeditions depended upon the apparatus employed. With good reason, the Prince was deeply interested in this question, as he often made clear: 'To accomplish the operations on which oceanography depends, it is necessary to create special equipment, the use of which demands both the experience of the seafarer and an exact knowledge of the questions which are to be studied. It is this aspect with which I have been most particularly concerned during my expeditions' (Albert I of Monaco 1905: 33).

He was fascinated by technological innovations such as the ciné-camera (Carpine-Lancre 1995, 1999), colour photography and aeronautics (Lécuyer 1981). He experimented with new procedures and new materials, applying them whenever he could both to his scientific activities and to the fitting out of his ships. Thus, following the construction of the first *Princesse Alice*, he installed a sea-water distillation unit on board, as well as electricity and refrigerators; the *Hirondelle II* was equipped with radio.

Either on his own or with the help of his colleagues, Prince Albert improved existing apparatus, inventing new instruments and introducing more reliable and effective methods. The trials were often carried out during a short spring voyage in waters close to Monaco. Almost all aspects of work at sea were dealt with. For the study of currents on board the

Hirondelle, the Prince invented several types of float (Carpine 1987). The traditional hemp rope (see Chapter 1) was replaced, first by steel wire and later by a cable, which proved more supple and durable. A sounding machine and several kinds of sounder were made to the Prince's specifications (Carpine 1996). Buchanan's tube sounder was modified to allow longer cores of sediment to be collected. The Prince's engineering adviser, Maurice Léger, produced a grab sounder while Dr Jules Richard invented a water-sampling bottle (Carpine 1993).

For the collection of animals the Prince devised a surface trawl, nets with curtains that could be opened and closed at a distance by messenger – in its simplest form, a metal ring that can be slid down the wire to induce a specific action in the apparatus at the end (Carpine 1998). One of his best known pieces of apparatus was the 'nasse triédrique' or three-cornered trap, regularly used from 1888 onwards to depths of several thousand metres. Jules Richard invented a small plankton net intended for collecting when the ship was under way at normal cruising speed. He also, with Paul Portier, produced a 'box for microbes', an apparatus for bringing up water for bacteriological studies. Wide-opening nets (up to 5m frame dimension) were employed to collect bathypelagic fauna. The first of these, made according to Dr Richard's instructions, was intended for vertical fishing (designed to sample marine fauna directly beneath the ship, that is downwards rather than horizontal), whereas that designed by Henry Bourée was for making horizontal hauls (Carpine 1991).

Cartography

Cartography was another area to which Prince Albert attached great importance. His publications and lectures were often accompanied by charts. Following his experiments on the currents of the North Atlantic, he plotted the probable courses followed by his floats. A preliminary version of this chart was displayed at the Monaco pavilion during the Paris Universal Exhibition of 1889. The finished work was presented to the Academy of Sciences in Paris and then at the annual meeting of the British Association for the Advancement of Science at Edinburgh in 1892.

The itineraries of his expeditions were recorded on successive sheets, with symbols to indicate the operations carried out at each station. The hydrographic and geographical surveys carried out in Spitsbergen, both by the Prince himself and by the expeditions of Isachsen and Bruce, which he supported, also led to the publication of maps.

However, the most important contribution of the Prince in this field was undoubtedly the General Bathymetric Chart of the Oceans. Following a decision taken by the International Geographical Congress at Berlin in 1899, a meeting of experts was held in Wiesbaden four years later. The proposals of Professor Julien Thoulet for the principal characteristics of the chart – scale, projection and bathymetry – were then discussed and adopted. The twenty-four sheets of GEBCO were published in the spring of 1905, all the drawing and printing costs having been borne by Prince Albert (see Chapter 5).

As well as cartography, the studies directed by the Prince and the results obtained were described in many publications designed for the scientific community. After each of his expeditions, Prince Albert presented a summary of the results as a communication to the Academy of Sciences in Paris. His collaborators published their preliminary accounts in specialist journals, followed by full reports that appeared in the series specially created for this purpose: the *Résultats des campagnes scientifiques accomplies sur son yacht par Albert Ier, Prince Souverain de Monaco*.

But the Prince regarded it as essential that the progress of oceanography should be made known to everyone. To this end he gave many lectures, especially to geographical societies, in Paris, London, Laon, Marseilles, Edinburgh and Glasgow. The entire social spectrum was represented among his audiences, from members of the 'popular' universities in Paris to the sovereigns and royal families of Belgium, Italy, Spain, Bavaria and Austria-Hungary. The Prince hoped that by this means he would convince heads of state and their governments of the usefulness of oceanographic research, not just for science but also for national economies.

It was in the same spirit that he displayed specimens collected during his expeditions, the apparatus used on board and publications based on his researches at the various international exhibitions then being organized throughout Europe. During the two universal exhibitions held at Paris in 1889 (Carpine-Lancre 1989) and 1900, half the Monaco pavilion was taken up by a presentation of his oceanographic work. When the second exhibition came to an end the collections were transferred to Monaco, where an oceanographic museum was under construction, its chief purpose being to conserve and display to the public the results of the expeditions. One of the museum's galleries, the hall of applied oceanography, put into effect Prince Albert's desire to convey to as wide an audience as possible the role played by the sea in many aspects of daily life, whether it was a question of food resources, communications, ethnography or art. In addition, laboratory accommodation was planned for the use of visiting oceanographers of all nationalities and specializations (Mills and Carpine-Lancre 1992).

The Musée océanographique was ceremonially opened on 29 March 1910. While still completing its building and equipment, the Prince decided to create an oceanographic institute in Paris, which he hoped would become a 'centre of diffusion' for oceanographic knowledge (Albert I of Monaco 1932a: 335).

Finally, Prince Albert did not dissociate his oceanographic activity from his efforts to promote peace between nations and individuals; he appears to have been a fervent believer in the scientific internationalism developing at that time. Whenever possible, he brought together men of learning of different nationalities on his cruises. He took advantage of both official and scientific visits to see laboratories and institutes in the host country. He became president of two commissions created on the recommendation of the International Geographical Congress held at Geneva in 1908. The intended commission for the Atlantic was a victim of World War I, but the International Commission for the Scientific Exploration of the Mediterranean Sea still functions, under the active presidency of Prince Rainier III.

Through the breadth and continuity of his expeditions, and through his understanding of the problems posed by the study of the marine environment, Prince Albert made an important and enduring contribution to the progress of oceanography. No field of research could have been better suited to someone of his temperament, as it was, in his own words, 'a science well fitted to seduce the imagination by the bond which it forms between poetry, philosophy, and pure science' (Albert I of Monaco 1898: 446).

KING CARLOS OF PORTUGAL

Having long been engaged in zoological studies, and having had since childhood a passion for the sea, in the month of August 1896 I resolved to devote my yacht to scientific research on our coasts, and after some preliminary trials, started work in earnest on 1 September 1896.

This is how Dom Carlos de Bragança, King of Portugal (1902: 11), described the start of his oceanographic work. However, it seems likely that other elements may have influenced his decision.

Dom Carlos was born in Lisbon in 1863. His father, Dom Luiz, had begun his career as an officer in the Portuguese Navy and spent several years at sea before becoming King of Portugal after the premature death of his elder brother. His father's influence undoubtedly contributed to Dom Carlos's youthful enthusiasm for the sea and for yachting.

Intelligent and energetic, he spoke several foreign languages and had a remarkable talent for painting and drawing. As a keen hunter and fisherman, he learned how to observe and identify the creatures he pursued and he produced a fine monograph on the birds of Portugal. The first indication of a more serious interest in marine matters was a study of the estuary of the Tagus, made while he was still Crown Prince.

When the *Princesse Alice* called at Lisbon in 1894 and 1895, Dom Carlos visited Prince Albert's yacht on several occasions and made a thorough inspection of her equipment and the work in progress. It was perhaps the discovery of the Princesse Alice Bank, in Portuguese waters in 1896, that was the deciding factor in his resolve to take up oceanographic research on his own account.

S. A. R. LE DUC DE BRAGANCE
Héritier du trône de Portugal. — Phot. Fillon, à Lisbonne.

Figure 3.5 Dom Carlos de Bragança (King Carlos of Portugal) as a young man.

Dom Carlos outlined the limited but carefully chosen objectives of his research very clearly: 'a methodical and purposeful study of the sea which bathes our coasts' (Carlos de Bragança 1902: 15). The King had neither the time nor the means at his disposal to plan distant or lengthy expeditions. In general he contented himself with cruises lasting no more than a day, but the characteristics of the Portuguese coast meant that considerable depths of water could be easily reached. Above all, the King wished that the exploration of his territorial waters should be carried out by a Portuguese ship and not only by foreign expeditions such as those of the *Porcupine* and the *Talisman* (Saldanha 1980).

Dom Carlos made use of four ships in succession, all named *Amelia* in honour of his wife, the Queen. They were all yachts intended primarily for transporting the royal family; the first two ships were not adapted in any way for the installation of equipment and scientific research. The officers and sailors assisted with the operations at sea, but the King had only one scientific collaborator, Albert Girard.

Between 1896 and 1907 they carried out 290 scientific stations. The soundings were intended to appear on a bathymetric chart prepared by Girard on a scale of 1:100,000, but this was never published. Most of their operations were directed towards the exploration of the fauna of Portuguese waters and the collection of observations that would explain the distribution of species, in order to improve the exploitation of marine living resources (Carlos de Bragança 1904: 9):

> It would be a great service to our fishing industry to publish a catalogue of the fish found in our seas; giving precise indications of their habitat, reproductive seasons, dates of arrival (or migration) and the methods of capture that experience has shown to be most successful. As it is, though the number of species of fish known on our coasts is already great, there is little or no information about their local and bathymetric distribution.

Instruments and working methods constitute one of the themes most frequently touched on in the correspondence exchanged between the two oceanographic sovereigns from 1894 onwards, either directly or through Jules Richard and Albert Girard as intermediaries. Dom Carlos adopted most of the apparatus recommended by Prince Albert, or invented by him, particularly the three-cornered trap, Richard's water-bottle and the Buchanan sounder. However, for the 'palangre', a long line used frequently and with much skill by Portuguese fishermen (Saldanha 1990), he was able in his turn to clarify how it was operated, to give advice on its use and provide hooks which were particularly effective.

To publicize the results he obtained, Dom Carlos followed the example set by Prince Albert. National and international exhibitions held at Lisbon, Oporto, Marseilles and Milan gave him the opportunity to present his work to a wider public. Two of his four oceanographic publications appeared in a series whose title was borrowed from the Monegasque publication *Resultados das investigações scientificas feitas a bordo do yacht* Amelia *e sob a direcção de D. Carlos de Bragança*. His two main works dealt with fish and fisheries.

In one important respect, however, Dom Carlos chose to work differently from Prince Albert; he himself examined and identified the species caught, especially the fish, with Girard as his only helper. In his last volume, dealing with the sharks of the Portuguese coast, he described a new species: *Odontaspis nasutus*.

Dom Carlos and his elder son were assassinated on 1 February 1908. The accumulated collections, the King's manuscripts and his scientific library were preserved in the museum which he had planned and which today forms part of the Vasco da Gama Aquarium in Lisbon.

Figure 3.6 King Carlos's last yacht, *Amelia IV.* (From a drawing by Christian Carpine.)

THE INFLUENCE OF THE *CHALLENGER* EXPEDITION

A story that has often been repeated says that Prince Albert was in Lisbon when the *Challenger* made her first landfall there early in 1873.[1] It seems likely that John Young Buchanan was responsible for this misunderstanding. In a lecture given to the Royal Institution in London in 1903 he described the early voyages of Prince Albert (Buchanan 1905: 358):

> In various very small vessels, he made adventurous cruises in the North Atlantic. It is remarkable that in the course of one of these early cruises he came to be lying in his yacht in the Tagus when the *Challenger* anchored at Lisbon in January 1873, being the first port of call in her long voyage.

However, Prince Albert was in Paris when the *Challenger* put in to Lisbon, and it was not until the following autumn that he acquired the *Hirondelle*.

The first undisputed evidence for contact between members of the scientific staff of the *Challenger* and the Prince dates from 1887 when Jules de Guerne wrote to Henry Moseley and asked for details of apparatus used for capturing creatures at great depths.[2] He later wrote to John Murray to ask how the collections made during the voyage round the world were being studied and preserved and what guidelines were being used in the publication of the results.[3] A request made through diplomatic channels enabled the Prince to obtain the volumes of the *Challenger Report*.[4]

These exchanges became increasingly frequent and informal: the Prince sent copies of his publications to Murray and Buchanan, who sent him their own works in return. The two British scientists were invited to the launch of the *Princesse Alice* on 12 February 1891. On behalf of the Royal Society of Edinburgh, Murray sent an invitation to the Prince to address one of its sessions and arranged the programme for the Prince's visit to Scotland when he came to give the lecture the following summer. The relationship between Prince Albert and Sir John Murray continued to be extremely cordial and was marked by exchanges of publications and letters, visits to Challenger Lodge when the Prince was in Scotland, and to the Palace in Monaco when Murray was on the continent.

The professional contact between the Prince and Buchanan, on the other hand, developed into friendship. Besides science, the two men had many interests in common: photography, bicycles and motor cars, and aeronautics. The Scottish scientist assisted the Prince efficiently and also tactfully on a wide variety of occasions, from the purchase and improvement of scientific instruments and the translation of articles to advice on how to obtain British support for the establishment of meteorological observatories in the Azores.

From 1892 onwards, Buchanan was invited on all the Prince's scientific expeditions. Circumstances allowed him to join only four of these, though several times he took part in the short trial cruises made in spring.

It was Buchanan who was chosen to represent the Royal Society of London at the ceremonies connected with the inauguration of the Musée océanographique at Monaco. When the Prince created his foundation, the Institut océanographique, in 1906 he made Buchanan vice-president and Sir John Murray a member of the Scientific Advisory Board.

Prince Albert never missed an occasion to express the importance that he attached to the *Challenger*'s circumnavigation in the development of modern oceanography. The name of the British ship was engraved on the façade of the Musée océanographique at Monaco (Carpine 1968) and painted in the frieze that decorates the library of the Institut océanographique in Paris. In no fewer than twenty-five of his publications (Albert I of Monaco 1932b) the Prince referred to work carried out in the famous voyage. He wrote: 'When . . . I began my oceanographic researches it was with precarious means and with no other guide than the expeditions of the *Challenger* and the *Talisman* already carried out through the initiative of British and French scientists' (Albert I of Monaco 1904: 162).

As for Dom Carlos, he was too young to have accompanied his father, Dom Luiz, when he visited the *Challenger* during her stay at Lisbon. However, he met Buchanan when the *Princesse Alice* visited Portugal in 1894. The collection of *Challenger Reports* held pride of place in his library and the King often referred to them, both in his publications and in his letters to Prince Albert.

The intellectual and scientific affinities between the oceanographic sovereigns and the scientists of the *Challenger* are indisputable. They help us to understand how the fame of this expedition spread around the world. In a letter to Prince Albert, Buchanan perfectly expressed the spirit of these endeavours:

> I learned . . . that you start on the *campagne* on Wednesday. I wish you every success in your work which is going to take you into some of the most interesting waters in the Atlantic. It is also indelibly connected in my mind with the early days of the voyage of the *Challenger* in the first half of the year 1873. Almost every day some part of a new world revealed itself to my enquiring eyes and mind. And the same process was at work in your mind and practically in the same waters.[5]

Acknowledgements

The author wishes to express her deep gratitude to Margaret Deacon and Anita McConnell for their help in preparing an English version of this chapter.

Notes

1 This statement was repeated in 1910 by Franz Doflein, in 1922 and 1923 by Sir William Herdman, and in 1958 by Mario Ruivo.
2 Letters of Jules de Guerne to Prince Albert I of Monaco, [Paris,] 31 March and 19 April 1887 (Archives du Musée océanographique, Monaco).
3 Letter from John Murray to Jules de Guerne, Edinburgh, 1 May 1888 (Archives du Musée océanographique, Monaco).
4 Copy of a letter from Lord Lytton to the Marquis de Maussabré, Paris, 9 February 1889 (Archives du Musée océanographique, Monaco).
5 Letter of John Young Buchanan to Prince Albert I of Monaco, [London,] 13 July 1914 (Archives du Musée océanographique, Monaco).

References

Albert I of Monaco (1898) 'Oceanography of the North Atlantic', *Geographical Journal* 12: 445–69.

Albert I of Monaco (1904) 'Les progrès de l'océanographie [Conférence à la Sorbonne]', *Revue scientifique (revue rose)* 5th series 1: 161–6.

Albert I of Monaco (1905) 'L'outillage moderne de l'océanographie', *La science au XXe siècle* 3e (26) 33–9.

Albert I of Monaco (1932a) 'Les progrès de l'océanographie [Conférence de Madrid]', *Résultats des campagnes scientifiques accomplies sur son yacht par Albert Ier, Prince Souverain de Monaco* 84: 323–36.

Albert I of Monaco (1932b) 'Recueil des travaux publiés sur ses campagnes scientifiques', *Résultats des campagnes scientifiques accomplies sur son yacht par Albert Ier, Prince Souverain de Monaco* 84: 369—[v].

Albert I of Monaco (1998) *Des œuvres de science, de lumière et de paix*, J. Carpine-Lancre (ed.). Monaco: Palais de S. A. S. le Prince.

Buchanan, J. Y. (1905) 'Historical remarks on some problems and methods of oceanic research', *Proceedings of the Royal Institution of Great Britain* 17: 357–74.

Carlos de Bragança (1902) 'Rapport préliminaire sur les campagnes de 1896 à 1900', *Bulletin des campagnes scientifiques accomplies sur le yacht* Amelia *par D. Carlos de Bragança* 1: 9–40.

Carlos de Bragança (1904) *Resultados das investigações scientificas feitas a bordo do yacht* Amelia *e sob a direcção de D. Carlos de Bragança. Ichthyologia II. Esqualos obtidos nas costas de Portugal durante as campanhas de 1896 a 1903*. Lisbon: Imprensa Nacional.

Carpine, C. (1968) 'Les navires océanographiques dont les noms ont été choisis par S.A.S. le Prince Albert Ier pour figurer sur la façade du Musée océanographique de Monaco', *Bulletin de l'Institut océanographique, Monaco*, special number 2: 627–38.

Carpine, C. (1987) 'Catalogue des appareils d'océanographie en collection au Musée océanographique de Monaco. 1. Photomètres. 2. Mesureurs de courant', *Bulletin de l'Institut océanographique, Monaco* 73: no. 1437.

Carpine, C. (1991) 'Catalogue des appareils d'océanographie en collection au Musée océanographique de Monaco. 3. Appareils de prélèvement biologique', *Bulletin de l'Institut océanographique, Monaco* 74: no. 1438.

Carpine, C. (1993) 'Catalogue des appareils d'océanographie en collection au Musée océanographique de Monaco. 4. Bouteilles de prélèvement d'eau', *Bulletin de l'Institut océanographique, Monaco* 75: no. 1440.

Carpine, C. (1996) 'Catalogue des appareils d'océanographie en collection au Musée océanographique de Monaco. 5. Instruments de sondage', *Bulletin de l'Institut océanographique, Monaco* 75: no. 1441.

Carpine, C. (1997) 'Catalogue des appareils d'océanographie en collection au Musée océanographique de Monaco. 6. Thermomètres', *Bulletin de l'Institut océanographique, Monaco* 76: no. 1442.

Carpine, C. (1998) 'Catalogue des appareils d'océanographie en collection au Musée océanographique de Monaco. 7. Instruments divers, matériel de pont, instruments de laboratoire', *Bulletin de l'Institut océanographique, Monaco* 76: no. 1443.

Carpine, C. (1999) 'Catalogue des appareils d'océanographie en collection au Musée océanographique de Monaco. 8. Suppléments, matériel de démonstration, météorologie. Additions et index cumulatifs', *Bulletin de l'Institut océanographique, Monaco* **76**: no. 1444.

Carpine-Lancre, J. (1989) 'Le Prince Albert de Monaco et l'Exposition universelle de 1889', *Annales monégasques* **13**: 7–42.

Carpine-Lancre, J. (1992) 'L'*Hirondelle* aux Açores', *Açoreana*, suplemento [Centenaire de la dernière campagne océanographique du Prince Albert de Monaco aux Açores à bord de l'*Hirondelle*. Communications, Açores, 1988, L. Saldanha, P. Ré and A. F. Martins (eds)], 22–49.

Carpine-Lancre, J. (1993) 'Le Prince Albert Ier de Monaco marin et océanographe: chronologie sommaire', *Océanis* **19**(4): 121–35.

Carpine-Lancre, J. (1995) 'Prince, océanographe et "cinématographiste": Albert Ier de Monaco', *1895* **18**: 84–95.

Carpine-Lancre, J. (1996) 'Préparatifs, déroulement et bilan de la campagne océanographique de 1896', in *La campagne de la* Princesse-Alice *en 1896*: 6–22. Monaco: Musée océanographique.

Carpine-Lancre, J. (1998) *Albert I Prince of Monaco*. Monaco: Éditions EGC.

Carpine-Lancre, J. (1999) 'Le Prince Albert Ier de Monaco et le cinématographe', in *L'aventure du Cinématographe, actes du Congrès Mondiale Lumière*. Lyon: Aléas, pp. 25–33.

Carpine-Lancre, J. and Saldanha, L. V. C. (eds) (1992) *Dom Carlos I, Roi de Portugal, Albert Ier, Prince de Monaco: souverains océanographes*. Lisbon: Fondation Calouste Gulbenkian.

Doflein, F. (1910) 'Das ozeanographische Museum in Monaco', *Naturwissenschaftlichen Wochenschrift* N.S. **9**: 481–95.

Herdman, W. A. (1922) 'H.S.H. Prince Albert of Monaco', *Nature* **110**: 156–8.

Herdman, W. A. (1923) 'The Prince of Monaco and the Oceanographic Museum', in *Founders of oceanography and their work: an introduction to the science of the sea*. London: Edward Arnold, pp. 119–33.

Lécuyer, R. (1981) 'Monaco, berceau de l'aviation moderne', *Annales monégasques* **5**: 45–86.

May, C. D. (1985) 'The ancestry of allergy: being an account of the original experimental induction of hypersensitivity recognizing the contribution of Paul Portier', *Journal of Allergy and Clinical Immunology* **75**: 485–95.

Mills, E. L. (1980) 'Alexander Agassiz, Carl Chun and the problem of the intermediate fauna', in M. Sears and D. Merriman (eds) *Oceanography: the past*. New York: Springer-Verlag, pp. 360–72.

Mills, E. L. (1983) 'Problems of deep-sea biology: an historical perspective', in G. T. Rowe (ed.) *The sea. 8.: Deep-sea biology*. New York: John Wiley, pp. 1–79.

Mills, E. L. and Carpine-Lancre, J. (1992) 'The Oceanographic Museum of Monaco', in E. M. Borgese (ed.) *Ocean frontiers, explorations by oceanographers on five continents*. New York: H. N. Abrams, pp. 120–35.

Novella, R. (1992) *Seigneurs et princes de Monaco*. Monaco: Éditions arts et couleurs.

Richard, J. (1934) 'Liste générale des stations (1885–1915)', *Résultats des campagnes scientifiques accomplies sur son yacht par Albert Ier, Prince Souverain de Monaco* **89**: 5–348.

Ruivo, M. (1958) *D. Carlos de Bragança, naturalista e oceanógrafo*. Lisbon: Fundaçao da Casa de Bragança.

Saldanha, L. (1980) 'King Carlos of Portugal, a pioneer in European oceanography', in M. Sears and D. Merriman (eds) *Oceanography: the past*. New York: Springer-Verlag, pp. 606–13.

Saldanha, L. (1990) 'The Forbes's azoic theory and the Portuguese zoologists of the 19th century', in W. Lenz and M. Deacon (eds) *Ocean sciences: their history and relation to man, Deutsche hydrographische Zeitschrift*, Ergänzungsheft, Series B **22**: 166–73.

Schadewaldt, H. (1965) 'La croisière du Prince Albert Ier de Monaco en 1901 et la découverte de l'anaphylaxie', *Vie et Milieu*, supplement **19**: 305–13.

4 Expedition to investigation
The work of the Discovery Committee

Rosalind Marsden

The *Challenger* Expedition has been identified as perhaps the world's first example of 'big science' (see Chapter 2). Certainly the £200,000 or so that it cost the British Government seems to have so shocked the Treasury that it would be forty years before an even remotely similar British-inspired venture would be undertaken again. Even then, the officials involved at the beginning of what became the Discovery Investigations had no idea that it would develop into one of the largest research programmes of the first half of the twentieth century.

For the main object of the original *Discovery* Expedition of 1925–7 was seemingly limited: to make the first serious attempt to place the southern whaling industry on a scientific basis. Secondary objectives were oceanographic observations, investigation of the possibility of creating a fishery industry in the Falkland Islands Dependencies, and hydrographic surveys. However, what was planned, like the *Challenger* Expedition, as a single expedition of limited geographical and scientific extent developed into a research programme that extended over many years and the entire Southern Ocean, and greatly increased understanding of both its biology and oceanography. In this chapter I look at how this work came to be undertaken and particularly at the body responsible for its administration, the Discovery Committee. I show how tensions could arise when scientific goals diverged from perceived institutional and economic needs, but that the impact of the latter was not necessarily negative. The Discovery Committee's activities extended over ten to twelve years of preparatory work (from 1913 onwards), followed by fifteen years of active practical oceanography and a further decade of largely chairborne work, nearly forty years in all from the first discussions until the bulk of the reports had been published, and the committee wound up. This period of time greatly exceeded that taken to plan and undertake the *Challenger* Expedition, and publish its report, a mere twenty-four years. This was partly because the work of the Discovery Committee and its forebears was twice brought to a standstill by world war, but also because it evolved into a far more ambitious and wide-ranging project than had been envisaged in the early stages, involving not one but a whole series of related expeditions to the Southern Ocean and adjacent seas. These voyages not only provided much new data on this then little known area, but also enabled valuable contributions to be made to the development of oceanography in general. Yet they attracted little attention at the time, except for reasons unconnected with their scientific work, and have been comparatively neglected by historians both of oceanography and of polar exploration.[1] Institutional rather than scientific reasons provide an explanation of how this came about.

Origins and aims of the Discovery Committee: the early years

It is not generally realized that concern about the southern whaling industry arose early in the twentieth century. Whaling began in South Georgia in 1904 and increased rapidly (Headland 1989). As a result, the Falkland Islands Dependencies came into formal existence in 1908 (Headland 1992). In 1910 Andreas Morch, a Norwegian concerned about the wastefulness of whaling, suggested to Dr Sidney Harmer FRS (1862–1950)[2] that 'a certain proportion of whaling dues should be set aside for scientific research' (Tønnessen and Johnsen 1982). Harmer, worried about the possible extermination of southern whales, passed Morch's ideas to E. R. Darnley (1875–1944)[3] at the Colonial Office in 1911. This resulted in a succession of three Interdepartmental Committees. The first,[4] under the Chairmanship of H. G. Maurice (1874–1950),[5] was to consider 'whether whales need protection owing to the present rate of fishing, and whether, if so, one can find a method by which they can be protected' (Savours 1992). The 1914–18 war stopped this committee's work.

The second committee[6] was convened in April 1918 'to consider what can be done . . . in regard to the preservation of the whaling industry and the development of other industries in the Falkland Islands'.[7] Members included J. O. Borley (1872–1938),[8] Darnley and Harmer, with H. T. Allen (1879–1950) of the Colonial Office as Secretary. Its report (Anon. 1919)[9] recommended tripling the export duty on whale oil from the Dependencies and the employment of two research vessels. As these were estimated to cost a prohibitive £300,000, the Colonial Office dallied.[10] Eventually Winston Churchill, then Secretary of State at the Colonial Office, set up the third Interdepartmental Committee[11] to consider what action was practicable regarding the special vessels recommended by the 1919 report. This committee recommended purchase of the *Discovery* and construction of a new vessel.

The Whaling Research Executive Committee, later to become the Discovery Committee,[12] was set up by the Colonial Office as a result of this proposal.[13] It first met on 19 April 1923,[14] and had seven members: Darnley (Chairman), Sir Sidney Harmer (Vice-Chairman),[15] Borley, Vice-Admiral R. W. Glennie (1868–1930), J. M. Wordie (1889–1962) representing the Royal Geographical Society,[16] Allen as Secretary, and Sir James Fortescue-Flannery Bt (1851–1943), the consulting naval architect of the Crown Agents.[17]

The individual interests and those of the departments represented are worth considering. Darnley and Borley were keen to save the whaling industry, but Darnley and Allen were also concerned about Anglo-Norwegian relations. Harmer was anxious about over-fishing of whales, and stressed the need for a reserve. The Admiralty representative, usually the Hydrographer, and Wordie, a geologist, were much less concerned with whales but heartily backed geographical exploration. The idea of carrying out research in such a remote part of the world was assisted to some degree by government policy concerning the establishment and maintenance of territorial rights in the area, in the Falkland Islands Dependencies in particular, and in the Antarctic generally, but as will be seen these did not necessarily lead to long-term support for science. Despite the stated objectives of the Committee, each individual member to some extent had his own agenda. These differences ultimately resulted in serious conflicts, not only within the Committee, but also between the Committee and the scientists and seafarers in the field.

At the beginning the influence of the *Challenger* Expedition was pervasive. In 1897 Harmer, aged 10 when the *Challenger* sailed, had written an appendix to the *Challenger Report* on *Cephalodiscus dodecalophus* by W. C. M'Intosh (1838–1931) (Calman 1951), while Vice-Admiral

J. D. Nares (1877–1957), an early replacement of Admiral Glennie[18] on the Committee, was a son of Vice-Admiral Sir G. S. Nares (1831–1915), Captain of the *Challenger* during the first part of her voyage. E. R. Gunther (1902–40), who was employed as a zoologist by the Committee, was a grandson of Albert C. L. G. Gunther (1830–1914), a member of the Circumnavigation Committee who had seen off the *Challenger* in 1872[19] (and also a great-nephew of M'Intosh). At a Challenger Society dinner in January 1925,[20] he sat next to Alfred Carpenter (1847–1925), who as a lieutenant had joined the *Challenger* in Hong Kong in January 1875. The members of the 1923 Committee saw themselves planning a one-off expedition, *Challenger* style, with no apparent expectation of an organization that was to last twenty-five years and employ more than 360 people, and also with little understanding of how oceanographic research had altered since the 1870s.[21]

It is therefore hardly surprising that some of the Committee's early decisions, or lack of them, created difficulties for smooth running scientific research many years later. The Committee's first action was to purchase the *Discovery*, which as events were to show was far from an ideal choice. In July 1923 they chose Commander J. R. Stenhouse[22] as Captain. Only after another eight months of advertising, interviewing and inviting people[23] to apply, was Dr Stanley Wells Kemp (1882–1945) appointed as Director of Science in February 1924.[24] Meantime, in September 1923, Stenhouse had advised the Committee that the 'surgeon might do duty as biologist on the scientific staff',[25] showing little understanding of what twentieth-century marine biology involves, and possibly presaging later problems between himself and Kemp.

The *Discovery* sailed south in September 1925 on a two-year commission, on which she was later joined by the Committee's purpose-built ship, the *William Scoresby*. The work of the expedition was centred on the island of South Georgia, and in particular on the whaling station at Grytviken. Some of the zoologists worked at a shore-based laboratory at Grytviken, which enabled them to examine the whales as they were brought into the station. Meanwhile, the two ships carried out a survey of the physical conditions and marine biology of the waters round South Georgia, and southwards to the Antarctic Peninsula. During this time, as in the *Challenger*, all the staff were on short-term contracts and had no pension provision. Once in the South, Kemp and Stenhouse were reminded that 'no compensation would be paid in the event of death from accident or any other cause as officers having dependents to provide for had been recommended to insure themselves against such a contingency'.[26]

When they returned home, Kemp's original wish was to allow the scientists to write up their own work rather than to farm it out to others. However, there was such a wide range of specimens that many of the collections were distributed among outside specialists,[27] again as in the case of the *Challenger*. Some of the scientists were employed working up their results at the Natural History Museum where, lacking even an adding machine for their statistics, they found the work tedious. Most returned to South Georgia or to sea on later commissions of the Committee's ships, but the insecurity of three-year contracts, a desire to practise zoology not just sort, label and preserve, the long periods of the commissions and salaries often too low for marriage, caused many of them to leave.[28]

Changing emphases in the 1930s

Though much was done during the 1925–7 commission, it soon became clear that even the limited objectives of the original commission could not be achieved that quickly. The observations on whales brought into the whaling station at Grytviken were continued and further

Figure 4.1 RRS *Discovery* at Port Lockroy (Wiencke Island, Palmer Archipelago, Arctic Peninsula) in
March 1927 (All photographs from the Discovery Collections, Southampton Oceanography
Centre.)

work at sea was planned, using the *William Scoresby* and a new purpose-built research vessel,
RRS *Discovery II*, to replace the old *Discovery*. Strongly built for Captain Scott's (1901–4)
Antarctic Expedition (Savours 1992), *Discovery* had proved an indifferent sailer.

Until 1931 the Committee limited the range of its work to the Falkland Islands Dependencies sector but three interrelated events, the introduction of pelagic whaling, the retirement of Darnley from the chairmanship in 1933, and Kemp's resignation three years later,
caused shifts of emphasis within the investigations. The first of these had the effect of
expanding the scientific and geographical parameters in which the work was taking place,
leading to its continuation up to the outbreak of World War II, but the other two contributed
from within to external pressures for the demise of the Committee and its work.

The expansion of the Committee's work in the 1930s was a direct consequence of the
introduction of pelagic whaling, which began with the Norwegian factory ship, *Lancing*, in
1926 (Mackintosh 1965). The Committee foresaw the problems ahead: 'If the practice of
whaling in this form increased it might easily diminish the number of whales so far as
seriously to injure or even destroy the whaling industry and consequently the Government
revenues derived from it.'[29] The resulting increase in Antarctic whale oil production, from
780,000 barrels in 1926 to 2,500,000 barrels in 1929–30, led to a world glut.

Figure 4.2 The whaling station at Grytviken, South Georgia.

Harmer, greatly concerned, produced a warning about the future of the whaling industry[30] and a confidential note on measures designed to preserve it. He recommended reserves, probably 'along the ice edge and in its vicinity in that portion of the Antarctic bordering on the Pacific Ocean'.[31] The Committee suggested sending the *Discovery II* to look at the area. Kemp thought the idea an important one, but did not want the *Discovery II* 'inspecting distant areas unless there is a good certainty that a reserve will be created. Otherwise we merely waste time that might be spent more profitably in our own sector'.[32] Nonetheless, *Discovery II*'s second commission in 1931–3 extended the geographical scope of the investigations, and included the first winter circumnavigation of Antarctica.

In 1933 the Under Secretary of State noted that the income of the Research and Development Fund was rapidly declining due to a dramatic drop in whale oil duties, and that it might not be possible 'to further commission either vessel'. However, following interviews with the Secretary of State,[33] and a personal appeal by Harmer[34] to the Prime Minister, Stanley Baldwin, the Secretary of State agreed to the recommissioning of the *Discovery II* for a third and final commission. Because of unforeseen interruptions to the scientific programme, permission was eventually given for a fourth, and then a fifth commission. Provision was also made for the scientific staff to complete work ashore after the conclusion of the investigations at sea.[35]

However, after Darnley retired at the end of 1933 the Committee was chaired not by civil servants but by politicians who had less commitment either to science or to the whaling industry and the Falkland Islands.[36] When Kemp was appointed Director of the Marine Biological Association's Plymouth Laboratory in 1936,[37] N. A. Mackintosh (1900–74), who replaced him as the Committee's Director of Research, was appointed on a five-year contract and at a salary smaller than Kemp had received in 1924. He had the difficult task of overseeing an organization that was suffering from official downsizing, and one that had

Figure 4.3 RRS *William Scoresby* leaving Grytviken for the Humboldt current cruise, 1931.

already experienced tensions between the Director of Research and the overseeing committee for some years.

Kemp's relations with the Committee

Kemp's job was often an unenviable one with many difficulties, not least those caused by the Committee. In October 1927, after the *Discovery's* return to the United Kingdom, Kemp told Darnley that he could not serve a second commission with Captain Stenhouse. 'Though we have managed to pull through the past two years, I have come to the conclusion that I could not undertake a second voyage under the same conditions.'[38] This must have come as something of an unpleasant surprise to the Committee.

However, the Committee's next move was hurtful to Kemp. They decided to institute a scientific post within the Committee. At a special meeting on 1 March 1928,[39] Borley read a Note[40] in which he expressed the view that a change of machinery was needed to bring the scientific work more thoroughly within the Committee. Borley said that 'in his opinion the task of preparing economic work for discussion made such an appointment imperative'.[41] The appointee would be called upon 'to approve or give reasoned dissent from suggested conclusions' of the research.[42] Allen[43] was responsible for originating this proposal and the suggestion that Borley should be appointed, a manoeuvre arranged without telling Harmer, who was both Vice-Chairman and Chairman of the Scientific Sub-Committee. Borley was offered full-time paid employment as a member of the Committee.[44] Kemp, interpreting the move as the installation of an 'overboss', submitted his resignation. The Committee itself was split: Wordie felt that Kemp was indispensable but Maurice[45] urged that his resignation should be accepted. Harmer was asked to intervene and he persuaded Kemp to stay on.

Figure 4.4 RRS *Discovery II* at Port Lockroy, off the Antarctic Peninsula, in January 1931.

The cause of the row was one of the basic dichotomies within the Committee's objectives. Borley's memo modified the criteria, stating that:

> The main aim of the Committee is to aid the Dependencies of the Falkland Islands by providing an accurate basis for the rational exploitation of the chief economic resources, their whales. The survey of whaling grounds and harbours in the interests of navigation, the exploration of the fishery possibilities of the region, the furtherance of general oceanography are included, but their main ultimate concern is economic.

Figure 4.5 Stanley Kemp on board the *Discovery*.

Research however is not the Committee's only function . . . results must be interpreted to give economic conclusions.[46]

The immediate trigger for the affair appears to have been the appointment of A. C. Hardy (1896–1985) to the chair of zoology at Hull.[47] Hardy, fourteen years Kemp's junior and the senior zoologist in the old *Discovery* on the 1925–7 commission, had initially been keenly supported by some members of the Committee for the position of Director of Research. They were loath to lose their plankton expert and fisheries man. Hardy had previously worked at Lowestoft with Borley, who possibly saw Kemp[48] as guiding the investigations too much towards basic research. Borley's wish to emphasize economic results is an example of the underlying conflict between the objectives of fisheries/economic and fundamental marine research described by Mills (1989) and Deacon (1993).

Extra-scientific activities

However, this conflict was hardly responsible for the extent to which the smooth running of Kemp's proposed research programmes was frequently modified by the Committee either deliberately or accidentally. Deliberate interruptions stemmed from decisions made by the Committee with what were little more than courtesy references to Kemp, as scientific director.

A typical example occurred with the Wilkins expedition. Sir Hubert Wilkins (1888–1958) was an opportunist Australian-American journalist with a taste for a frenetic life-style of adventure. At the end of 1928 he and his pilot made the first flight in the Antarctic (Grierson 1960). In May 1929 he asked the Colonial Office for £10,000[49] towards a further expedition and for their research vessel, the *William Scoresby*, to be on standby from November to the end of December 1929.[50] Kemp was told that the ship would only be required for a few days (he estimated two weeks). Having loaded his equipment at Deception Island, Wilkins sailed south in early December. Except for a refuelling voyage to Stanley and Grytviken, he monopolized the *William Scoresby* for two months.[51] The Committee received no scientific report from Wilkins as the American Geographical Society had been promised all publishing rights.[52] The vessel's intended scientific programme for the season had to be abandoned. Another such occasion was a last-minute request[53] by the Committee that the *Scoresby* should survey the Humboldt Current off the coast of Peru in 1931 (Marsden, in press).

Accidental adjustments were usually the result of ships leaving home long after the date suggested by the Director of Research, or modifications to the route he recommended. Other interruptions were attributable to the temptation to make use of government vessels operating in remote waters for a variety of purposes. These ranged from state occasions, such as delivering a picture of the King to Tristan da Cunha,[54] to rescue missions. The first of these, in February 1934, was when Rear-Admiral Richard E. Byrd (1888–1957) radio-grammed that his doctor had dangerously high blood pressure and was unfit to spend the winter in his Antarctic base at Little America: 'must get Doctor somehow please move heaven and earth to get competent Doctor to help us out'. The *Discovery II* went to the rescue. To the doctor was eventually added large quantities of supplies both for the base and for Byrd's vessel, the *Bear of Oakland*.[55]

In December 1935 it was suggested that the *Discovery II* should participate in search and rescue for Lincoln Ellsworth (1880–1951) and his pilot Hollick Kenyon, who had made the first flight across Antarctica, and were presumed lost. The story is a lengthy one. Suffice it to say that the *Discovery II* lost a season's research although Ellsworth had already arranged to be picked up by Wilkins in the *Wyatt Earp* and considered himself in little need of rescuing.[56] (For Wilkins and oceanography in the Arctic, see Chapter 9.)

The Committee and the public

These high-profile events consorted oddly with the degree of secrecy, almost, surrounding the real, scientific, work of the Committee. It is sometimes suggested that the lack of awareness by the general public of the *Discovery* expedition's activities was due to Kemp's intense dislike of publicity. However, this dislike was for personal publicity, now called the cult of personality, rather than public relations for the achievements of the expedition as a whole. It was exemplified in his comments on Wilkins: 'I thought he was really anxious to do serious exploration. It seems however, that his main purpose is publicity and press notoriety . . . . Every small incident in the work is recorded by cinema and arrangements were made on some occasions that the ship should pass through the more picturesque sections . . . during daylight.'[57]

In spite of Darnley's view, expressed to Harmer, that it was 'undesirable for the Committee as a Committee to seek to place any restrictions on the public utterance of its members',[58] the Committee was reluctant to get involved with the press or to allow their staff to do so. The Committee had total control of communication with the press. All photographs were their property. Kemp had to persuade the Committee that the staff must publish their scientific results to motivate them as scientists[59] and Hardy told the Committee that the lack of recognition of their achievements was hurtful.[60]

As Director of Research, Kemp had to seek the Committee's consent for every scientific article or lecture. In 1930 he attempted to get it to promote public awareness[61] but a further seven years elapsed before a booklet on the investigations appeared (Anon. 1937).[62] Only two books about the investigations were published by staff before 1939 and another three after 1944.[63]

Photography was all in black and white and, as in *Challenger* days, watercolours were used to record colours. The waters of the Peru Current (Gunther 1936) and the birds of South Georgia (Matthews 1929) are two examples. During the old *Discovery*'s voyage of 1925–7 Dr Marshall, the surgeon, became a semi-official photographer and made a very short film. Alfred Saunders, the laboratory technician, exhibited at the Royal Photographic Society in 1933[64] – after the Committee had inspected his prints. He was refused support towards going on a 'cinematographic course' and towards purchase of a second-hand cinema camera because Mackintosh thought that cinematography was not of much commercial value to the Committee.

In 1937 Gaumont were allowed to make a film of *Discovery II*'s activities and the BBC asked for a short television talk by a member of the marine staff of the *Discovery II*. This was agreed after supervision of the script.[65] Scientific results appeared in learned journals and in the thirty-seven volumes of *Discovery Reports*. These were modelled on the *Challenger Reports* and are large and heavy to handle. It has been suggested that if they had been in another format the work of the *Discovery* would have become better known (Roberts 1975).

Ships and apparatus

Harmer had as his first priority the marking of whales and emphasized this at every opportunity. The Committee's second ship, the *William Scoresby*, was designed with that in mind, although she was also to carry out oceanographic and trawling surveys. Britain's first purpose-built scientific research ship, she was described as too 'clumsy, made too much noise and her bows were dangerously low in the water'. Any potential marksman got dangerously cold. Although built for either coal or oil with auxiliary sails, she used oil from the beginning. The *Scoresby* had a very large bunker capacity and was designed to have a steaming distance of 5500 nautical miles at 10 knots;[66] nevertheless, fuelling was frequently a problem and her suitability for exploration was limited.

The *Discovery*, like the *Challenger* a wooden sailing vessel with auxiliary engines, was never going to be suitable as an oceanographic research vessel, even though she carried what was then the largest ever plankton net (4.5m in diameter)[67] and the prototype of Hardy's continuous plankton recorder. Coaling stations were disappearing during the 1920s, and to make the distance the *Discovery* had her decks piled high with sacks of coal. The engines, only intended by the Committee for auxiliary use, were necessary for reasonable progress. To save fuel on the outward voyage in 1925 the dynamo was frequently turned off so that scientists and crew were without electricity.[68] Both Kemp and Stenhouse wrote advising the Committee of her unsuitability.[69] Their letters were ignored and it was seriously proposed to send her south again. This decision was strongly deplored by Kemp.

Permission for building the Committee's third ship, the *Discovery II*, came about from a combination of circumstances. Reinforcing his previous letter Kemp wrote: 'Taking a general view of the whole commission it can only be said that it has been extremely disappointing . . . any work that has been done has been accomplished with very great difficulty. In my view it is not fair that those on board should be asked to sail again under the same depressing conditions.'[70] Here government policy worked to the scientists' advantage – an indirect effect of the Imperial Conference of 1926 and the intention that 'The process of bringing the Antarctic regions under British sovereignty must be gradual'.[71] The Discovery Committee, keen to acquire a new research vessel, agreed to the charter of *Discovery* to the BANZARE (British, Australia and New Zealand Antarctic Research) Expedition (Savours 1992) provided that the Treasury agreed to make funding for a new ship available. The *Discovery II*, an oil-fired steam vessel, was ordered in February 1929 and sailed on her first commission in December 1929. Markedly more comfortable than the *Discovery*, she also had a range of 7000 miles, which enabled her to undertake the circumpolar voyages of the 1930s.

In 1926 Roald Amundsen (1872–1928) told Byrd: 'Aircraft is the new vehicle for exploration' (Byrd 1930). The consultant naval architects in 1923 expected the *Discovery* to carry a seaplane.[72] This plan was never implemented but Wilkins used the *Scoresby* for transporting an aircraft. It made a total of six flights, but problems occurred with landing sites and the weather. Kemp wrote to Darnley:

> The result is that he has been unable, and probably always will be unable to make landings and hoist the flag. Since as I understand the Antarctic Committee recommended a grant to Wilkins mainly for the purpose of effecting such landings, I find it difficult to understand why they wish him to be given further facilities now that it is clear that landings cannot be made. It is evident that the committee has lacked information . . .[73]

Although Wilkins flew further south in 1929 than he had before, his report that Graham Land was made up of one or more islands was wrong, as was proved by the British Graham Land Expedition of 1934–7.[74] The *Discovery II* was specially adapted to carry aircraft in the search and rescue of Ellsworth and Kenyon.

The refit of Scott's *Discovery* had included the provision of a wireless room, but sometimes those on the *Discovery* experienced the *Challenger*'s isolation as radio communication was sparse. The *Scoresby* was fitted with a low-powered transmitter, at times inadequate. Wilkins broadcast to the American press regularly. When, in February 1930, there was no communication from him for nine days, lurid disaster headlines appeared in the American and British press. By the time of the Ellsworth and Kenyon rescue, easy radio communication had been established.

Technological advance was also reflected in the scientific work. In 1923, before any staff had been appointed, D. J. Matthews (1873–1956) of the Admiralty Research Laboratory suggested that the *Discovery* should be fitted with echo-sounding gear.[75] The Americans had an echo-sounder in the USS *Stewart* in June 1922 (Schlee 1973) and the Admiralty Research Laboratory was developing British versions. Within a month Captain Stenhouse visited the laboratory and reported to the Committee 'that the shallow sounding gear operates satisfactorily. The deep-sounding gear is still in the experimental stage'.[76] He recommended fitting both to the *Discovery*, but an experimental model had to be hung overboard to avoid cutting holes in the ship. Rolling caused excessive noises in the hydrophone and the unusual position of the hammer prevented it from working properly so it was removed in 1926. From 1926 to 1929 only a Lucas-type sounding-machine was used (Herdman 1932). The Committee discussed the equipment frequently, prompted by reports such as Captain Shannon's from the *Scoresby* that the echo-sounding gear was out of action throughout the Wilkins expedition,[77] or Commander Carey's from *Discovery II* in 1932 that: 'It would be a very good reason to recommend ships using these waters where the weather is usually thick, to supply themselves with a Deep Echo machine were it not for the amount of work which had to be done by the ship's staff to make it efficient since its overhaul by the makers in London.'[78] An automatic deep-water set was fitted to the *Discovery II* in 1933, though it was not often effective until 1934, and a supersonic shallow-water set in 1936. But despite the shortcomings of the early echo-sounders, whereas the *Challenger* achieved less than 300 soundings exceeding 1000 fathoms in three-and-a-half years, *Discovery II* took 1432 soundings in only twelve days in 1930–1. By 1936 they managed 120 hours of continuous recording (Herdman 1948).[79]

Whale marking, in order to determine their migration routes and estimate populations, was Harmer's first priority. The original marks, shaped like collar studs, were unsuccessful. They were replaced first by aluminium and then by 'Staybrite' steel marks (tubes 23cm long) which became known as '*Discovery* marks'. Of the 4988 whales marked in the Antarctic before 1939, 3.8 per cent had been recovered by 1939 and 7.9 per cent by 1975. However, the marking system relied on the death of the whale to retrieve the mark and with the end of the whaling industry the return of marks diminished (Brown 1977).

The end of the Committee and its work

The outbreak of World War II postponed the expected demise of the Committee, but work came almost to a standstill. During the war the *Discovery II* was on charter to the Admiralty for Trinity House work. The *William Scoresby*, also requisitioned, served around the Falkland Islands and the Falklands Islands Dependencies. A nucleus of staff completed a few reports, gave advice to government departments and made preparation for the postwar regulation of whaling (Tønnessen and Johnsen 1982).[80] Mackintosh, arguing for the continuation of the investigations, seemed to recognize a change in its objectives, saying: 'The main object of the committee's work is the exploration of the Antarctic'.[81] But changes were afoot that would, at least in the short term, ally further research on whaling with other branches of marine science.

In January 1944, at a meeting of the Scientific Advisory Committee of the War Cabinet, Vice-Admiral J. A. Edgell FRS (1880–1962)[82] suggested that a British Oceanographical Institute should be set up.[83] This led to a Royal Society Committee, chaired by Edgell, but with much input from Dr George Deacon (1906–84). Their report,[84] in August 1944, proposed the setting up of a National Oceanographical Institute for Great Britain,[85] the purpose of which would be to 'advance the Science of Physical Oceanography in all its aspects'.[86] This remit was later widened.

On D-Day, 6 June 1944, the Discovery Committee met, for the first time since 1939,[87] to consider the future of the Committee's work. Mackintosh spent much time over the next few years attempting to show that the Discovery Committee should not be swallowed up by the proposed Oceanographic Institute.[88] Meanwhile, the Admiralty asked if they could keep *Discovery II* and 1948 drifted by in uncertainty. Finally, on 22 March 1949, the Scientific

Sub-Committee held its final meeting. It noted that from 1 April the work would be transferred from the Colonial Office to the Admiralty. Financial provision had now been made for the National Institute of Oceanography, which was 'to advance the science of oceanography in all its aspects and which it is intended will take over the function of the *Discovery* Committee. . . . Members of Staff will join the Royal Naval Scientific Service. It is expected that they will in due course be seconded to the NIO.'[89]

Achievements

Throughout the life of the Committee there were tensions between its members and the organizations that they represented, and between the Committee and the Scientific Director and his staff. Though the primary objectives were biological, the scientific staff were under constant pressure from the Admiralty to survey, from the Royal Geographical Society representative to explore, and from various government offices not to upset Norway, Argentina, or the whaling companies. From the beginning, Kemp coordinated the work of the scientists with individual specialisms, always stressing that it was the overall picture that mattered. The *Discovery Reports* cover every aspect of marine science, from plankton to large mammals and birds, and from geology to ocean circulation. Meteorological observations did not appear in the reports, but were published in the *Marine Observer*. Working for the Committee or Kemp was good training for future distinction. Of the first ten scientists recruited, three became Fellows of the Royal Society,[90] three were directors of national institutions[91] and one an Oxford professor.[92] Deacon, appointed later, was also elected FRS and became the founding Director of the National Institute of Oceanography.

Although this work failed to save the whale, much else was achieved. The study of whale biology was carried to the limits of available technology, and from a starting point unaware even of what whales ate! In the long term this contributed to a scientific basis for present-day control of whaling, and of fishing for krill. Hydrographic surveys led to greatly increased knowledge of the Southern Ocean so that it became (temporarily) one of the best known, while observations of the Scotia Arc went some way towards understanding the underwater geology of the area. On land, some 240 features were named by or after *Discovery* personnel (Hattersley-Smith 1980). But perhaps the most important and enduring results emerged as the research progressed, vindicating the original decision to study the marine environment and the whales' place in it, as well as whales themselves. Plankton studies revealed the hitherto unknown life-cycle of *Euphausia superba* (krill), the principal food of the baleen whales, and its distribution (Marr 1962), while hydrological work, charting the salinity and nutrient[93] content and temperature of the sea water, advanced knowledge of the ocean circulation system in general, and of features such as the Antarctic Convergence in particular, as well as helping to understand the Southern Ocean as an ecosystem. The idea of a biological expedition with 'collections' had become full-scale oceanographic and ecological investigations. At no one moment was the name formally altered, the change from Expedition to Investigations being introduced gradually from 1928 onwards.[94]

No member of the Committee achieved all his personal objectives, for few of the original aims of the 1920 report were accomplished. Those wanting great explorations or the enhancement of the nation's image abroad were probably happy with the Wilkins, Byrd and Ellsworth episodes, and the support given to the British Graham Land Expedition. Much research work of economic as well as scientific importance was done, but it failed to carry decisive weight at the level of international negotiations. In 1932, in the midst of major cuts in the Committee's funding, Wordie unwittingly revealed that exploration was his real objective and that he had failed to understand the value of the scientific work. In trying to persuade the Committee to fund an expedition by Gino Watkins to Graham Land, he said: 'When the Discovery Committee finishes its work it will have little apart from Oceanography to show for the expenditure of half a million.'[95] He was probably right, but most oceanographers today would claim that it was money well spent.

Notes

Abbreviations

IOS   Institute of Oceanographic Sciences (see note below)
PRO   Public Record Office (Kew)
SOC   Southampton Oceanography Centre (University of Southampton)
SPRI   Scott Polar Research Institute (University of Cambridge)
NB: Papers relating to the work of the Discovery Committee, National Institute of Oceanography and Institute of Oceanographic Sciences are now located at the Southampton Oceanography Centre.

1   Five books have been written by the scientific staff that give accounts of some of the voyages and work done under the auspices of the Discovery Committee (Deacon 1984, Gunther 1928, Hardy 1967, Ommanney 1938, Saunders 1950). There is also a history written by a journalist (Coleman-Cooke 1963). This chapter is the first comprehensive account of the Committee's activities based on a reading of all their minutes and some of their papers.
2   Dr Sidney Harmer was then Keeper of Zoology at the British Museum (Natural History Department). He was created KBE in 1920.
3   Ernest Rowland Darnley was officially concerned with Falkland Island affairs at the Colonial Office (Wordie 1945).
4   The Whaling Committtee. It first met 10 December 1913.
5   Maurice was Fisheries Secretary at the Board of Agriculture and Fisheries from 1912 to 1938, Vice-President of the International Council for the Exploration of the Sea from 1912 and President from 1920 to 1938 (Went 1972).
6   The Interdepartmental Committee on 'Research and Development in the Dependencies of the Falkland Islands'.
7   PRO MAF41/139, Minute 8 May 1918.
8   PRO MAF41/139, Maurice to Under Secretary of State, 25 March 1918. Borley was Superintendent Naturalist Inspector of the Fisheries Division of the Board of Agriculture and Fisheries. His name was put forward in a letter from Maurice.
9   PRO MAF41/39 Minute, Borley 20 June 1919. Borley records that 'Dr Harmer took a very pronounced view indeed of the dangers to the stock'. Harmer's more debatable statements were relegated to Appendix VIII of the report.
10   PRO MAF/41/140 Letter, Colonial Office to Board of Agriculture and Fisheries, 19 January 1922. It begins: 'With reference to your letter of the 26 February 1920, and the connected correspondence . . .'.
11   SPRI /1284/4/9/945, Discovery Committee Minutes: Notes regarding the constitution of the Committee. This Committee was usually referred to as The Research Ships Committee.
12   SPRI MS 1284/4/1/6 Discovery Committee Minutes: Whaling Research Expedition Executive Committee Minutes, 19 April 1923.
13   The first executive committee; its predecessors were purely advisory.
14   SPRI MS 1284/4/1/16, Minutes of 3rd Meeting, 8 June 1923. Name 'Discovery' Committee to be retained.
15   SPRI 1284/4/1/22. Harmer represented the Royal Society informally as well as the British Museum (Natural History).
16   Mr (later Sir James) Wordie was a veteran of Shackleton's *Endurance* expedition. See Savours (1992).
17   Sir Fortescue Flannery was co-opted initially for one year, but remained on the Committee for many years.

18 Admiral Glennie, representing the Admiralty, was followed on the Committee by Admirals Nares, Douglas and Edgell.
19 R. T. Gunther, letter to E. R. Gunther, 12 June 1925, in private hands.
20 E. R. Gunther, letter to R. T. Gunther, 29 January 1925, in private hands.
21 Since the voyage of the *Challenger* there had been little British activity in oceanographic research, due mainly to lack of funds, so this was not really surprising.
22 SPRI 1284/4/1/29, Discovery Committee Minutes. Commander Stenhouse DSO, DSC, OBE, Croix de Guerre, RD, RNR had been Master of Shackleton's *Aurora* 1914–17.
23 SPRI 1284/4/1/24, Discovery Committee Minutes. Minutes of 24 June 1923 and 23 July 1923. The Committee approached J. S. Huxley, Mr Carr Saunders, Stanley Kemp, Mr Hardy and Mr Matthews when seeking applicants. Hardy may have been the biological secretary of the Royal Society, William Bate Hardy (1864–1933), or possibly Alister Clavering Hardy, who was one of the candidates for the post (see page 76).
24 SPRI MS 1284/4/2/85, Discovery Committee Minutes. Lists of candidates for zoologists. Kemp was Superintendent of the Zoological Survey of India but before that he had worked in the Irish Fisheries Department. It is worth noting, in view of his later work in contributing to the design of *Discovery II*, that (with G. H. Fowler) he wrote on 'Yacht equipment' and (with E. J. Allen) on 'Dredging and trawling' in *Science of the sea* (Fowler 1912).
25 SPRI 1284/4/1/34, Stenhouse, submission to Committee, 22 August 1923.
26 SPRI 1284/4/5/364, Discovery Committee Minutes: discussion of 1st scientific report of the Director, 6 January 1926.
27 SPRI 1284/6/1–6, Minutes of the Scientific Sub-Committee from October 1927 until 1949.
28 SOC Discovery Committee File 2460, Provident Fund. After pressure from Kemp a Provident Fund was started, but contributions were taken from pay for two years before any was invested and later it was found that inappropriate rules meant that the fund was not tax exempt. In the late 1930s the Inland Revenue claimed seven years' arrears of tax so that benefit was reduced for the few remaining members.
29 SPRI 1284/4/5/398, Discovery Committee Minutes. Minutes, 24 February 1926.
30 SPRI 1284/4/13/1360, Discovery Committee Minutes. Harmer: Warning as to the effect of the growth of Antarctic whaling on the future of the whaling industry. Sent to Darnley on 20 August 1930.
31 SPRI 1284/4/13. Confidential note as to measures designed for the preservation of the whaling industry.
32 SPRI 1508/3/12, Letter from Kemp to D. D. John, 3 July 1930.
33 SPRI 1284/4/19/1977, Discovery Committee Minutes. Note of a Meeting held in Secretary of State's room, 30 May 1933. The Secretary of State was Mr Cunliffe Lister.
34 SPRI 1284/4/19/1985, Discovery Committee Minutes. Letter from Harmer to Baldwin, 28 June 1933.
35 SPRI 1284/5/10, letter from the Under Secretary of State, Colonial Office, to Secretary of Discovery Committee, 16 June 1937.
36 They had poor attendance figures, Lord Plymouth attending 75 per cent of possible meetings, the Earl de la Warr 55 per cent and the Marquess of Dufferin and Ava 61 per cent. As vice-chairman, Harmer deputized in their absence until his retirement in 1942.
37 SPRI 1284/5/7–11, Discovery Committee Minutes. 153rd Meeting, 7 July 1936.
38 SPRI 1284/4/5–7/710, Discovery Committee Minutes. Letter from Kemp to Darnley, 10 October 1927.
39 SPRI/1284/4/8/806, Discovery Committee Minutes. Minutes of a Special Meeting of Discovery Committee, 1 March 1928.
40 SPRI/1284/4/8/807, Discovery Committee Minutes. Borley, Note on the decision to institute a scientific post within the Committee, 1 March 1928.
41 SPRI/1284/4/8/806, Discovery Committee Minutes. Minutes of a Special Meeting of Discovery Committee, 1 March 1928.
42 SPRI/1284/4/8/807, Discovery Committee Minutes. Borley, Note on the decision to institute a scientific post within the Committee, 1 March 1928.
43 H.T. Allen had originally been firmly opposed to Kemp's appointment, saying that he thought he would be difficult.

44 SPRI 1284/4/8/818, Discovery Committee Minutes. Letter from Discovery Committee to Kemp, 16 March 1928.

45 Maurice was not a member of the Committee at that time, but had been invited to attend.

46 SPRI 1284/4/8/807, Discovery Committee Minutes. Borley, Note on the decision to institute a scientific post within the Committee, 1 March 1928.

47 SPRI 1284/4/8/784, Discovery Committee Minutes. Letter from Hardy to Kemp, 6 January 1928.

48 At the time of his appointment Kemp was strongly supported by E. J. Allen, the Director at Plymouth Marine Laboratory, who attended the selection committee in an advisory capacity. SPRI MS 1284/4/1/45, Letter from E. J. Allen to Harmer, 13 October 1923.

49 Wilkins wanted £7,500 to be made available by 15 June.

50 As an afterthought Wilkins said he would like to borrow an air camera from the Air Ministry.

51 SOC Discovery Committee Files 6252. Proposed Assistance to Sir H. Wilkins. On 2 January 1930, Wilkins requested that he be allowed *Scoresby* until 19 February. London telegraphed Kemp: 'Work contemplated by Wilkins outweighs regrettable interference with scientific work involved trust you will be able to authorise employment of *Scoresby* as proposed.'

52 Ibid.

53 In January 1931 when the *William Scoresby* was nearing South Georgia on her way south, Admiral Douglas suggested a hydrographic survey in the Pacific. Kemp was 'consulted' while in *Discovery II*. On 28 March, Gunther was given the Admiralty's instructions and the *Scoresby* sailed from Grytviken for Chilean, Peruvian and Ecuadorian waters on 11 April.

54 SPRI 1284/4/2/357, Discovery Committee Minutes. Telegrams from Kemp to Committee, 18 December 1925, and from Secretary to Kemp, 29 December 1925.

55 These included drums of gasoline (initially 30 but later increased to 60), 7 cases of soup extract, 60 cases of wheatna cereals, 17 cases of oatmeal, 105 cases of canned vegetables, 3 cases of tractor parts, 1 crate of generator armature, 1 case of hardware, 17 boxes of film, 5cwt. of iron and steel and a large quantity of mail, to which was later added a request for 'any lines or cordage' and food supplies. The *Discovery II* also handed over 500lb. fresh meat, 200lb. bacon, 500 eggs, 10cwt. potatoes, 50lb. fish, 6 hams, 112lb. butter and 2cwt. of onions to the *Bear of Oakland*. (SOC Sir Sidney Harmer's papers. Discovery Expedition Minutes 21.)

56 Ommanney (1938) and Saunders (1950) both contain narrative accounts of this rescue.

57 SPRI 1284/4/12/1244, Confidential letter, Kemp to Darnley, 26 January 1930.

58 SPRI 1284/4/8/808A, Discovery Committee Minutes. Letter, Darnley to Harmer (n.d.).

59 SPRI 1284/4/5–7/527, Discovery Committee Minutes. Letter from Kemp to Secretary, 23 October 1926.

60 SPRI 1284/4/8/757, Discovery Committee Minutes. Special Meeting of Discovery Committee, 1 December 1927.

61 SPRI 1508/3/18, Letter from Kemp to D. D. John, 19 August 1930, says: 'What on earth has happened about the intended publicity re S. Sandwich work?'

62 SPRI 1284/5/8. This was first suggested in December 1935, and was to be written by Borley, Kemp and Admiral Edgell. In March 1937 (SPRI 1284/5/10) Sir Fortescue Flannery was anxious that mention of future surveying must give no impression that survey work was anything but incidental to whaling investigations, for which the whaling dues had originally been levied. It was to go to the Colonial Office for approval before printing.

63 See note 1.

64 SOC Discovery Committee file no. 15033. Exhibition of Committee's photographs, Dominion Office, 1932. Royal Photographic Society, 1933.

65 SPRI 1284/5/10, Discovery Committee Minutes. Minutes of 164th Meeting, 1 June 1937 (item K).

66 SPRI 1284/4/5–7/411, Discovery Committee Minutes. Notes of a Meeting held at the Admiralty, 11 April 1926.

67 E. R. Gunther, 28 November 1925. Diary in private hands. Johan Hjort's net of 3 metres was its largest predecessor.

68 E. R. Gunther, 28 November 1925. Diary in private hands.

69 SPRI 1284/4/5–7/660, Discovery Committee Minutes. Communication from Kemp. See also SPRI 1284/4/5–7/455. Letter from Stenhouse to Discovery Committee, 29 April 1926.

70 SPRI 1284/4/8/782, Discovery Committee Minutes. Kemp: Memorandum on proposal to recommission the RRS *Discovery*, 12 January 1928.

71 PRO CAB32/51, Report of Committee on British Policy in the Antarctic, 19 November 1926.
72 SPRI 1284/4/1/3, Discovery Committee Minutes. Letter, Flannery Baggaley & Johnson to Crown Agents, 21 March 1923.
73 SPRI 1284/4/12/1244, Confidential letter, Kemp to Darnley, 26 January 1930.
74 SOC Discovery Expedition Minutes 21, 9 January 1934 to 31 August 1934. Sir Sidney Harmer's Papers 2090–2180: Paper 2114. In 1934 the British Graham Land Expedition, which received £10,000 of support from the Falkland Islands Dependencies Research and Development Fund (i.e. Discovery Committee money) showed that Graham Land was not made up of islands (Rymill 1938).
75 SPRI 1284/4/1/28, Discovery Committee Minutes. Letter from D. J. Matthews to Harmer, 28 April 1923. Matthews, then at the Admiralty, had previously worked at the Marine Biological Association's Laboratory, Plymouth. Harmer described him as 'an experienced hydrographer (in the zoological sense)'.
76 SPRI 1284/4/2/102, Discovery Committee Minutes. Letter from Captain Stenhouse to E. Baynes, Secretary of the Committee, 28 April 1924. At the ARL he had discussions with C. V. Drysdale and B. S. Smith.
77 SPRI 1284/4/12/1259, Discovery Committee Minutes. Letter from R. L. V. Shannon to Secretary, Discovery Committee, 8 March 1930.
78 SPRI 1284/4/17/1778, Discovery Committee Minutes. Commander W. M. Carey, 20th Report of the *Discovery II*, 9 March 1932.
79 It is interesting to note that Herdman felt the need to explain the meaning of 'supersonic' in the text.
80 At the end of 1943, there were already plans for postwar large-scale whaling. The British–Norwegian Joint Committee first met on 29 June 1943 to plan reconstruction of their whaling fleets.
81 IOS GERD B8/1(5), Mackintosh, Memorandum on the future prospects of the Committee's work, 11 October 1943 (George Deacon Papers).
82 Vice-Admiral Edgell was then Hydrographer of the Navy.
83 PRO DO35/1171/W152/1, in paper 45 (sheets 20/21).
84 PRO DO35/1171/W152/1, Report of the Royal Society's National Committee for Geodesy and Geophysics (Sub-Committee for Oceanography). Deacon had joined the Discovery Investigations in 1927 as a hydrographer (which in this context meant not nautical surveying but studying the distribution of salinity and temperature, important for learning about ocean circulation and marine productivity). He acted as chief scientist on the fourth commission of *Discovery II* but went to work for the Admiralty on the outbreak of war. He was subsequently appointed Director of the National Institute of Oceanography.
85 The Royal Society Sub-Committee's Report recommended that the Institute should be in Liverpool and absorb the Tidal Institute at Bidston.
86 PRO DO 35/1171/W152/1, Report of the Sub-Committee for Oceanography. Appendix A: The Work of the Proposed Institute of Physical Oceanography. Eventually the Scientific Advisory Committee of the Cabinet recommended that 'the creation of a National Institute for Oceanography should be recommended to the favourable consideration of His Majesty's Government'.
87 The new Chairman was the Duke of Devonshire.
88 IOS GERD B6/6, Mackintosh, N. Notes given to Barton (George Deacon Papers).
89 SPRI 1284/6/6, Discovery Committee. Minutes of the Scientific Sub-Committee.
90 Kemp, Hardy and Harrison Matthews.
91 Kemp became Director of the Marine Biological Association's Laboratory in Plymouth, L. Harrison Matthews Director of the Zoological Society, and D. D. John Director of the National Museum of Wales, Cardiff.
92 Hardy became Linacre Professor of Zoology in 1946.
93 See Chapter 13.
94 The first printed use of the word appears to be in 1929, in *Discovery Investigations Second Annual Report* (HMSO).
95 SPRI MS 1284/4/17/1800, Minutes of Special Meeting, 31 May 1932.

References

Anon. (1919) *Report of the Interdepartmental Committee on Research and Development in the Dependencies of the Falkland Islands with appendices, maps &c.* London: Discovery Committee, Colonial Office.

Anon. (1924) [Unsigned communication by staff of the Director of Scientific Research, Admiralty 29 March, 1924]. 'The acoustic method of depth sounding for navigation purposes', *Nature* 113: 463–5.

Anon. (1937) *Report on the progress of the Discovery Committee's investigations 1937.* London: Discovery Committee Colonial Office.

Brown, S. G. (1977) 'Whale marking: a short review', in M. Angel (ed.) *A voyage of discovery.* Oxford: Pergamon, pp. 569–81.

Byrd, R. E. (1930) *Little America.* New York: Putnam.

Calman, W. T. (1951) 'Sir Sidney Frederic Harmer', in *Obituary notices of Fellows of the Royal Society* 7: 361–71.

Coleman-Cooke, J. (1963) *Discovery II in the Antarctic.* London: Odhams.

Deacon, G. E. R. (1984) *The Antarctic Circumpolar Ocean.* Cambridge: Cambridge University Press.

Deacon, M. B. (1993) 'Crisis and compromise: the foundation of marine stations in Britain during the late 19th century', *Earth Sciences History* 12: 19–47.

Fowler, G. H. (ed.) (1912) *Science of the sea.* London: John Murray.

Grierson, J. (1960) *Sir Hubert Wilkins.* London: Robert Hale.

Gunther, E. R. (1928) *Notes and sketches.* Oxford: reprinted from the *Draconian.*

Gunther, E. R. (1936) 'A report on oceanographical investigations in the Peru Coastal Current', *Discovery Reports* 13: 107–276. Cambridge: Cambridge University Press.

Hardy, A. C. (1967) *Great waters.* London: Collins.

Hattersley-Smith, G. (1980) *The history of place-names in the Falkland Islands Dependencies (South Georgia and the South Sandwich Islands).* Cambridge: British Antarctic Survey.

Headland, R. K. (1989) *Chronological list of Antarctic expeditions and related historical events.* Cambridge: Cambridge University Press.

Headland, R. K. (1992) 'Delimitation and administration of British dependent territories in Antarctic regions', *Polar Record* 28: 167.

Herdman, H. F. P. (1932) 'Report on soundings taken during the *Discovery* investigation, 1926–32', *Discovery Reports* 6: 205–36. Cambridge: Cambridge University Press.

Herdman, H. F. P. (1948) 'Soundings taken during the *Discovery* investigations 1932–39', *Discovery Reports* 25: 39–106. Cambridge: Cambridge University Press.

Mackintosh, N. A. (1965) *The stocks of whales.* London: Fishing News (Books).

Marr, J. W. S. (1962) 'The natural history and geography of the Antarctic krill (*Euphausia superba* Dana)', *Discovery Reports* 33: 33–464. Cambridge: Cambridge University Press.

Marsden, R. R. G. (in press) 'Investigations on the Humboldt Current following a long series of misadventures: the voyage of the *William Scoresby*, May–August 1931', in K. R. Benson and P. F. Rehbock (eds) *Proceedings of Fifth International Congress on the History of Oceanography.*

Matthews, L. H. (1929) 'The birds of South Georgia', *Discovery Reports* 1: 561–92.

Mills, E. L. (1989) *Biological oceanography.* Ithaca, N.Y.: Cornell University Press.

Ommanney, F. D. (1938) *South latitude.* London: Longmans.

Roberts, B. (1975) 'Obituary of Neil Alison Mackintosh', *Polar Record* 17: 422–8.

Rymill, J. (1938) *Southern lights.* London: Chatto and Windus.

Saunders, A. (1950) *A camera in Antarctica.* London: Winchester Publications.

Savours, A. (1992) *The voyages of the Discovery.* London: Virgin Books.

Schlee, S. (1973) *A history of oceanography.* London: Robert Hale.

Tønnessen, J. N. and Johnsen A. O. (1982) *The history of modern whaling.* Berkeley and Los Angeles: University of California Press.

Went, A. E. J. (1972) *Seventy years agrowing: a history of ICES, 1902–1972.* Copenhagen: International Council for the Exploration of the Sea.

Wordie, J. M. (1945) 'Obituary of E. R. Darnley', *Polar Record* 4: 287.

Part 2
Ocean basins

Introduction
The exploration of the sea floor

Colin Summerhayes

Technology has been the key to mankind's understanding of the ocean floor since earliest times, in the same way that it has to our knowledge of the surface of other planets. Essentially primitive technologies, such as the lead-line sounding used in HMS *Challenger* (see Chapter 1), have in the twentieth century given way to widespread use, first of single-beam echo-sounding, then of multi-beam echo-sounding, and now of interpretations based on satellite altimetry. Our knowledge of the topography of two-thirds of our planet's surface more than one mile deep has increased exponentially as a result and is now truly global, though questions remain about the veracity of the altimeter interpretations of bathymetry in areas poorly charted by ships. Even so, the detail we have is nowehere near as fine as it is for the surfaces of Mars and the Moon, which have been photographed by space missions. Since shape is the first clue to process, this rise in application of technology, and a concomitant rise in topographic resolution, have revolutionized our understanding of seabed processes in ways undreamed of even thirty years ago, as addressed by Tony Laughton in Chapter 5.

Advances in our knowledge of topography provided one of the first important leads to the hypothesis of sea-floor spreading, the necessary precursor to the plate tectonic theory. Application of another technology clinched the matter. The arrival of the towed, continuously recording precession magnetometer in World War II complemented the fast growing topographic database by providing information on the properties of the seabed, demonstrating that it was magnetically striped. Not until the early 1960s did the evolving bathymetric and magnetic data all make sense. Now we take for granted the use of reversals in the polarity of the Earth's magnetic field as a guide to the motions of the Earth's tectonic plates away from the mid-ocean ridges where they are created.

Our *Challenger* forebears were also denied the close-up visualization of the seabed that we now take for granted through the manipulation of deep-sea cameras, both still and video, and through the portholes of manned submersibles. These technologies, too, have changed forever our knowledge base and our perceptions of processes. With visualization, combined with manipulator arms on manned and robot submarines, marine geologists are much better off than they used to be in getting to grips with the geology of the seabed. They are still worse off than land geologists, who can drive up to an outcrop in a Land Rover and lean out of the window to hack off a sample with a hammer, but in a much better position than lunar geologists who can only do it with rockets.

Some technologies have changed little in the 125 or so years since the *Challenger* set sail. Dredging is still done in much the same way, although by putting a pinger on the wire to show the approach of the seabed, and taking advantage of GPS (Global Positioning System) navigation and excellent seabed maps derived from echo-sounding, we can now direct our dredge to more or less precisely the right spot, and know more or less precisely where it has been when we haul it up again. Precision sampling has come of age.

The same goes for coring, where the basic technology of dropping a weighted pipe on the seabed has not changed much since *Challenger* days. But now we can get much longer cores, and sample precisely where we want to. When we want even more penetration we can, at present, drill for it with the Ocean Drilling Program. Our reach into the past would be the envy of our *Challenger* predecessors. From this has come a whole new science, palaeoceanography, and with it an amazing window into the climates of the past, a window clearer and wider than anything available on the continents, which promises to illuminate the story of climate change in ways undreamed of, as Brian Funnell illustrates in Chapter 7. Climate modellers are now turning to the fast growing palaeoceanographic record to validate the models they will use to tell our futures. Advanced coring and drilling is also an essential aid to documenting the sedimentary processes filling the ocean basins, as Dorrik Stow explains in Chapter 6, with surprising spin-offs for geologists trying to model the petroleum potential of structures such as deep-sea fans.

In recent years, as advances in technology have improved our knowledge of how ocean basins work, more focused questions have been prompted about the nature of fundamental Earth processes, such as 'How does sea-floor spreading take place at the crest of the mid-ocean ridge?', spawning the international inter-RIDGE programme and its national counterparts like RIDGE in the USA and BRIDGE in the United Kingdom. To answer these questions multi-beam echo-sounders have been used to map in unprecedented detail the topography of the 50km-wide rift valley in the crest of the Mid-Atlantic Ridge; advanced side-scan sonars such as the UK's TOBI (Towed Ocean Bottom Instrument) have proved essential in understanding the geological evolution of the rift and tools like CTDs (measuring conductivity, which enables a salinity reading to be made, temperature and density of sea water), nephelometers, transmissometers, manned submersibles and manganese sensors have been used to detect and map hydrothermal vents and plumes within the rift. How envious our *Challenger* colleagues would have been of our abilities!

As Chris German tells us in Chapter 8, new efforts with increasingly sophisticated technologies have disclosed in recent years an ever growing number of such vents and plumes, which may turn out to be quite common. Their presence has been shown to modulate the chemistry of the ocean. Their exhalations create vast mineral deposits and feed communities of bizarre organisms. These comparatively recently discovered phenomena have come to be seen as fundamental and important parts of the Earth's biogeochemical system, and may even contain important clues to the story of evolution.

Technology is so vital a part of our exploration of the surface of the planet beneath the sea that it is surprising that it carries such a low profile, or has done until recent years. Wunsch (1989) underlines the importance of investment in technology for deep ocean science and argues for special means for funding it, echoed by Woods (1991). Clark (1989) sets out the need for the National Science Foundation in the United States to set aside funds especially devoted to the support of such technologies, and Summerhayes and Girard (1992) make the case for a collective approach to funding the more costly items not affordable by individual nations. In the United Kingdom a growing realization that technology is important to all our futures is manifest in the change of the names of the science advisory boards of the Natural Environment Research Council to Science and Technology boards, though as yet the change of name has not resulted in more investment in technology. In addition, the government has created a Marine Technology Foresight Panel to help decide what ought to be invested in. Chapter 16 outlines some of the ways in which oceanographers are working to develop new methods for studying the sea.

To solve the progressively more complex problems thrown up by our continuing investigation of the surface of our planet and its resource potential demands the application of progressively more sophisticated, and hence much more expensive technologies. No one nation can do it all, and there is much to be said for banding together to pool resources to accomplish what is necessary in any strategy for progressing the planetary-scale research now demanded of us if we are to see in the next 100 to 125 years as much change as we have seen since *Challenger* days. We must not only think global, but act global.

References

Clark, H. L. (1989) 'Ocean science instrumentation development at the National Science Foundation: a status report', *Oceanography* **2**(2): 22–5.

Summerhayes, C. P. and Girard, D. (1992) 'Oceanographic equipment and instrumentation: needs, trends and priorities', *Underwater Technology* **18**: 16–31.

Woods, J. D. (1991) *New technology for ocean science* (Mike Adye Lecture). London: Marine Technology Directorate.

Wunsch, C. (1989) 'Comments on oceanographic instrument development', *Oceanography* **2**(2): 26–7, 64.

5 Shape as a key to understanding the geology of the oceans

A. S. Laughton

Seeing is believing, or so they say. The geologist on land can readily believe the existence of the mountains, the rivers and their valleys and the strata exposed on fault scarps, and can set about explaining how they were formed and what forces of nature were at work. Marine geologists have no such advantage and have to create in their own minds, and by means of maps and photographs, the scenery of the ocean floor before they can get to work. They have to devise means to measure the ocean depths, to interpolate and correlate these into charts that even now are of very variable quality, and to lower cameras and sampling equipment on long wires to obtain more detailed data. And yet marine geologists have led the revolution of the past forty years in the understanding of global geology through the concepts of plate tectonics. The shape of the ocean floor has played a significant role in initiating this revolution.

The opening of the Southampton Oceanography Centre in 1995 marked the start of a new era in British oceanography. It seemed appropriate at the same time to celebrate the hundred years since the completion of the reports of the great *Challenger* Expedition of the 1870s.

It is timely, then, to review the advances made in the field of marine geology and the impact that the increasing knowledge of ocean morphology has had on them in the past century. These advances are very often marked by major improvements in instrumentation and technique, so I plan to highlight some of the most important technical developments and to discuss their impact on geological thought.

Early charts

From the time of the very earliest seafarers, there was a need to measure the depth of the sea in order to avoid grounding, and the simplest method was to lower a weighted line to see when it hit the bottom. But outside the hazards of the coastal zone, the depth of the ocean was of only academic interest – to be argued about but never measured.

In the sixteenth century, soundings had become useful not only to avoid running aground but also as an aid to navigation. Charts, such as those of the *Mariners Mirrour* of 1584, showed soundings on the continental shelf, which gave seamen advanced notice of the proximity of land after long ocean voyages (Putman 1983) (see Fig. 5.1). But it was scientific curiosity that led Count Luigi Marsigli, in 1706, to run a line of soundings perpendicular to the southern coast of France, and to note that, after a long and gradual slope of the sea floor away from the coast, the depths then rapidly increased (Deacon 1971: 177–8). Although he did not know it, he had discovered the continental shelf and upper continental slope, which were defined and described globally by Bruce Heezen two-and-a-half centuries later. Marsigli

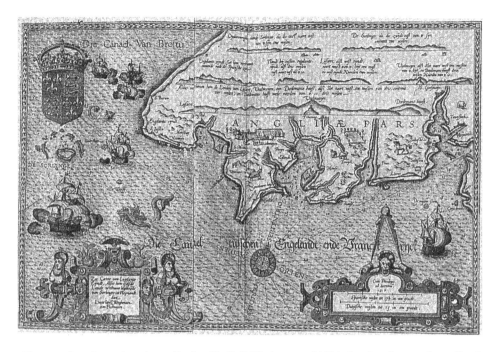

Figure 5.1 Sea chart of southwest England by L. J. Waghenaer of 1584, showing the use of sounding for navigational purposes, from the pilot guide *The Mariners Mirrour*. (Putman 1983: 60.) (Reproduced by kind permission of Abbeville Press Inc., 488 Madison Avenue, New York, USA.)

recognized the importance of the sea floor in understanding continental geology, but surmised that its geology was the same as that of the continents and that the depths compensated for the mountains.

Marsigli's sounding equipment was relatively simple and could not cope with the greater ocean depths. Many methods were proposed during the eighteenth century for measuring ocean depths, including pressure measurements, but the weighted line was still the only practicable option. The first successful ocean sounding is usually attributed to Captain Constantine Phipps, in command of HMS *Racehorse*, who made a sounding of 683 fathoms in 1773 in the Norwegian Sea during his attempt to reach the North Pole. It was not for another sixty-seven years that Sir James Clark Ross, on his voyage with HMS *Erebus* and HMS *Terror* in 1840 to make magnetic measurements in the southern hemisphere, successfully sounded a depth of 2425 fathoms in the South Atlantic (Ritchie 1967: 256).

It is interesting to record how he had to do this in order to minimize the effect of the wind drift of the ship. He had 3000 fathoms of rope, fitted at intervals with swivels, wound onto a huge reel mounted on a ship's boat (see Fig. 5.2). The 76-pound sinker took the rope down and the time was recorded for the passage of every 100 fathom mark. When the rate of fall decreased, it was presumed that the weight had touched the bottom. The whole operation took many hours and clearly was very dependent on the weather. A subsequent echo-sounding survey of the same spot showed that Ross's sounding was too deep by about 300 fathoms.

During the next few decades, others used the same method and soundings in the Atlantic began to accumulate. Many were grossly in error because of wind drift and unknown surface

Figure 5.2 Method of sounding from an open boat used by Sir James Clark Ross (1847, vol. 2, p. 355) in 1840 to obtain the first deep-sea sounding, of 2425 fathoms.

and subsurface currents, and attempts to make sense of them in terms of the shape of the ocean floor were fraught with difficulty.

Midshipman Brooke, under the influence of Matthew Fontaine Maury (Director of the US Naval Observatory), developed a more efficient method of deep sounding (McConnell 1982: 51–2) by which the sounding weights were discarded after touching the bottom, a sample of the bottom was taken and the line recovered by a small engine. By this means the US Naval Survey collected many more soundings in the Atlantic and Maury was able, by the vigorous pruning of erroneous data, to prepare the first contoured chart of a whole ocean. In 1855 he published this chart of the North Atlantic in his book, *The physical geography of the sea* (Maury 1855) and was able to demonstrate for the first time the continuity of an extensive shoal area running north and southwest from the Azores and dividing the North Atlantic into two parts.

By this time the telegraph had been invented and plans were being made to link Europe and America by cable. This stimulated the collection of more soundings across the North Atlantic so that by the time the eighth edition of Maury's *The physical geography* came out in 1861 (reprinted 1963) the central shoal area of the North Atlantic was better defined and was named 'Telegraphic Plateau' (see Fig. 5.3).

So what was the state of knowledge of submarine morphology when the *Challenger* Expedition was being planned? Nothing was known about the world encircling mid-ocean ridges, nor of the deep trenches. The deep-sea floor was thought to be a region of passivity, where little if anything ever happened and no life could exist. Indeed, there were still those who thought that the water density increased with depth such that sunken ships and drowned

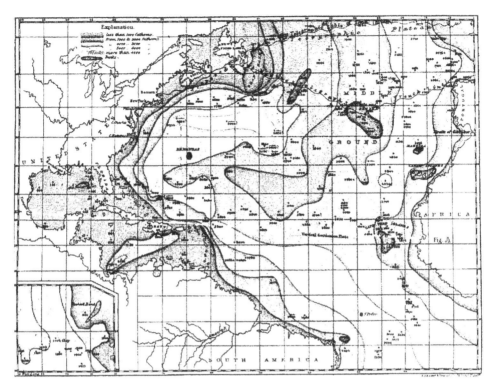

Figure 5.3 Basin of the North Atlantic Ocean. The 1861 version of the chart published by Matthew
Fontaine Maury. (Reprinted in Maury 1963: pl. VII.)

sailors never reached the bottom but remained suspended in mid-water forever (Deacon
1971: 285). Apart from Charles Darwin's ideas about the origin of atolls, there was little
speculation about the geology of the ocean floor.

As well as a wealth of biological and physical data, the *Challenger* Expedition brought back
a large number of geological samples, which were worked on by Sir John Murray. He wrote
(Murray 1886: 354):

> In the abysmal areas, we find here and there small volcanic islands rising as great cones
> from the bottom of the sea, sometimes capped with coral, and forming atolls; but we do
> not find in these areas any traces of continental rocks. Indeed it is extremely unlikely
> that any continental land ever existed in these abysmal areas during past ages, and the
> deposits now forming in these regions, far from our present continental lands, have, so
> far as we know, no analogues in the geological series of rocks.

But there was great argument about the permanence or otherwise of the ocean basins and
no generally accepted theory. G. H. Darwin had suggested that the oceans were the result of
the gap left after the breakaway of the moon, an idea that persisted for decades. Others
supposed that the canyons, which were found to be incised into the continental shelf and slope
(terms that were coined by H. R. Mill in 1888), were formed by subaerial erosion at a time
when the ocean basins were virtually empty, and sea level was 10,000 feet lower than today.

As a result of temperature measurements made of the deep oceanic water in the South Atlantic on the *Challenger* Expedition, Thomson and Tizard deduced that there was a barrier between the east and west basins that stopped the waters mixing and that there were branches from this barrier reaching towards both South America and Africa, defining distinct ocean basins. In their diagram of the deep basins of the Atlantic Ocean (see Fig. 5.4), published in a preliminary report of the *Challenger* Expedition (Tizard 1876), they show the Dolphin Ridge in the North Atlantic (essentially Maury's ridge) and the Challenger Ridge in the South Atlantic connected by Connecting Ridge to form a continuous feature stretching the entire length of the Atlantic. It was not until the 1950s that this was recognized as part of the mid-ocean ridge system.

Macromorphology

A key player in oceanography at the turn of the century was Prince Albert I of Monaco, who used his private yachts for extensive research in the Mediterranean and the Atlantic. In 1899, the 7th International Geographic Congress, held in Berlin, set up a Commission to

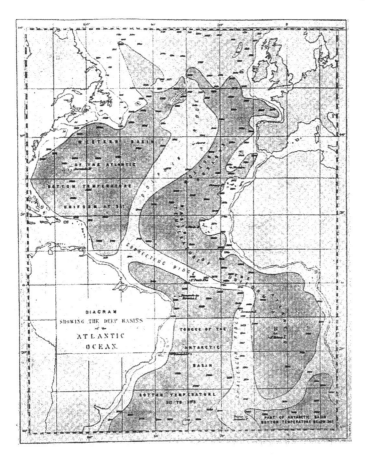

Figure 5.4 Chart of the Atlantic Ocean, by Thomson and Tizard, in which the east and west Atlantic basins are shown to be separated by a continuous ridge; its continuity was deduced from the difference in bottom water temperature on the two sides. (Tizard 1876: pl. 6.) (See also Deacon 1971: 355.)

draw up a general chart of the oceans to bring together and contour on a global basis all the soundings that were by then being obtained (see Chapter 3). Prince Albert chaired the Commission and organized and financed a series of charts to be known as the General Bathymetric Chart of the Oceans, known now as GEBCO. The series of 16 Mercator and eight polar charts were published in 1905 and provided for the first time a global perspective of submarine morphology that could be studied and interpreted by geologists (Kapoor and Scott 1984). Inevitably the first edition was rapidly out of date as more soundings were collected and earlier soundings were found to be erroneous, and new editions were brought out throughout the next eighty years.

The shape of the oceans includes the shape of the edges of the continents. Many geologists have pondered over the similarity in shape between the coastlines of South America and Africa. As long ago as 1858, Antonio Snider in Paris explained the similarity of the Coal Measure formations in Europe and in North America by proposing the outrageous suggestion that in Carboniferous times the two continents were adjacent (Holmes 1965: 1197).

However, these ideas were not taken seriously until Tayler in America in 1908, and later Wegener in Germany in 1915, collected evidence to show that continents might have moved around on the face of the Earth (Wegener 1924). Whereas Tayler had proposed that the creation of the moon from the Pacific had torn the continents apart, Wegener suggested that forces within the Earth drove the continents as blocks drifting through the weaker oceanic crust.

Wegener's theory of continental drift held sway for several decades and received strong support from the South African, Du Toit (1937), who had collected considerable data on the matching of fossils and facies on either side of the South Atlantic, which indicated earlier contiguity of coastlines. But geophysicists were by now studying the physical properties of the interior of the Earth using seismic waves from earthquakes, measurements of the gravitational field, the study of ocean and Earth tides, together with an increasing body of experimental data on the behaviour of rocks under pressure and high temperature. Professor Harold Jeffreys, the *éminence grise* of the period, in his various editions of *The Earth* from 1924 to 1952 and beyond, categorically denied the possibility of continental drift on the grounds that there were no forces large enough to drive continents through the strong oceanic crust.

Sverdrup, Johnson and Fleming (1942), in summing up the state of knowledge in *The oceans*, stated that 'there is yet no agreement concerning the processes involved in the geological history of the ocean basins'. So by the time of World War II there were reasonable, if somewhat inaccurate and undetailed, charts of the shape of the ocean floor, but very little understanding of how it was created.

The technology developed during the war changed all that. In the field of submarine warfare, sonar played a vital role, magnetic detection methods were perfected, seaborne gravity measurements became important and an understanding of the propagation of ultrasound and low-frequency sound waves in the water and the sediments and rocks below was essential. Data were collected by the navies for operational purposes, many of which, alas, were not available to science for decades. In particular, echo-sounders, which had replaced rope and wire sounding in the 1920s, enabled vast quantities of new sounding data to be collected and profiles of the sea floor to be generated. Since in the USA the production of contoured charts using high-quality navigation was classified, scientists at the Lamont Geological Observatory (Columbia University) compiled diagrams of the shape of the sea floor from profiles drawn as oblique views and using the growing knowledge of geological processes as a guide to interpolation.

Bruce Heezen, Marie Tharp and Maurice Ewing (1959) published the first (and only) volume of *The floors of the oceans: 1. The North Atlantic*, together with a large Physiographic Diagram, which was the forerunner of the physiographic diagrams seen on the walls of most oceanographic and geological laboratories and which has been used extensively as a location slide in lectures. In this volume, the authors formally categorize the physiographic provinces of the ocean floors (see Fig. 5.5), and summarize the outputs of Lamont geologists in these productive postwar years.

In compiling these diagrams Heezen & Tharp recognized the existence of a world-encircling mid-ocean ridge that linked all oceans and penetrated the continents in California and in the Gulf of Aden. They discovered that on the crest of the ridge on nearly all the profiles there was a significant cleft, which coincided with the epicentral belt of mid-ocean earthquakes, giving rise to the idea that this was an active boundary between more rigid domains.

It is worth remembering that in 1953 John Swallow, who was at that time working with a later HMS *Challenger* in the Atlantic, had discovered a median valley at about 46°N in the Atlantic, which scientists from the UK National Institute of Oceanography (later Institute of Oceanographic Sciences) later investigated in 1954 in RRS *Discovery*. They found it to extend at least 75 nautical miles along the axis of the ridge (see Fig. 5.6) and an account was published by Maurice Hill (1960).

The story of the development of sea-floor spreading and of plate tectonics is too well known to be repeated here, but it depended critically on the improving knowledge of the sea-floor morphology. Mapping revealed transverse valleys crossing the mid-ocean ridges, sometimes from coast to coast, which were later interpreted as fracture zones and transform faults by Tuzo Wilson (1965) and which provided strict constraints on the relative movement of plates. The Pacific trenches became better mapped and associated with the subducting zones of the plates. Information on the morphology of the continental shelves and the associated slopes was combined with gravitational and seismic refraction data to demonstrate that the transition from continent to ocean occurred usually at the foot of the slope, thereby constraining the location of the closest fit of continents in reconstructions.

The introduction of precision echo-sounders, in which the timing of echoes was made by crystal clocks not subject to the vagaries of a governor-controlled electric motor, enabled the small and subtle changes of gradient of sediment-filled basins to be accurately measured, and led to the recognition of abyssal plains whose slope merged with the foot of the continental rise. Mapping also showed that there were underwater channels that meandered thousands of miles across the deep ocean basins. The whole story of slumps, debris flows, and density or turbidity currents grew out of the attempts to explain the morphology of these sediment basins and sedimentary features (see Chapter 6).

Micromorphology

All of the foregoing processes are macroprocesses affecting the large-scale morphology. New technologies of deep-sea photography, remote-controlled television and direct observation from manned submersibles gave new insights into the micromorphology and microprocesses (Laughton 1963).

Underwater photographs showed that the deep-sea sediments were being constantly reworked by the benthic fauna living below the surface or crawling across it (Hersey 1967). Even the tails of large mammals such as whales leave their marks on the sediment. Direct evidence of the existence and strength of bottom currents was shown by ripple marks and

PHYSIOGRAPHIC PROVINCES, ATLANTIC OCEAN

Figure 5.5 Diagram of the physiographic provinces of the North Atlantic as defined by Heezen et al. (1959: pl. 20) from analyses of bathymetric profiles. (Reproduced by permission of the Geological Society of America.)

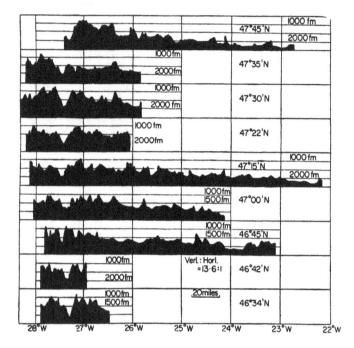

Figure 5.6 East–west profiles of the central part of the Mid-Atlantic Ridge at about 47°N, showing the continuity of a median valley first discovered by HMS *Challenger* in 1953. (Hill 1960: 195.) (Reproduced with kind permission from Elsevier Science Ltd, The Boulevard, Langford Lane, Kidlington, OX5 1GB, UK.)

sand waves. In the deep and still environment of red clay deposition, manganese nodules were seen lying as cobbled pavements over the surface.

On the mid-ocean ridges and on the volcanic sea-mounts, exposed rock outcrops and scree slopes were visible and it was observed that the older rocks, far from the spreading centres, were smoothed by a manganese oxide layer many centimetres thick. Major fault scarps, such as that of the south side of Palmer Ridge in the northeast Atlantic, exhibited the entire oceanic crust sequence from ultrabasic rocks to recent sediments.

Many photographs showed the way in which volcanic lavas arrived at the sea floor. The more viscous were extruded like toothpaste before solidifying. Others, more fluid, flowed into lava lakes, which filled, then broke and drained, only to be filled again, leaving solidifying feeder pipes ringed with the remains of the former crusts of the lake. Most recently, hydrothermal vents were identified first by camera and later visited by manned submersibles. Visual observation of their morphology was crucial to understanding their physics, chemistry and biology, and to deducing their geological origins and ultimate fate (see Chapter 8).

Intermediate scale morphology

In the postwar development of the field, we have seen how the major morphological features of the ocean basins, in the scale range of 10,000 to 0.1km, were revealed by echo-sounding and contouring, and how photographs or direct visual observations gave detail in the scale range of 10m to 1cm. However, there was no technique was available to look at the intermediate scale range of metres to kilometres. This need was identified in the 1969 Report of the Intergovernmental Oceanographic Commission, 'Global Ocean Research'. It was met in several ways as follows.

The US Navy had a requirement for much higher-quality bathymetry that gave coverage perpendicular to the track equal to that along it so that interpolation was unnecessary. In the 1960s it developed the multi-narrow beam sounder known as the Harris array (Glenn 1970). Although the system and its results were classified for many years, it spawned a whole range of swath bathymetry sounders that are made today for the scientific and surveying community.

The first US multi-beam data released for science were made available to Project FAMOUS, the French–American Mid-Ocean Undersea Study, carried out in 1971 (Phillips and Fleming 1978). An area of the mid-ocean ridge southwest of the Azores was chosen for an intensive study by all geological and geophysical means, including submersible exploration (see Fig. 5.7a).

The Navy data, the so-called P charts, enabled the precise measurements and sampling to be carried out in relation to the detailed topography. This had shown that there was an elongated volcanic ridge running down the axis of the median valley, and that there were overlapping spreading centres that were not explicable in simple plate tectonic terms. Some of the so-called fracture zones appeared to have no offset.

The swath mapping technique had many other advantages. Since the data were all handled digitally they could be processed and manipulated to produce instant contour charts, profiles, oblique views of topography with simulated shadows and other tricks of the digital age, thereby giving the geologist a better basis for planning and executing further work.

A complementary approach to higher-resolution bathymetry was taken by the Institute of Oceanographic Sciences using side-scan sonar (Belderson, Kenyon, Stride and Stubbs 1972). Already successful at showing geological features on the continental shelf at short range, the side-scan sonar technique was translated into a deep-ocean survey tool.

The GLORIA (Geological Long Range Inclined Asdic) system sends sound to the deep ocean floor from near the sea surface out to a range of 30km both sides of the ship's track, at a normal cruising speed (Laughton 1981). Since, for much of the ground covered, the insonification is near to grazing angle, the data obtained are complementary to those from swath sounding. Steep slopes can be measured from the shadow geometry, variations in backscattering correlated between adjacent sweeps can build up a visual image of the topography, and textural differences can be seen where there is no apparent relief.

GLORIA was used to great effect, as a contribution to the FAMOUS project, when a survey was made covering the whole area mapped also by the swath bathymetry (Laughton and Rusby 1976). This revealed that the area was dominated by faults, up to 30km long and separated by only a few kilometres, which ran parallel to the spreading axis. Shadow analysis showed that the majority of the fault scarps were facing inwards, in contrast to the fault pattern at the faster-spreading East Pacific Rise. The faulting, associated with the stresses on the young crust close to the spreading centre, could explain the existence of a median valley in slow-spreading ridges (see Figs. 5.7b and 5.7c).

GLORIA has been used extensively throughout the world since the mid-1970s, giving detailed morphology of the continental slope and its canyons, of sediment slumps and debris flows, of meandering deep-sea channels (see Fig. 5.8), of massive landslides, of the fine

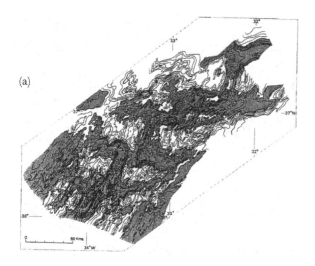

(a)

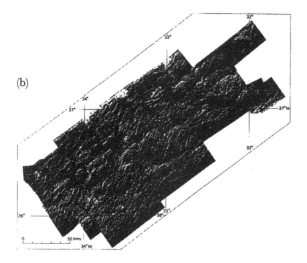

(b)

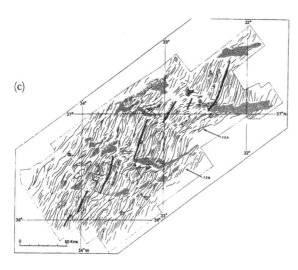

(c)

Figure 5.7 (a) Detailed bathymetry of the FAMOUS area of the Mid-Atlantic Ridge southwest of the Azores, obtained by multi-beam sounders and made available to science by the US Navy; (b) a comprehensive side-scan sonar mosaic of the same area made by GLORIA; (c) a tectonic interpretation based on both sets of data. (From Whitmarsh and Laughton 1976: figs 4, 3 and 6.) (Reproduced with kind permission from Elsevier Science Ltd, The Boulevard, Langford Lane, Kidlington, OX5 1GB, UK.)

structure of transform faults, such as the Charlie-Gibbs Fracture Zone, and of propagating rifts. In a study of the Easter microplate in the southeast Pacific, the complex story of the rotation of the plate by the interaction of propagating rifts from adjacent plates could never have been unravelled without the GLORIA imagery (Searle et al. 1989).

Although GLORIA provided a fast survey tool that could cover enormous areas, the geologist is always looking for more detail and higher resolution. Another approach to using side-scan sonar was started at the Scripps Institution of Oceanography with the development of a near-bottom towed sonar (Mudie, Normark and Cray, 1970). 'Deeptow' was the forerunnner of a series of deep-tow sonars developed at different laboratories – Seamark, Epaulard, TOBI and others.

These tools add to the growing armoury of techniques available to study sea-floor geology. A common pattern is to survey initially by long-range side-scan sonar, such as GLORIA, identify interesting target problems, which are mapped by swath echo-sounding, and then to home in with near-bottom sonar before making direct observations and sampling from a submersible. The studies of hydrothermal vents have exploited all these techniques.

The marine geologist now has almost as much freedom of action as the land geologist, but within the confines of a very inhospitable environment. A consequence of this is that new discoveries are being made all the time and a new understanding achieved of the marine geological processes, which often have an impact on geological interpretations on land.

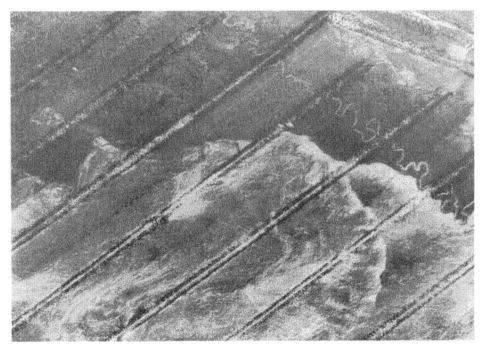

Figure 5.8 GLORIA II sonograph of meandering deep-sea channels of the Mississippi Fan. (For explanation, see Twitchell *et al.* 1991: 351, 353.)

Satellite altimetry

The oceans, however, are vast. Although there are several hundred ships measuring ocean depths during surveys and on passage, there are huge areas of the oceans as yet unvisited. In the South Pacific and in the Southern Indian Ocean there are regions many hundreds of kilometres wide where there are no data. The demand for these data exist, not only from geologists, but also from physical oceanographers who need accurate bathymetry on a global basis to provide the boundary conditions for ocean models. Luckily a new tool has come from an unexpected source.

The measurement of the height of a satellite above sea level can now be done with an accuracy of several centimetres using radar altimetry. The undulations of the level of the sea surface relative to the centre of the Earth are determined by the local gravity field, which in turn is controlled by the distribution of mass in the Earth beneath. With suitable filtering and using the crossover values of intersecting tracks, global maps of the geopotential field at the sea surface have been produced that principally reflect those parts of the sea-floor topography that are not isostatically compensated by buried density contrasts (Haxby 1987).

First Seasat and later GEOS and ERS-I satellites carried radar altimeters, and several global maps have been produced that give indications of topography in areas never sounded by ships. Although these resemble topographic charts, they are in fact gravity charts, reflecting the horizontal density contrasts of the sea-floor topography and of that buried beneath sediment basins.

Work currently in progress on ERS-I data is combining the scattered shipboard sounding data with the gravity data to provide what has been called 'predicted bathymetry' (Smith and Sandwell 1994) (see Fig. 5.9). The tremendous value of these data is that there is a relatively uniform global coverage at a track spacing of 8km, allowing features to be delineated that hitherto were not known at all. Although the data do not give depths as such, they indicate where there are features of interest and where future surveys should be made.

Data availability

The quantity of data on the shape of the ocean floor now available at a wide variety of scales is considerable, but it is relatively difficult to access since it comes in many different formats and from different sources. Contours of global bathymetry derived from shipboard soundings (published as the 5th edition of GEBCO) are now available in digital form in the *GEBCO digital atlas* (Jones et al. 1994), or GDA, on a CD-ROM. Although much of this is relatively old now, the GDA is being updated with new surveys.

A greater problem is how to generate a gridded database that reflects the personal input that is essential when interpolating sparse sounding data, and also includes the information provided by sonar and satellite radar altimetry. Modellers like gridded databases that can easily be manipulated by computer and can generate visual images. Some are available, such as DBDB5 from the US Defense Mapping Agency, but there is no way of assessing the quality of the product. The GDA hopes to rectify this.

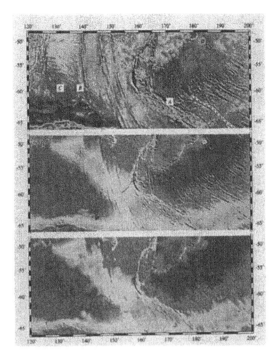

Figure 5.9 (top) Sea-surface gravity measurements in the southwest Pacific made by radar altimetry from satellite; (middle) 'predicted bathymetry' made by combining gravity field
with known but sparse bathymetric data; and (bottom) for comparison, the ETOPO-5
bathymetry (from the DMA database of ocean soundings). The gravity field shows
fracture zones not revealed on ETOPO-5. (From Smith and Sandwell 1994: 21805, pl.
1.) (Copyright 1994 by the American Geophysical Union.)

The future

We have travelled far from the single deep ocean sounding of Sir James Clark Ross 150 years
ago, and from the almost total ignorance about oceanic geology. The technological developments have each brought new understandings, but at no time more dramatic than the
eruption of new technology in the postwar years and the revolution of plate tectonics.

What will the future bring? The need is there for eliminating the interpolation and
interpretation that arises from inadequate data. Unlike the dynamic ocean, the sea floor
remains relatively stable over centuries, at least on the large scale, and so, once determined,
the topography can provide the base maps for all oceanography. There are now a variety of
efficient tools to map the oceans systematically and completely, and position fixing, with the
advent of GPS (global positioning system) giving continuous and accurate fixes, is no longer
the problem that used to beset surveys. Over a hundred ships are now fitted with swath
bathymetry sounders.

I believe the time has come when there should be an international effort to crack this
problem once and for all and to make a systematic survey of all the world's oceans. At the
18th Assembly of the Intergovernmental Oceanographic Commission in June 1995, Resolution XVIII–10 on 'Support for the Joint IOC–IHO Ocean Mapping Programme' was
adopted in which it 'Instructs the Executive Secretary IOC':

to initiate discussions on how to establish scientific priorities for bathymetric surveys of the world's oceans and then, in collaboration with the International Hydrographic Organisation, to establish a well coordinated and comprehensive plan for the coming decade.

We have the tools. Do we have the will?

References

Belderson, R. H., Kenyon, N. H., Stride, A. H. and Stubbs, A. R. (1972) *Sonographs of the sea floor.* Amsterdam: Elsevier.

Deacon, M. B. (1971) *Scientists and the sea, 1650–1900.* London: Academic Press; reprinted 1997, Aldershot: Ashgate.

Du Toit, A. L. (1937) *Our wandering continents.* Edinburgh: Oliver and Boyd.

Glenn, M. F. (1970) 'Introducing an operational multi-beam array sonar', *International Hydrographic Review* **47**: 35–9.

Haxby, W. F. (1987) 'Gravity field of the world's oceans. A portrayal of gridded geophysical data derived from SEASAT radar altimeter measurements of the shape of the ocean surface (scale 1:40,000,000 at equator)', World Data Center A for Marine Geology and Geophysics, *Report MGG–3.*

Heezen, B. C., Tharp, M. and Ewing, M. (1959) *The floors of the oceans, I: The North Atlantic.* Special Paper 65. New York: Geological Society of America.

Hersey, J. B. (1967) *Deep-sea photography.* Baltimore: Johns Hopkins University Press.

Hill, M. N. (1960) 'A median valley of the Mid-Atlantic Ridge', *Deep-Sea Research* **6**: 193–205.

Holmes, A. (1965) *Principles of physical geology.* London: Thomas Nelson.

Jeffreys, H. (1952) *The earth.* Cambridge: Cambridge University Press.

Jones, M. T., Tabor, A. R. and Weatherall, P. (1994) *GEBCO digital atlas.* Bidston: British Oceanographic Data Centre.

Kapoor, D. C. and Scott, D. P. D. (1984) 'The General Bathymetric Chart of the Oceans (GEBCO)', in IHO/IOC/CHS 5th edition GEBCO, pp. 5–13. Ottawa: Department of Fisheries and Ocean.

Laughton, A. S. (1963) 'Microtopography', in M. N. Hill (ed.) *The sea* 3: 437–72. New York: Interscience.

Laughton, A. S. (1981) 'The first decade of GLORIA', *Journal of Geophysical Research* **86**(B12): 11511–34.

Laughton, A. S. and Rusby, J. S. M. (1976) 'Long-range sonar and photographic studies of the median valley in the FAMOUS area of the Mid-Atlantic Ridge near 37°N', *Deep-Sea Research* **22**: 279–98.

McConnell, A. (1982) *No sea too deep: the history of oceanographic instruments.* Bristol: Adam Hilger.

Maury, M. F. (1855) *The physical geography of the sea.* New York: Harper. London: Sampson Low.

Maury, M. F. (1963) *The physical geography of the sea, and its meteorology,* ed. J. Leighly. Cambridge, Mass.: Harvard University Press (Belknap).

Mill, H. R. (1888) 'Sea temperatures on the continental shelf', *Scottish Geographical Magazine* **4**: 544–9.

Mudie, J. D., Normark, W. R. and Cray, E. J. (1970) 'Direct mapping of the sea floor using side-scanning sonar and transponder navigation', *Geological Society of America Bulletin* **81**: 1547–53.

Murray, J. (1886) 'The physical and biological conditions of the seas and estuaries about North Britain', *Scottish Geographical Magazine* **2**: 354–7.

Phillips, J. D. and Fleming, H. S. (1978) 'Multi-beam sonar study of the Mid-Atlantic Ridge rift valley, 36°–37°N', Geological Society of America, Map Chart Series MC–19.

Putman, R. (1983) *Early sea charts.* New York: Abbeville Press.

Ritchie, G. S. (1967) *The Admiralty chart: British naval hydrography in the nineteenth century.* London: Hollis & Carter; reprinted 1995. Edinburgh: Pentland Press.

Ross, J. C. (1847) *A voyage of discovery and research in the Southern and Antarctic regions, 1839–43*. London: John Murray.

Searle, R. C., Rusby, R. I., Engeln, J., Hey, R. N., Zukin, J., Hunter, P. M., LeBas, T. P., Hoffman, H.-J. and Livermore, R. (1989) 'Comprehensive sonar imaging of the Easter microplate', *Nature* **341**: 701–5.

Smith, W. H. F. and Sandwell, D. T. (1994) 'Bathymetric prediction from dense satellite altimetry and sparse shipboard bathymetry', *Journal of Geophysical Research* **99**(B11): 21803–4.

Sverdrup, H. U., Johnson, M. W. and Fleming, R. H. (1942) *The oceans*. New York: Prentice Hall.

Tizard, T. H. (1876) 'Report on temperatures' and 'General summary of Atlantic Ocean temperatures', HMS *Challenger*, No. 7. *Report on ocean soundings and temperatures*. London: Admiralty.

Twichell, D. C., Kenyon, N. H., Parson, L. M. and McGregor, B. A. (1991) 'Depositional patterns of the Mississippi Fan surface: evidence from GLORIA II and high-resolution seismic profiles', in P. Weimer and M. H. Link (eds) *Seismic facies and sedimentary processes of submarine fans and turbidite systems*. New York: Springer, pp. 349–63.

Wegener, A. (1924) *The origin of continents and oceans*. London: Methuen.

Whitmarsh, R. B. and Laughton, A. S. (1976) 'A long-range sonar study of the Mid-Atlantic Ridge crest near 37°N (FAMOUS area) and its tectonic implications', *Deep-Sea Research* **23**: 1005–23.

Wilson, J. T. (1965) 'A new class of faults and their bearing on continental drift', *Nature* **207**: 343–7.

6 Silent, strong and deep
The mystery of how basins fill

Dorrik A. V. Stow

INTRODUCTION

There are many different types of basin – continental or marine, deep or shallow, large or small, long-lived or ephemeral. All represent regions of topographic or bathymetric low and all, therefore, will receive an input of sediment from the surrounding basin catchment area and, with continued sedimentation over time, become sedimentary basins. Only some basins will eventually fill completely, whereas many others will remain only partially filled. Some are preserved in the ancient rock series, even though they may have become fragmented in the process; others are subducted and lost.

In this chapter I focus on deep-sea basins and on the several different processes responsible for the input and accumulation of sediment. These processes and their resultant deposits have been the subject of study since HMS *Challenger* circumnavigated the world between 1872 and 1876 and returned with many sediment samples from the ocean depths (Stow, Reading and Collinson 1996). However, because of the nature of the samples recovered (red clays and biogenic oozes), the main sedimentary process active in all deep-sea basins was believed to be the slow, continuous, vertical fall-out of material by pelagic settling. Only much later and after many more years of research on both modern and ancient series were turbidity currents, one of a family of downslope processes first recognized in 1950, accepted as a principal agent of basin fill. In the mid-1960s, contour-following bottom currents were recognized as important in shaping the fill. Hemipelagic advection, volcanic fall-out, glacial input, chemical precipitation and hemiturbiditic processes are all now known to contribute to basin fill, the last of these being recognized as late as 1990 (Stow and Wetzel 1990).

The nature of the sediment fill is illustrated by examining case studies of existing and ancient basins on which the author has worked. This brief survey is no more than a very eclectic view of a subject of increasing importance. The study of sedimentary basins as a separate discipline has received great impetus in recent years because of their economic importance for oil, gas and mineral deposits, and several important publications on the topic have appeared in the 1990s (Allen and Allen 1990, Miall 1990, Busby and Ingersoll 1995).

SEDIMENT SUPPLY

Many different types of material find their way eventually into deep-sea basins. Terrigenous material is derived primarily from the physical and chemical weathering and erosion of pre-existing rocks – igneous, metamorphic and sedimentary alike. Collectively, the world's rivers transport the bulk of this detritus across vast areas of continental drainage basin to the sea. The suspended loads are deposited first in coastal and shelf areas, only slowly moving seaward to the brink of the abyss under the constant influence of currents, tides and waves. Dissolved loads add their individual chemistry to that of the global ocean and may remain in solution for many aeons before being fixed into the sediment pile by organic or inorganic precipitation.

At high latitudes, the processes of melting and calving from glaciers and floating ice add much particulate material directly to both shelf and deep-sea deposition, but introduce far less dissolved solute. Wind-blown dust from arid and semi-arid low-latitude land masses, as well as from explosive volcanic eruptions, may travel far into the neighbouring ocean basin. The wave-pounding, undercutting and erosion of coastal cliffs, particularly effective during major storms or hurricanes, are a further source of detritus that may find its way ultimately to an abyssal sink. The relative input from these different sources is only poorly constrained at present (Table 6.1) and will have varied markedly in the past.

Material of biological origin (biogenic), including siliceous and calcareous tests of plank-tonic plants and animals as well as soft organic tissue, is supplied to the deep sea following primary productivity in the surface waters. Other important sources are organic growth in shoal areas with subsequent erosion and resedimentation, and run-off from biogenic-rich areas of the continents and coastline (e.g. mangrove or other swamps and rainforests, river weed and associated fauna, shoreface molluscs, and so on). The figures estimated for the flux of this material into the ocean basins (Table 6.1) are subject to a relatively wide margin of error as so much is dissolved and recycled, often many times over, before reaching the basin floor.

The ocean waters above any sedimentary basin have an enormous range of dissolved chemicals but a relatively constant ionic composition. This has been built up from the dissolved loads of riverine and groundwater input, wind dispersal of evaporitic salts from arid lands, dissolution of biogenic tests and organic matter, progressive corrosion of the more unstable terrigenous and volcaniclastic components such as ferromagnesian minerals, feldspars, glass and amorphous compounds, and by primary volcanic and hydrothermal emissions from mid-ocean ridges and active plate margins. A crude steady-state chemical balance is maintained within the oceans over long periods of geological time, subject always to large material flux in and out of solution, together with minor local variability between basins and between different layers or water masses within the oceans (Table 6.2).

Table 6.1 Sediment flux to the ocean basins (From Stow 1994.)

Source	Supply $\times 10^9$ t year^{-1}
Rivers (suspended load)	18.3
Rivers (solution load)	4.2
Groundwater (solution load)	0.48
Ice (ice shelves and bergs)	3.0
Coastal erosion	0.25
Wind-blown dust	0.6
Volcanic ejecta	0.15
Biogenic carbonate	1.4
Biogenic silica	0.49

Data from Goldberg (1974) and Open University (1984)

Table 6.2 Principal properties of sea water (From Stow 1994.)

Salinity	
General average	35‰
Surface waters	32.4–39.8‰
Deep water (>1000m)	34.5–35‰
Temperature	
General average	3.52 °C
Surface range	− 1.87°–30.0 °C
Deep water (>1000m)	− 3.56°–8.9 °C
Density	
(expressed as specific gravity)	1.024–1.029
pH	
General average	8.0

SEDIMENTARY PROCESSES

Once material has been supplied to a basin or its margins, the range of sedimentary processes that operate within that basin (Fig. 6.1) can be divided into four main groups. These are: (1) downslope (or resedimentation) processes, (2) bottom current processes, (3) pelagic and hemipelagic settling, and (4) chemogenic processes. These affect different basins or parts of basins, involve very variable amounts of sediment, may be sudden and catastrophic or slow and continuous, and are subject to varying allogenic and autogenic controls. The brief review below draws heavily on previous work by the author (Stow 1985, 1986, Stow et al. 1996).

Downslope processes

Downslope (or resedimentation) processes are those that move material from the basin margins down a slope under the influence of gravity. They are mainly episodic, short-duration events that are typically dramatic in their nature and effect and generally introduce by far the greatest volume of material into the receiving basin. Distinct processes can be recognized within the downslope family, although in nature there is overlap between these end-members, such that a process continuum of mechanical behaviour exists from elastic through plastic deformation to viscous flow (Fig. 6.2).

Two very different end-members that are generally of only minor importance in terms of volume of material moved are rock falls and sediment creep. Rock falls are sudden, rapid, freefall events that can occur only on steep slopes, triggered by undercutting and erosion or by earthquake shocks. Single displaced clasts may be very large (> 10m) and bounce or roll downslope for several tens or hundreds of metres, in some cases dislodging other loose debris to create a submarine avalanche. Talus slopes of rockfall debris are common off coral reef mounds and volcanic sea-mounts. Sediment creep is an extremely slow, continuous process analogous to soil creep on land, that occurs over wide areas of even very gentle basin-margin slopes, wrinkling the surface and gradually displacing the upper few tens of metres of sediment. It may be an important precursor to slides and debris flows on unstable slopes.

Sliding and slumping are extremely common on most submarine slopes and involve the

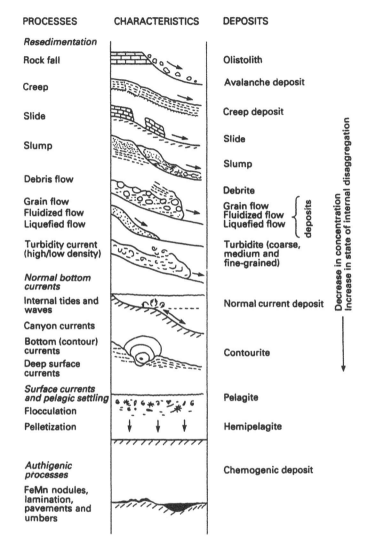

Figure 6.1 The range of processes that operate in deep-sea basins and their deposits. (From Stow 1994.)

sudden downslope displacement of the upper layers of sediment along a basal shear plane. In slides the internal disturbance of these upper layers is minimal, whereas slumps show more internal disruption. Single slide masses range from very small (< 1m³) localized displacements to very large (> 100km³) and often complex bodies that may be as much as several hundred metres in thickness. Catastrophic slides in human terms have involved the complete loss of a drilling platform with all hands on board off the Mississippi Delta on the northern slope of the Caribbean Basin, the removal of half the new runway at Nice Airport in southern France and its disappearance downslope towards the Alboran Basin, and the overnight vanishing of a Ukrainian village from the Crimean Peninsula as it slid towards a watery grave in the Black Sea Basin. Such large-scale slides can lead to the generation of major tsunamis and further disastrous effects around the periphery of semi-enclosed basins.

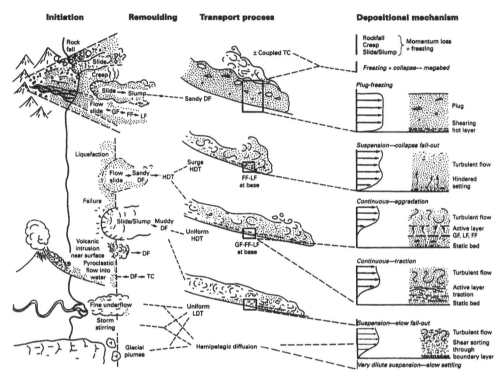

Figure 6.2 Downslope processes responsible for basin filling: initiation, remoulding, transport and deposition. (From Stow et al. 1996.)

Destruction of the Minoan civilization on Crete some 3500 years ago is believed to have resulted from a combination of explosive volcanism and associated seismicity on the island of Santorini, followed by submarine sliding around the flanks of the caldera and the generation of devastating tidal waves (tsunamis).

Sediment instability on basin slopes is affected by a combination of interacting variables. These include: (1) the slope angle, (2) high rates of sedimentation leading to high water content and low shear strength, (3) seismic shock, (4) repeated cyclic stress (minor tremors, tides, and so on), (5) high primary productivity leading to high organic carbon flux, or deposition on the seabed of organic matter derived from plants and animals in the water column, (6) diapiric activity, or subsurface intrusions (usually sedimentary) causing doming in the overlying strata, and mud volcanoes, and (7) generation of gas in the sediment due to clathrate[1] decomposition. Interestingly, recent data suggest that gas-triggered slides on the Bermudan slope remove sufficient of the upper sediment layers to decapitate large areas of buried clathrates and hence release giant bubbles of methane gas. Further speculation links these natural events with the countless and mysterious disappearances of ships and even planes in the Bermuda Triangle (Anon. 1992).

Debris flows are highly concentrated slurry-like flows that advance down slopes in excess of only 0.5°, either rapidly and continuously (for at least a short time) or slowly and intermittently. Typically they have a mud-rich matrix that can support large boulders or slabs the size of a small car, potentially moving at speeds of 20m per second and eventually 'freezing' to leave deposits standing up to 30m or more above the surrounding sea floor, after downslope transport of tens or even hundreds of kilometres. Their chaotic bouldery mud deposits, known as debrites, are well documented from both modern and ancient systems throughout the world, although they may in places be confused with debris-flow deposits on alluvial fans, glacial tillites and sediment injection or mud-volcano bodies.

Turbidity currents and their deposits, turbidites, are one of the most common and best known features of deep-sea basins, although a full-size prototype has never yet been observed in nature. They are a type of density current in which the denser fluid is a relatively dilute suspension of sediment in sea water generated in the first instance in one of four main ways: (1) from the transformation of slumps and debris flows by excessive mixing with sea water; (2) from sand-spillover, grain flows and rip currents feeding sediments into the heads of submarine canyons; (3) from concentrated shelf or canyon nepheloid layers[2] built up by storm-stirring and biogenic suspension of unconsolidated sediment; (4) directly from suspended sediments delivered to the sea by flooding rivers, glacial meltwaters or volcanic plumes; and (5) from the transformation of pyroclastic flows and falls into or onto the sea.

Depending on the way in which the flow has been initiated and on the subsequent supply of sediment, two main types of turbidity current are recognized: short-lived surge-type flows and relatively long-lived steady or uniform flows. Both can occur on even the gentlest of slopes (<1°), but above a critical density the flow will 'ignite', increasing in density and velocity and achieving a state of autosuspension or flow self-maintenance. In this mode, a turbidity current can travel several thousands of kilometres downslope, across almost flat abyssal plains, and even upslope and over minor topographical obstacles for short distances. Truly large turbidity currents, generated perhaps from gigantic basin-margin slides, may reach over 500m in thickness, 10km in length and speeds of over 70km per hour, and eventually deposit a 2m thick layer of sand and mud over an area the size of England! If the receiving basin is small and confined, or if the flow does not reach ignition, then megaturbidites up to 25m thick, grading from coarse gravel and even small boulders at the base to fine silt and clay at the top, may be deposited by single flows. Such large-scale events are relatively rare, perhaps occurring once every 1000–3000 years, interspersed with more frequent small-scale flows. They occur more commonly during times of lowered sea level and in areas of high tectonic activity. All, however, form geologically instantaneous deposits, gravel and sand depositing within a matter of minutes to hours and even the very finest material settling over a period of a few days.

Bottom currents

The second main group of processes that operate in deep-water basins, actively eroding, transporting and depositing sediment, are collectively known as bottom currents. These include all types of deep current driven by normal oceanographic forces such as winds, tides, waves and thermohaline circulation. Four main groups are recognized: (1) major surface currents, (2) internal tides and waves, (3) canyon currents, and (4) bottom (contour) currents. The last of these is the most important in shaping basin-fill and so forms the main focus for our discussion below. They tend to be semi-permanent in nature rather than episodic, although they typically display periods of greater and lesser intensity.

Bottom (contour) currents

Ocean waters, shallow and deep, can be compartmentalized into different water masses formed in different places, each having distinctive salinity/temperature characteristics. The deepest water masses in each basin are formed by the cooling and sinking of surface waters at high latitudes and their subsequent slow thermohaline circulation, that is, gravitational exchanges of sea water throughout the world oceans caused by local variations in density, which is a function of both salinity and temperature (Figs 6.3a and 6.3b). Upward mixing leads to the formation of intermediate water masses, which may impinge directly on the sea floor on basin margin slopes. Highly saline but warm water also flows out of the Mediterranean Sea as an intermediate water mass.

Any slowly moving water mass is affected by the Coriolis force, caused by the Earth's spin, which deflects moving bodies to the right in the northern hemisphere and to the left in the southern hemisphere. The result of this is that the once slowly moving flows are banked up against the western slopes of ocean basins, unable to move upslope against gravity and so become restricted and hence intensified, forming more powerful currents up to several kilometres in width and tens to hundreds of metres in thickness, flowing at different levels in the water column. Bottom currents are also locally intensified by flow restriction through narrow passages in the deep sea, typically linking one deep basin with another, such as those through fracture zone gaps in the mid-ocean ridge system. The Mediterranean outflow forms a contour current and contourites on the eastern margin of the North Atlantic, off the Portuguese continental slope.

Whereas much of the ocean is swept by very slow currents (less than 2cm/s) that have very little effect on basin fill, the western boundary currents commonly attain velocities of 10–20cm/s and up to more than 100cm/s where the flow is particularly restricted. Clearly these currents are sufficiently competent to erode, transport and deposit material, although their variability in both velocity and direction leads to a similar complexity in their deposits – contourites. Both tidal and seasonal periodicities have been noted from direct measurements of bottom-current flow, whereas longer-term variations, perhaps linked to Milankovitch climatic cycles,[3] can be inferred from the sedimentary record. It is now also known that temporary very high surface-energy conditions, related to tropical storm build-up for example, can propagate downwards and induce high energy over the deep-sea floor. These lead to benthic storms that stir and erode sediment for periods of a few days at a time, interspersed with longer and quieter periods of contourite deposition (Gardner and Sullivan 1981).

Contourite deposits or drifts develop in small irregular patches commonly associated with deep-sea channels and moats, in very broad only slightly mounded sheets often covered by extensive sediment wave fields, and as distinct mounded elongate drifts typically aligned parallel or subparallel to the basin margin. Individual drifts may grow to over 1000km long, 150–200km wide and several hundreds of metres in thickness, at mean accumulation rates of 5–10cm per 1000 years for several million years. Some of the giant drifts of the North Atlantic basins, for example, have been established for over 25 million years, bearing witness not only to periods of slow semi-continuous accumulation but also to basin-wide episodes of much more vigorous bottom currents causing regional hiatuses to develop.

Pelagic and hemipelagic processes

Slow vertical settling of microscopic biogenic and non-biogenic particles that occurs through the water column to any depth as a basin-wide continuous background sedimentation is known as pelagic settling. It involves the hard tests of calcareous and siliceous planktonic organisms and their associated soft body tissues, which have been biosynthesized in the surface layers of the oceans. These are mixed with small amounts of the very finest terrigenous, volcaniclastic and even cosmic materials, carried by surface currents or windblown and

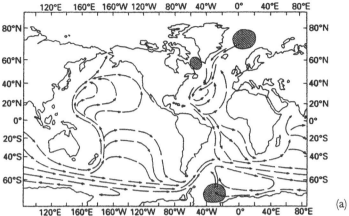

Contourite drift models

TYPE I Elongate drifts

Separated drift

Detached drift

TYPE II Contourite sheets

Plastered drift   Drift sheet

TYPE III Channel-related drifts

Contourite 'fan'

Lateral and axial patch drifts

(b)

Figure 6.3 Bottom current processes that shape the basin fill: (a) global pattern of deep thermohaline circulation in the ocean basins; (b) the nature of contourite drift deposits. (From Stow et al. 1996.)

then subject to the same very slow rates of settling, yielding mean accumulation rates for pelagic sediments of < 1cm per 1000 years.

Normal settling rates are enhanced by physical mechanisms of flocculation, that is particle aggregation, and biogenic processes (e.g. faecal pellets), whereas overall sedimentation rates are increased by as much as a factor of ten in areas of high surface productivity associated with oceanic upwelling. Pelagites[4] in these regions have a high potential for preserving organic matter, particularly where bottom circulation is restricted and oxygen levels very low. In the central parts of deep oceanic basins, by contrast, not only is all the organic matter destroyed by oxidation during settling, but calcareous skeletons dissolve in the lower pH waters found below a level known as the carbonate compensation depth (CCD). Where siliceous organisms make a significant part of the plankton then the resulting pelagite is a siliceous ooze but, in the absence of this material, pelagic red clay is deposited. Very fine terrigenous material blown by winds or transported in surface ocean currents forms a large part of these red clays.

On the continental margins of large basins and in many smaller basins, silty terrigenous material can form a significant proportion of the settling material. In this case the resulting deposit is termed hemipelagic. At high latitudes, glaciogenic hemipelagites[5] are very common, whereas in active margin basins volcaniclastic material may be dominant. Hemipelagic processes typically involve a component of slow lateral advection in mid and bottom waters via suspension cascading, lutite flows of fine-grain material such as silt or clay, and canyon currents, as well as vertical settling. Rates of accumulation are typically between 5 and 10cm per 1000 years.

Chemogenic processes

Although chemogenic processes and their deposits cannot be considered volumetrically important in terms of basin fill, they do serve to indicate significant environmental niches or events in basin history. Ferromanganese nodules and pavements grow very slowly by direct precipitation from sea water enriched in iron and manganese, together with the co-precipitation of a range of other metallic elements. For their growth there must be a complete absence of other sedimentary materials, perhaps in areas swept clean by slow bottom currents or far removed from continental input beneath areas of ocean desert, such as the Sargasso Sea in the central North Atlantic.

Metalliferous sediments or umbers are formed much more rapidly in localized ponds and patches associated with the upwelling of hot fluids into cold bottom waters primarily along mid-ocean ridge systems. These deep-seated hot fluids vent through dark encrusted chimney stacks known as black smokers, which have been constructed by earlier precipitates. They then play host to a weird and wonderful world of bizarre life forms that thrive in these lightless depths of high temperatures and enormous pressures (see Chapter 8).

Even the deepest basins will eventually close and become shallow when caught between the inexorable forces that weld plates together and force up mountain chains. In low latitude areas of low precipitation and high evaporation, an enclosed basin may evaporate to near dryness and natural sea salts begin to precipitate out from the concentrating brine solution. Chemogenic evaporites, such as gypsum, anhydrite and halite, are formed in this way. They are well known from the basal sections of many Atlantic margin basins, having formed during the incipient stages of basin development, and from the upper sections of Tethyan basins, formed during final isolation of the Mediterranean region from the global ocean that led to a major salinity crisis between 6.0 and 5.5 million years ago.

BASIN DEVELOPMENT: CASE STUDIES

There is such a variety of deep-sea basins, both present-day and preserved in the ancient rock record, that it is hard to do justice in an article of this length to the many and varied styles of basin fill that we know to exist. The following summary case studies, therefore, are selected from those basins in which the author has worked himself, and comparisons are drawn with other systems as appropriate.

Angola Basin, southeast Atlantic

The Angola Basin is one of the four main basins in the South Atlantic, having an area of some 1 million square kilometres lying below the 4000m isobath (marine depth contour) and a maximum depth in excess of 5500m (Fig. 6.4). In many ways it is typical of a modern ocean basin with a flat-lying central abyssal plain, bordered on the west by a mid-ocean ridge system and on the east by a major continental land mass. To the north and south the topographic barriers are now submerged but have been at least partially emergent in the past. The South Atlantic has been opening since the early Cretaceous over 110 million years

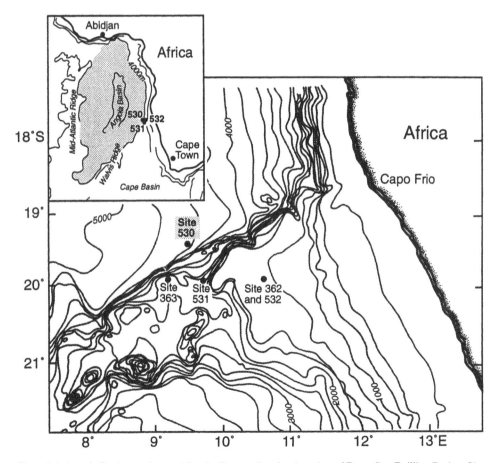

Figure 6.4 Angola Basin, southeast Atlantic Ocean, showing location of Deep Sea Drilling Project Site 530. (After Stow 1984.)

ago, with the Angola Basin deepening progressively as the oceanic crust on which it formed cools and subsides, but also being slowly but steadily filled with sediment shed from its various margins.

At its thickest the sediment fill is little over 2km, but it is best known from the Deep Sea Drilling Project (DSDP) Site 530 in the southeast portion of the basin where 1.1km of sedimentary section was drilled, overlying ocean floor basalts of mid-Albian age (approximately 100 million years ago). Five main phases of basin fill can be recognized from the earliest sediments upwards (Figs 6.5a and 6.5b).

(1) The lowermost 200m of section comprises fine-grain mainly terrigenous sediments characterized by over 250 distinct beds of black shale having a mean organic carbon content of around 5 per cent. These were deposited by normal turbiditic and hemipelagic processes in a semi-restricted, narrow elongate Angola Basin, in which the bottom waters periodically became anoxic. Interestingly, the same mid-Cretaceous black-shale event can be recognized clearly throughout the Atlantic-Tethyan realm (including the Mediterranean and its predecessor, the Tethys) and, to a lesser extent, in other ocean basins.

(2) The overlying 250m is distinguished by an influx of greenish-coloured volcaniclastic turbidites derived from the Walvis Ridge volcanic chain to the south, which was active and partly emergent during late Cretaceous time. There is a progression upwards from

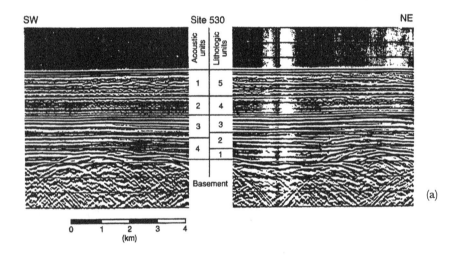

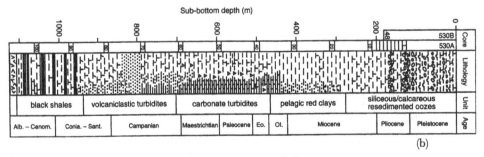

Figure 6.5 Sections through Angola Basin fill: (a) seismic reflection profile showing 1100m sediment fill (horizontal reflectors) over ocean crust basalts (strong hyperbolic reflectors); (b) graphical representation of lithological section drilled at Site 530. Units 1–5 refer to the sediment fill units described in the text. (After Stow 1984.)

thin-bedded fine-grained turbidites and interbedded hemipelagites, through a thick-bedded sand-dominated section, and then a return to thinner beds but of mixed vocaniclastic–carbonate composition, intercalated (layered) with very thin quartzo-feldspathic turbidites derived from the African mainland.

(3) Upwards, the mixed composition beds give way to purely carbonate turbidites made up of the fragmented remains of shallow-water organisms reworked from reefs and platforms that surrounded the now subsiding and inactive chain of volcanic sea-mounts. Individual calciturbidites are well separated by calcareous pelagites, the whole section extending for some 200m across the Cretaceous–Tertiary boundary. Careful micropalaeontological studies across this boundary reveal no particular lithological change and a rapid but phased transition from Mesozoic to Tertiary faunas. This is not consistent with a catastrophic end to the dinosaur era.

(4) The sudden end to this period of carbonate sedimentation occurred within the Oligocene period, when oceanographic changes, perhaps forced by climatic deterioration in Antarctica or major sea-level oscillation (or both), led to a dramatic shallowing of the carbonate compensation depth (CCD) in the South Atlantic and hence the non-preservation of any calcareous pelagites. One or two thin calciturbidites persisted, but the source area was clearly subsiding or exhausted and so these too disappeared. Meanwhile, the African continent bordering the Angola Basin had drifted northwards into the desert belt, so that little fluvial terrigenous material was available for redistribution. These events are reflected in the slow accumulation of some 200m of pelagic red clay.

(5) Then, in the late Miocene, calcareous background pelagites returned as suddenly as they had disappeared. However, they are interbedded with an intriguing 250m thick re-sedimented pile of mud-rich turbidites and giant debrites containing abundant siliceous material of biological origin, together with high concentrations of organic carbon. These were derived from the Walvis Ridge and mark the onset and development of the cold water Benguela Current at the surface, which led to upwelling and enhanced primary productivity along the African margin, and hence increased deposition and preservation of biogenic material and organic matter. This system is still in existence today.

Comparisons

Although there will be regional differences according to the different sources of sediment available for supply into the Angola Basin, the sort of history of sediment fill outlined above is believed to be characteristic. It also has many close similarities with the fill or partial fill of other open ocean basins worldwide, although some specific differences can be highlighted.

The northeast Indian Ocean basin, for example, is dominated by the world's largest submarine fan, the Bengal Fan, which has been receiving vast quantities of terrigenous sediment for the past 40–50 million years as a direct result of collision between the Indian and Eurasian plates and consequent uplift of the Himalayas. The fan itself covers over 1 million square kilometres and is up to 16km or more thick at its northern end. Most of the material is brought into the basin by turbidity currents, with slides and debris flows being more important across the northern slope, and pelagic interbeds thicker on distal parts of the fan. These typically occur as alternating layers of turbidites and pelagites.

In the North Atlantic Ocean basins there is no single dominant source of sediment and the varied input of material is swept along the margins in deep-water bottom currents constructing giant elongate drifts and contourite sheets. At certain periods, climatic and oceanographic changes have led to much intensified bottom currents and the occurrence of basin-wide hiatuses in the sediment record.

Sorbas–Tabernas Basin, southeast Spain

The Sorbas–Tabernas Basin (here referred to as the Sorbas Basin for convenience) is an example of a relatively small, fully filled, ancient basin now exposed on land. It lies within the Betic Cordillera, an Alpine mountain chain resulting from the collision of the European and African plates, which began in late Mesozoic times. The Sorbas Basin formed as an east–west orientated transtensional half-graben from the mid-Miocene onwards, when there was a switch from compressional to strike slip tectonics. At its maximum extent it was some 15–20km wide and 30–40km long, and was partly interconnected with the Vera Basin to the east and the Andarax Basin to the west (Fig. 6.6). The chief phases of basin fill can be defined as follows (Fig. 6.7).

(1) Terrestrial coarse clastic deposits first marked the onset of basin development, deposited in alluvial fan to fan delta settings by terrestrial and then submarine debris flows and associated processes. Rapid basin deepening at this stage flooded the area with fully marine deep-water sediments.

(2) The succeeding 500–700m section comprises resedimented slides, slumps, debrites and turbidites, varying from very coarse-grained megabeds to thin silt and mud turbidites, deposited both as a slope-apron fringe and in a series of submarine fans extending across the full width and partly along the axis of the basin system. They are interbedded with and pass distally into hemipelagic marls, or calcareous mudstones, and limestones. Together, these represent the thickest and deepest marine phase of basin fill that is everywhere capped by an unconformity related to continued tectonic activity and basin uplift.

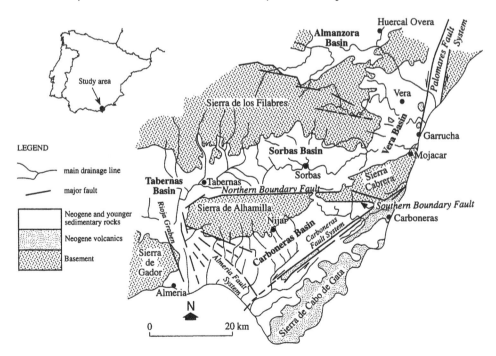

Figure 6.6 Sorbas–Tabernas Basin, Andalucia, Spain. (After Harvey 1990.)

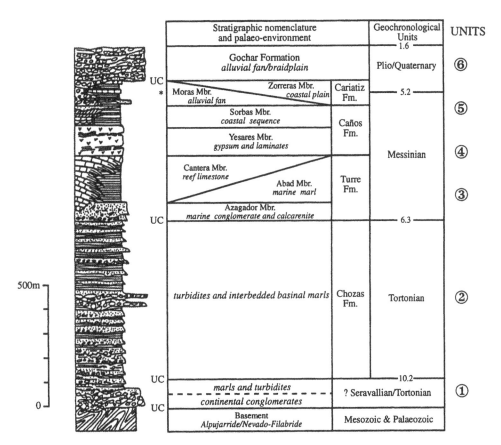

Stratigraphic nomenclature and palaeo-environment			Geochronological Units	UNITS
			— 1.6 —	
Gochar Formation *alluvial fan/braidplain*			Plio/Quaternary	⑥
UC				
* Moras Mbr. *alluvial fan*	Zorreras Mbr. *coastal plain*	Cariatiz Fm.	— 5.2 —	
Sorbas Mbr. *coastal sequence*		Caños Fm.		⑤
Yesares Mbr. *gypsum and laminates*			Messinian	④
Cantera Mbr. *reef limestone*	Abad Mbr. *marine marl*	Turre Fm.		③
Azagador Mbr. *marine conglomerate and calcarenite*			— 6.3 —	
UC				
turbidites and interbedded basinal marls		Chozas Fm.	Tortonian	②
UC			—10.2—	
marls and turbidites			? Seravallian/Tortonian	①
continental conglomerates				
UC				
Basement *Alpujarride/Nevado-Filabride*			Mesozoic & Palaeozoic	

Figure 6.7 Graphical representation of composite lithological section through the Sorbas Basin fill succession. Units 1–6 refer to the sediment fill units described in the text. (After Mather 1993.)

(3) The overlying series is represented by shallow-water reef limestones around the basin margins, fringed with resedimented coarse-grained reef talus that passes distally into marine marls of the basin centre.

(4) Continued tectonic activity led to isolation of the Sorbas Basin, together with a progressive shallowing, and the coincident isolation of the whole Mediterranean system. The result was a major phase of basin evaporation and the precipitation of gypsum–anhydrite evaporites in repeated cycles that indicate alternate evaporation and flooding events. This latest Miocene series is generally termed the Messinian salinity crisis throughout the Tethyan realm.

(5) A relatively thin (< 100m) shallowing upwards coastal sequence marks the end of marine conditions in the Sorbas Basin. Shallow marine sands, tidal sand barriers, and back-barrier lagoon sediments with birdfoot trails and fossil insects mark an intriguing conclusion of marine fill.

(6) In late Pliocene times the tectonic regime changed, with strike-slip transtension giving way to transpression and compressional tectonics so that regional uplift ensued. The former Sorbas Basin became a terrestrial basin for a brief period and was locally swamped by

alluvial fan and fluvial sediments from the uplifting basement areas to both north and south. Although the Sorbas Basin retains some topographic expression today and is clearly demarcated by the outcrops of metamorphic basement rocks in the bounding Sierras, it is essentially a filled basin that has moved into a phase of dissection and erosion.

The style of basin fill exemplified by the Sorbas Basin has many analogues in both modern and ancient systems worldwide. On the Pacific side of the San Andreas Fault system in North America, for example, there are numerous, relatively small, lozenge-shaped, fault-controlled basins in various stages of sediment fill. The exact nature of the sediment is variable, but all tend to pass from deep water through stages of progressive shallowing, followed by eventual uplift and emplacement on land. In island arc settings on compressive plate margins, small fault-controlled basins are formed and filled within a timespan of 5–15 million years. The Pohang Basin in southeastern Korea and the Miura Basin in south-central Japan are both arc-related basins of Miocene–Pliocene age (approximately 15–30 million years old) that filled with a mainly volcanic-rich suite of deep-water, shallow-water and finally terrestrial sediments.

CONCLUSIONS

Although many different processes conspire to introduce much material over millions of years to the deep-sea floor, it is not possible to fill an open ocean basin by sedimentary means alone. Episodic turbidity currents can deliver 100 billion tons of detritus in a single event and construct submarine fans that extend over 2500km across the ocean floor. Semi-permanent bottom currents winnow, erode and reconstruct the ocean margins and can build giant contourite drifts over 1000 km in length. The ever-present slow pelagic rain of material onto the sea floor may be enhanced 100 times or more beneath areas of rich nutrient upwelling and high primary productivity. Where melting ice, strong winds or volcanic eruptions spread large amounts of terrigenous material across the ocean surface, rates of hemipelagic sedimentation are still greater. However, even given all these conditions, deep open ocean basins will not fill.

It is only where tectonic forces along active plate margins create small enclosed and semi-enclosed basins that the combined processes of sedimentation acting over a long enough period of geological time completely fill the basin. It is the record of this type of basin fill that is most readily preserved and easily recognized in ancient series on land.

Notes

1 Clathrate: a gas hydrate deposit, mostly of methane trapped in frozen water molecules in the pore spaces of sediments.
2 Nepheloid layer: near-bottom water containing abundant suspended sediment.
3 This theory, put forward in the twentieth century by the Yugoslav scientist Milutin Milankovitch, ascribes variations in global climate and the onset of ice ages to cyclical changes in the Earth's orbit. An earlier version of this theory, by the nineteenth-century Scottish physicist James Croll, underlay a heated controversy that played a major part in the dispatch of the *Challenger* Expedition.
4 Pelagite: a sediment made up from the remains of dead pelagic (floating) planktonic organisms.
5 Hemipelagite: a sediment composed of a significant mixture of terrigenous detritus from land and the remains of pelagic (planktonic) organisms.

References

Allen, P. A. and Allen, J. R. L. (1990) *Basin analysis: principles and applications*. Oxford: Blackwell.

Anon. (1992) *The Bermuda triangle*. Shown on Channel Four (Equinox), 20 September 1992. Oxford: Geofilms.

Busby, C. J. and Ingersoll, R. V. (eds) (1995) *Tectonics of sedimentary basins*. Oxford: Blackwell.

Gardner, W. D. and Sullivan, L. G. (1981) 'Benthic storms: temporal variability in a deep ocean nepheloid layer', *Science* **213**: 329–31.

Goldberg, E. D. (ed.) (1974) *The sea*, vol. 5: *Marine chemistry*. New York: Wiley-Interscience.

Harvey, A. M. (1990) 'Factors influencing Quaternary alluvial fan development in southeast Spain', in A. H. Rachocki and M. Church (eds) *Alluvial fans: a field approach*. Chichester: Wiley, pp. 247–69.

Mather, A. E. (1993) 'Basin inversion: some consequences for drainage evolution and alluvial architecture', *Sedimentology* **40**: 1069–89.

Miall, A. D. (1990) *Principles of sedimentary basin analysis*, 2nd edn. New York: Springer-Verlag.

Open University (1984) *Oceanography, units 11 and 12. Sediments*. Milton Keynes: Open University Press.

Stow, D. A. V. (1984) 'Cretaceous to recent fans in the southeast Angola Basin', in W. W. Hay, J.-C. Sibuet et al. (eds) *Initial Reports of the Deep-Sea Drilling Project*, **75**(2), 771–84. Washington, D.C. : US Government Printing Office.

Stow, D. A. V. (1985) 'Deep-sea clastics: where are we and where are we going?', in P. J. Brenchley and B. P. J. Williams (eds) *Sedimentology: recent developments and applied aspects*, Geological Society Special Publication **18**: 67–93. Oxford: Blackwell.

Stow, D. A. V. (1986) 'Deep clastic seas', in H. G. Reading (ed.) *Sedimentary environments and facies*. Oxford: Blackwell, pp. 399–444.

Stow, D. A. V. (1994) 'Deep-sea processes of sediment transport and deposition', in K. Pye (ed.) *Sediment transport and depositional processes*. Oxford: Blackwell, pp. 257–91.

Stow, D. A. V., Reading, H. G. and Collinson, J. D. (1996) 'Deep seas', in H. G. Reading (ed.) *Sedimentary environments*. Oxford: Blackwell, pp. 395–453.

Stow, D. A. V. and Wetzel, A. (1990) 'Hemiturbidite: a new type of deep-water sediment', in J. R. Cochran, D. A. V. Stow et al. (eds) *Proceedings of the Ocean Drilling Program: Scientific Results* **116**: 25–34. Washington, D.C. : US Government Printing Office.

7 Palaeoceanography
Tapping the ocean's long-term memory

Brian M. Funnell

Introduction

Ocean sediments comprise the following: the remains of organisms that live on and in the deep-sea bottom; biological and mineral materials that are contributed by bottom currents; biological and mineral materials that sink from the ocean's surface waters, including those materials that reach those surface waters by atmospheric transport; and the products of chemical changes that take place in the sediment after deposition. Although erosion and redeposition do take place on the deep-sea floor, in many places deposition has continued without significant interruption for many hundreds of thousands, up to many millions, of years. The different components of the sediment provide evidence of past conditions, both at the ocean surface and at the ocean floor, so that past variations in temperatures, salinities, currents, chemical conditions and aeolian inputs can be inferred for very long periods of time, and over most of the oceans. Realization that such a storehouse of information on past conditions on earth existed in the oceans developed only very slowly between the *Challenger* and Swedish Deep-Sea Expeditions but has expanded exponentially during the past half century.

The *Challenger* Expedition (1872–6) only scratched the surface of ocean history, but it comprehensively set the scene for the later development and deployment of deep-sea coring by the Swedish Deep-Sea Expedition (1947–8), which extended our knowledge of ocean history several hundreds of thousands of years back into the past. Since then the Deep-Sea Drilling Project (DSDP) and Ocean Drilling Program (ODP) have taken the story back over 150 million years, to the limits of the ocean's preserved sedimentary record. The collaborative application of micropalaeontology, stable isotope analysis, palaeomagnetics and geochemistry has enabled a revolution in our understanding of both planetary tectonics and the processes that control climatic and chemical global environmental change. The recently instituted IMAGES project will concentrate on what the ocean's memory can tell us about changes in the global ocean during the past 300,000 years, in order to inform us better about possible future climatic changes.

The *Challenger* Expedition (1872–6)

When HMS *Challenger* sailed in 1872, she was equipped with only relatively primitive equipment for sampling the rocks and sediments of the deep-sea floor. However, one of those items was the tube (corer is perhaps too sophisticated a name for it), projecting 45cm beyond the end of the weights of the Baillie sounding machine (Fig. 7.1). In spite of its lack of sophistication, because the sounding weight was then the only way of determining the depth of the ocean, it was deployed many times. As a result some 400 bottom sediment samples were obtained during the expedition. From the examination of these sediment samples, a picture of the general distribution of sediment types in the deep sea emerged (Murray and Renard 1891), which has survived until the present day.

After a lively dispute between the expedition leader Wyville Thomson and John Murray, it was ultimately determined that most of the biological remains in the sediments were the skeletons of organisms that live in the surface waters of the oceans (Wyville Thomson 1877). Depending partly on their distribution in the surface waters, and partly on their susceptibility to solution on the ocean floor, they gave rise to *Globigerina* ooze, the radiolarian and diatomaceous oozes, and, with almost total absence of fossil remains, the Red Clay of Murray's classification of deep-sea sediments.

The shortness of the sounding weight 'corer' meant that little more than the top 30cm of sediment was ever sampled. We now know that that thickness probably represents no more than the past 10,000 years of sedimentation, i.e. the time period subsequent to the Last Glacial period.[1] The 'memory' of the ocean sediments that was tapped by the *Challenger* Expedition was therefore very short, and confined to a period when conditions were very similar to those of the present day. There is just a hint from some of the microfossils in the samples that some of their components had originated earlier in the ocean's history. Whether they were there because the 'corer' sampled older sediments, because

Figure 7.1 Baillie sounding machine, as used by HMS *Challenger*. The tube, which acted as a coring tube, projected 45cm below the weights, and usually collected 0.5–1 litres of bottom sediment. (See Chapter 1)

recent sediment had been swept clear by bottom currents, or simply because they had been eroded and redeposited in later sediments is not clear, but peeping out from among the coccoliths of Station 338 (21°15'S, 14°2'W; Murray and Renard 1891: plate xı, fig. 4) are some star-like discoasters that became extinct in the oceans some 2 million years ago (Fig. 7.2).

Whether or not the *Challenger* realistically sampled the earlier history of the ocean basins, it nevertheless provided a comprehensive record of the microfossils contributing to ocean sediments in the beautifully illustrated *Reports* of the *Challenger* Expedition by Brady (1884), Haeckel (1887) and others.

The Swedish Deep-Sea Expedition (1947–8)

For many years after the *Challenger* Expedition the techniques for recovering deep-sea sediments remained essentially the same. Longer coring tubes were used, and evidence was obtained showing that different microfossil assemblages, indicating different surface water conditions, existed deeper in the sediment and further into the past (Schott 1935). However, it was not until the invention of the piston corer (Kullenberg 1947, 1955), and its application during 1947–8 by the Swedish Deep-Sea Expedition, that a whole new range of insights into the past history of the oceans became possible.

Cores taken by the Kullenberg piston corer averaged 10m in length, and sometimes approached 20m long. At accumulation rates of 10mm per thousand years this carried the penetration of the sediment record back to 1 to 2 million years before the present. The same

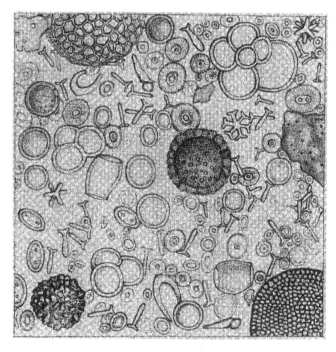

Figure 7.2 Sea-floor sediment sample obtained by HMS *Challenger* in the South Atlantic. Among the coccoliths and other microfossils seen in this illustration, from one of the original *Challenger Reports* (Murray and Renard 1891: plate xi, fig. 4) are some star-like microfossils, called discoasters, that became extinct some 2 million years ago.

methodology was taken up by the Lamont Geological Observatory, where Maurice Ewing established a systematic programme of piston-coring as the *Atlantis* and *Vema* travelled the world ocean.[2] A widespread awareness of changes in ocean sedimentation extending back hundreds of thousands of years was developed (Arrhenius 1952, Emiliani and Milliman 1966, Ericson 1961, Phleger, Parker and Peirson 1953, Schott 1966, Wiseman 1965). Bulk sediment geochemistry, planktonic foraminifera and early determinations of stable isotopes in both planktonic and benthic foraminifera featured in these studies. Evidence of alternating glacial and interglacial states in the world's oceans became well documented (Fig. 7.3), and preliminary indications of substantial changes in the ocean's chemistry and temperature became clear. After the discovery of the coincidence of palaeomagnetic reversals and micropalaeontological extinctions in ocean cores, the parallel use of palaeomagnetic reversals and micropalaeontological extinctions and evolutions to date deep-sea sediments

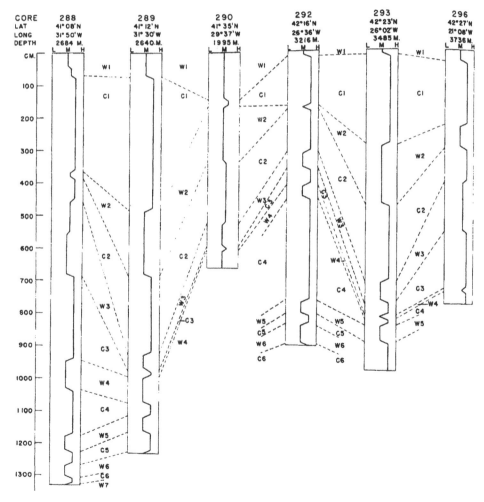

Figure 7.3 An early attempt to correlate the alternations of cold and warm climate inferred from planktonic foraminiferal assemblages recovered from deep-sea piston cores (after Phleger et al. 1953: fig. 26). Lack of an independent time-scale, and absence of a quantitative relationship to sea-surface temperatures, limited the usefulness of these interpretations.

was greatly extended by the Lamont Geological Observatory (Hays and Opdyke 1967), and Woods Hole Oceanographic Institution (Berggren 1971).

Whereas stable isotope investigations tended to concentrate on good Quaternary records, recovered by piston-coring, micropalaeontological investigations generally began to seek evidence of earlier ocean history recorded in core or dredge samples that had sampled sea-floor exposures of earlier ocean sediments. In many places these were found to have been relieved of their Quaternary sediment cover by erosion, non-deposition or faulting. All these investigations were based essentially on the technology already available at the time of the Swedish Deep-Sea Expedition, and by 1968 a first micropalaeontological demonstration of the reality of ocean-floor spreading had been obtained, and the Quaternary climate had already been probed back to 2 million years before present.

CLIMAP (1971–6)

One of the most seminal advances in understanding of the ocean and planetary system, which depended on the technology of piston-coring ocean sediments, was the Climate Long-Range Investigation Mapping and Prediction (CLIMAP) Project. Its success depended on a combination of several new scientific developments. First, there was a new approach to the analysis of the skeletons of planktonic organisms in terms of the surface water temperatures and salinities in which they lived. This was based not only on the development of a discriminating statistical procedure for interpreting the specific quantitative composition of faunal or floral assemblages in terms of past temperatures and salinities (Imbrie and Kipp 1971), but also on the availability of many core top samples for analysis from ocean cruises after World War II. CLIMAP also developed a new and uniquely reliable way of identifying particular time horizons in deep-sea sediments. This followed the recognition that the $\delta^{18}O$ values found in benthic (bottom-living) foraminifera were mainly dependent on changes in the global volume of glacier ice (Shackleton 1967).[3] Cyclical changes in $\delta^{18}O$ (Fig. 7.4) from benthic foraminifera (and global ice volumes) form a characteristic pattern over time (Shackleton and Opdyke 1973), and the maximum of the Last Glacial period, showing the high $\delta^{18}O$ values, is particularly distinctive. The CLIMAP Project, led by John Imbrie, coordinated the inputs of specialists from a variety of institutions and, having established calibrations relating to present-day conditions, inferred sea-surface temperatures at the maximum of the Last Glacial period and for various intervals during the late Quaternary. Papers resulting from this collaboration, representing the use of a variety of planktonic microfossil

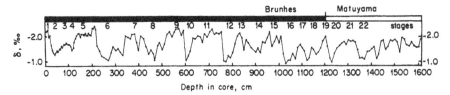

Figure 7.4 Interpretation of the $\delta^{18}O$ values, obtained from deep-sea benthic foraminifera, as reflecting global (glacier) ice volumes, allowed the independent identification of a sequence of glacial and interglacial stages in the deep-sea record. Even numbers indicate glacial (cold) stages, odd numbers interglacial (warm) stages. At intervals these stages can be cross-linked to the palaeomagnetic reversals recorded in the sediments. The Brunhes–Matuyama reversal occurred 780,000 years ago. (From Shackleton and Opdyke 1973: 48, fig. 9.) (Reproduced by kind permission of Academic Press Inc., 6277 Sea Harbor Drive, Orlando, Florida, USA, and the authors.)

groups, and their application to the interpretation of past ocean temperatures over a number of time-scales and for various parts of the ocean, were reported in Cline and Hays (1976), with a more general summary by CLIMAP Project Members (1976). In 1981 the CLIMAP Project Members published a complete set of sea-surface temperature and salinity maps for the maximum of the Last Glacial period (Fig. 7.5), covering all the world's oceans (CLIMAP Project Members 1981), and in 1984 they published another, less complete account of conditions at the maximum of the Last Interglacial period (CLIMAP Project Members 1984). Finally, arising out of the CLIMAP programme, it was realized that the climatic pattern found in the oceans could be correlated with the changes in solar insolation received by the Earth in response to cyclic changes in the earth's orbital parameters (Hays, Imbrie and Shackleton 1976). The story of the search for an explanation of the origin and periodicity of the Ice Ages and the main achievements of the CLIMAP Project is attractively told by Imbrie and Imbrie (1979).

The Deep-sea Drilling Project and Ocean Drilling Program

In the 1950s the pursuit of petroleum reservoirs into deeper and deeper waters, particularly in the Gulf of Mexico, led to increasing interest in the offshore potential and technology of submarine drilling for oil and gas. By that time it had been recognized that a major seismic discontinuity – the Mohorovicic discontinuity, which defines the boundary between the Earth's crust and mantle – occurred at much shallower depths under the oceans than under the continents. So an idea arose to drill to that discontinuity beneath the ocean. This project was called the Mohole. An experimental hole was actually drilled off Guadalupe Island, Mexico, in 1961, sampling the sedimentary section and terminating in basalt. Subsequently the original motivation to drill to the Moho, at more than 10km below the sea floor, was

Figure 7.5 Difference between inferred sea-surface temperatures for August at the maximum of the Last Glacial period (18,000 years ago) and those of the present day. (From CLIMAP Project Members 1981: map 5a.) (Reproduced by the kind permission of the Geological Society of America.)

transformed into a project with more achievable objectives, namely the drilling (mainly) of deep-sea sediments. A consortium of US oceanographic institutes, JOIDES (Joint Oceanographic Institutions for Deep Earth Sampling), put forward a coordinated programme, which was funded by the US National Science Foundation. Operations were managed from Scripps Institution of Oceanography, and the Deep-Sea Drilling Project began operations in 1968 (Cullen 1994). The vessel employed was the *Glomar Challenger*. At 120m long it was capable of drilling at oceanic depths to sub-bottom depths exceeding 1km. Drilling started in 1968 and continued until 1983. In all, 1112 holes were drilled at 635 sites, and more than 90km of core was collected. Between 1974 and 1976 five other nations joined the project, and from 1975 it continued under the title International Phase of Ocean Drilling (IPOD).

It is simply impossible to summarize the results of this project in a chapter of this length. They are recorded in ninety-six thick and heavy volumes, many running to two parts. Together with the bulky index, they occupy almost 6m of shelf space. *Glomar Challenger*'s early drilling explored a number of problems in different parts of the world. Rotary drilling in the ocean floor was not without its problems, and continuous sampling was not always achieved, but the hypothesis of ocean-floor spreading was soon confirmed, as the predicted age of the underlying igneous basement rocks was verified by dating the immediately over-lying sediments. The dating (biostratigraphy) of deep-sea sediments using the skeletal remains of different groups of planktonic organisms (planktonic foraminifera, radiolaria, calcareous nannoplankton and diatoms) developed by leaps and bounds, providing a framework not only for deep-sea sediments, but also for use in the rapidly expanding field of offshore exploration for oil and gas resources. Calibration with the reversals of the Earth's magnetic field, also recorded in the sediments, provided an excellent opportunity for applying absolute dating to the biostratigraphic record. A long-term view of oceanic sedimentation developed, with evidence of systematic major changes in chemical equilibria in the oceans, especially in relation to the carbon cycle and the production and dissolution of carbonate over many tens of millions of years. There were occasional spectacular surprises, like the evidence that the Mediterranean had completely dried out 6 million years ago, and extensive evidence of anoxic oceanic bottom waters over wide areas in the Early Cretaceous period, prior to 100 million years ago. As all deep-ocean floor is ultimately doomed to be destroyed by subduction into trenches at continental margins or under island arcs, nowhere is the sedimentary record *of* the oceans preserved *in* the oceans for ever. In fact the oldest oceanic sediments from the oceans, dating back to the Late Jurassic period (160 to 150 million years ago), were finally retrieved from the western Pacific, underlying thick basalt flows, and from near the margins of the Atlantic Ocean, dating back to a time when it was still a very young ocean.

During the Deep Sea Drilling Project, including its IPOD phase, ocean drilling techniques were progressively developed. Re-entry into holes was an essential requirement for drilling deep into igneous rocks, and continuous coring comparable to that achieved by piston-coring was essential for high-resolution stratigraphic study of climatic and palaeoceanographic change in older sediments. The later stages of the Deep Sea Drilling Project had not only achieved re-entry into igneous basement, but also developed a downhole hydraulic piston corer (APC), which enabled almost continuous sediment recovery (Storms 1990).

The year 1983 marked the end of the Deep Sea Drilling Project, and after a short take-over period, new objectives and a new ship inaugurated the Ocean Drilling Program. This new programme has been based on the College Station campus of Texas A & M University. The new drilling vessel JOIDES *Resolution* is larger than *Glomar Challenger*, at 144m long, and has significantly greater capability for working in deeper waters, keeping station to higher precision and operating safely at higher latitudes. It does not, however, have a riser system, which would allow holes to be sealed during drilling against escaping oil or gas, so drilling continues to be strictly scheduled to avoid any sequences that might contain significant quantities of oil or gas. Nevertheless, new ground has been broken, by drilling through tectonic thrusts, and into basalts near black smokers, as well as close to the Antarctic ice and in the Arctic. Downhole logging has been developed and combined with shipboard core-logging and analysis to produce a sophisticated and comprehensive multidisciplinary understanding of sediment sequences. An outstanding recent result has been the multidimensional description of the palaeoceanography of the eastern equatorial Pacific, which extends back over 10 million years (Pisias, Mayer, Jarecek, Palmer-Julson and van Andel 1995). On 1 February 1996 JOIDES *Resolution* (Fig. 7.6) completed work at Ocean Drilling Program Site 1000, having added 375 sites and some 980 holes to those previously drilled by *Glomar Challenger*.

Orbital chronology

The realization that ocean climate and resultant ocean sediment patterns appeared to exhibit a systematic relationship to the Earth's orbital cycles around the sun, as mediated via their effect on insolation (Hays et al. 1976), opened the prospect for dating deep-sea sedimentation in relation to those orbital cycles. Back calculation of the Earth's orbital cycles and the resultant effects on insolation by André Berger and others (Berger and Loutre 1991) can potentially be matched to any ocean sediment parameter that is responding directly to

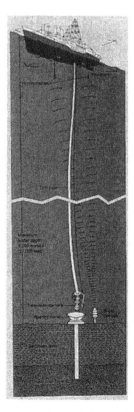

Figure 7.6 Artist's impression of the JOIDES *Resolution* drilling in deep water. The vertical dimension of the drill string is of course very much reduced compared with reality.

insolation. In 1990 Shackleton, Berger and Peltier published a correlation of the benthic $\delta^{18}O$ values from an eastern equatorial Pacific ODP Site 677, drilled by JOIDES *Resolution* in 1986, with such back calculations of insolation variations. The orbital variations and the resultant effects on insolation were correlated with the benthic $\delta^{18}O$ (glacier ice volume) signal back to 2.5 million years ago. A similar correlation between the astronomically driven insolation cycles and the occurrence of organic-rich, sapropel sediments in and around the Mediterranean was established by Hilgen (1991a, 1991b) back to around 5 million years ago. In both cases a knowledge of the timing of magnetic field reversals in relation to biostrati-graphic markers enabled an independent check to be made on the age of the sediments at fixed points in these correlations. The successful application of these correlations has in effect established a new and independent method of determining the precise age of the palaeomagnetic reversals recorded in deep-sea sediments (Shackleton et al. 1990), and generated the potential to carry out very high resolution, time-dependent studies of deep-sea sediments. By relating the calculated insolation variations with either the GRAPE[4] (Fig. 7.7) (mainly carbonate productivity related) density, or the $\delta^{18}O$ values recorded by the sediments in the eastern equatorial sediments sampled during Leg 138 of the Ocean Drilling Program, a correlation to orbital insolation cycles and absolute age has been extended back to 6 million years ago, with partial correlations back to 10 million years ago (Shackleton et al. 1995b). In addition to providing a close check on the timing of variations in magnetic intensity and reversals (Valet and Meynadier 1993), this correlation has also demonstrated a

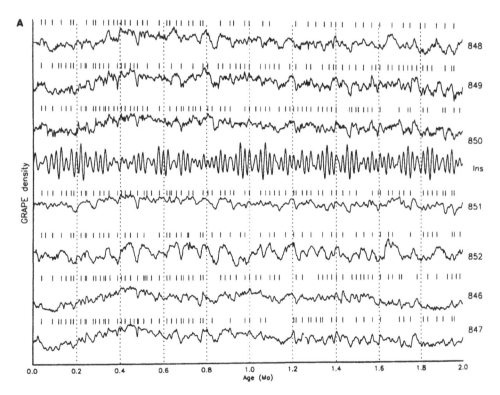

Figure 7.7 Correlation of GRAPE density records (which approximate carbonate content of the sediment and ultimately variations in calcareous planktonic organism productivity over time), with orbitally modulated insolation variations – for the past 2 million years. (After Shackleton et al. 1995b: 82, fig. 1.)

very close relationship between biological production and sedimentation and the orbital cycles (Pisias et al. 1995b) over the past 6 million years of the Earth's history, and allowed sediment samples and the dates of arrival and departure of fossil plankton in the fossil record to be determined to within 2000 years (Shackleton et al. 1995a). This high degree of accuracy provides a new perspective to resolving the problems of determining the rates of processes and the contemporaneity of changes in the whole ocean system over the longer term.

The future (IMAGES)

In spite of the exceptional successes of deep-sea drilling for understanding the long- and very long-term history of the ocean (only the sediment-derived aspects of which have been summarized above), increasing importance has recently become focused on changes during the past few hundred thousand years. This is the period when the boundary conditions of global climate and ocean circulation must have been similar to those of the present day. As a result, the focus for studies of the deep-sea sediment record is changing towards geographic- ally greater coverage of the top 50m or so of deep-sea sediments, rather than more localized investigation of the 1km or 2km thick sequences accessible to deep-sea drilling.

In 1987 the International Council of Scientific Unions initiated its International Geo- sphere–Biosphere Programme: a Study of Global Change (IGBP). IGBP is, at least in part, a response to increasing concern about potential changes in global climate caused by anthropogenic activities. It involves probably the most complex and wide-ranging collabor- ation between international scientists ever attempted. Although large sections of the programme are directed towards understanding contemporary physical, chemical and bio- logical processes in the atmosphere, hydrosphere and biosphere of planet Earth, a modest but very significant component is directed towards greater understanding of past global changes in those systems (PAGES). We already know, from previous research both in the oceans and at the land's surface, that climatic changes have been a continuing characteristic of our planet, as far back as we can determine. In the past 2.5 million years of the Ice Age in particular, extreme fluctuations of climate have affected the Earth. This climatic change was not caused by human activity. A clear understanding of the processes of natural climatic change in the past is an essential prerequisite for understanding the potential effects of anthropogenic inputs on future climatic change. Clearly, much data has already been collected on this subject, but targeting the collection of data in such a way that it can be directly related to the modelling of processes being investigated by other IGBP projects presents a new challenge.

The PAGES core project of IGBP is attempting to produce exactly that kind of palaeo- climate data. Initially it concerned itself mainly with the record that can be obtained regard- ing the past few thousand years or tens of thousands of years of climatic change, as recorded in glacier ice, tree rings, and other terrestrial data sources. In 1994, however, a new PAGES initiative in marine palaeoclimate was launched. IMAGES (International Marine Global Change Study) seeks to generate worldwide coverage of oceanic climatic change data for the past 300,000 years (Pisias et al. 1994). The IMAGES objectives are particularly suitable for investigation using giant piston corers (GPCs) (Fig. 7.8), whose depth of penetration (40–50m) usually probes sediment over the appropriate time-span, and whose diameter (110mm) provides almost three times more material for multidisciplinary investigations than the diameters of 60–70mm that have previously been used, both for most earlier piston corers and for the DSDP/ODP APC hydraulic piston cores (Weaver and Schultheiss 1990).

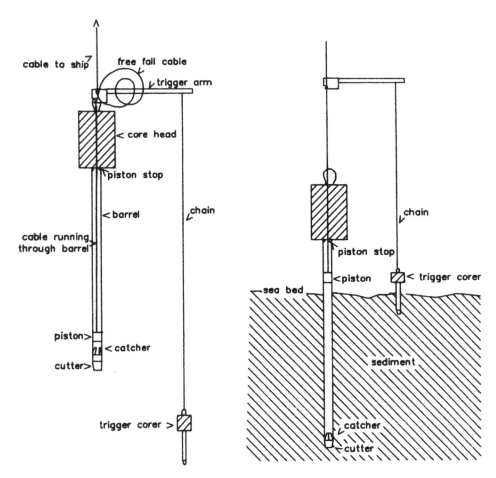

Figure 7.8 The operation of a typical piston corer. Giant piston corers differ only in their size, and the ancillary equipment required to ensure safe triggering and recovery of such large items of equipment. (After Weaver and Schultheiss 1990: 89, fig. 2.)

Although development of giant piston corers began as early as 1967, they have been consistently used for programmes of research only in recent years, most notably by the French vessel *Marion Dufresne* in connection with European marine Quaternary research programmes. The use of GPCs is now *de rigueur* in connection with the IMAGES programme. Some investigations using this technology have already been completed and others are under way. A lot more are planned, from which a detailed understanding of the late Quaternary, or past 300,000 years of the ocean's response to climatic change, can be expected.

Notes

1 Sedimentation rates in the oceans range from a maximum of 100m per million years, i.e.100mm or 10cm per thousand years, to a minimum of about 1m per million years, i.e. 1mm per thousand years, with the majority of ocean sediments accumulating at from 30 to 10m per million years, 30 to 10mm or 3 to 1cm per thousand years. In general, therefore, a 1cm thickness of ocean sediment may represent anything between 100 and 10,000 years of accumulation, and only at relatively high

sedimentation rates can millennial, let alone centennial scale changes in ocean chemistry and climate be studied. (See also Chapter 6.)

2 Similar piston-coring was adopted by the Scripps Institution of Oceanography and the National Institute of Oceanography in the UK, but, whether through lack of commitment or expertise, the results from their piston-coring often achieved little better penetration than simple gravity coring.

3 The lighter ^{16}O isotope is preferentially incorporated into glacier ice, and therefore the heavier ^{18}O is concentrated in the oceans, and incorporated into the skeletons of benthic foraminifera, in proportion to the global volume of glacier ice.

4 Gamma-Ray Attenuation Porosity Evaluator.

References and further reading

General

Crowley, T. J. and North, G. R. (1991) *Paleoclimatology*. New York: Oxford University Press.

Kennett, J. P. (1982) *Marine geology*. Englewood Cliffs, N.J.: Prentice-Hall.

Seibold, E. and Berger, W. H. (1993) *The sea floor: an introduction to marine geology*, 2nd edn. Berlin: Springer.

Thorpe, S. A. and Summerhayes, C. P. (eds) (1996) *Oceanography: an illustrated guide*. London: Manson.

Challenger *Expedition*

Brady, H. B. (1884) 'Report on the Foraminifera dredged by HMS *Challenger* during the years 1873–1876', *Report on the scientific results of the voyage of HMS* Challenger *(1873–1876)*, **9** (Zoology). London: HMSO.

Haeckel, E. (1887) 'Report on the Radiolaria collected by HMS *Challenger* during the years 1873–1876', *Report on the scientific results of the voyage of HMS* Challenger (1873–1876) **18** (Zoology). London: HMSO.

Murray, J. and Hjort, J. (1912) *The depths of the oceans*. London: Macmillan.

Murray, J. and Renard, A. F. (1891) 'Deep-sea deposits', *Report on the scientific results of the voyage of HMS* Challenger *(1873–1876)*. London: HMSO.

Wyville Thomson, C. (1877) *The voyage of the* Challenger*: the Atlantic*, 2 vols. London: Macmillan.

Swedish Deep-Sea Expedition

Arrhenius, G. (1952) 'Sediment cores from the East Pacific', *Report of the Swedish Deep-Sea Expedition (1947–1948)* 5 (1). Göteborg: Elanders.

Berggren, W. A. (1971) 'Oceanographic micropalaeontology', *EOS: Transactions, American Geophysical Union* **52**: IUGG 249–56.

Emiliani, C. and Milliman, J. D. (1966) 'Deep-sea sediments and their geological record', *Earth-Science Reviews* **1**: 105–32.

Ericson, D. B. (1961) 'Pleistocene climate record in some deep-sea sediment cores', *Annals of the New York Academy of Sciences* **95**: 537–41.

Hays, J. D. and Opdyke, N. D. (1967) 'Antarctic Radiolaria, magnetic reversals and climatic change', *Science* **158**: 1001–11.

Kullenberg, B. (1947) 'The piston core sampler', *Svenska Hydrografisk–Biologiska Kommissionens Skrifter*, 3rd series **1** (2): 1–46.

Kullenberg, B. (1955) 'Deep-sea coring', *Report of the Swedish Deep-Sea Expedition (1947–1948)* **4**(2). Göteborg: Elanders.

Phleger, F. B, Parker, F. L. and Peirson J. F. (1953) 'North Atlantic Foraminifera', *Report of the Swedish Deep-Sea Expedition (1947–1948)* **7**(1). Göteborg: Elanders.

Schott, W. (1935) 'Die Foraminiferen in dem äquatorialen Teil des Atlantischen Ozeans', *Wissenschaftliche Ergebnisse der Deutschen Atlantischen Expedition (1925–1927)* (Meteor) **3**: 43–134.

Schott, W. (1966) 'Foraminiferenfauna und Stratigraphie der Tiefsee-Sedimente im Nordatlantischen Ozean', *Report of the Swedish Deep-Sea Expedition (1947–1948)* **7**(8). Göteborg: Elanders.

Wiseman, J. D. H. (1965) 'The changing rate of calcium carbonate sedimentation on the equatorial Atlantic floor and its relation to continental late Quaternary stratigraphy', *Report of the Swedish Deep-Sea Expedition (1947–1948)* **7**(7). Göteborg: Elanders.

CLIMAP

CLIMAP Project Members (1976) 'The surface of the ice-age Earth', *Science* **191**: 1131–7.

CLIMAP Project Members (1981) *Seasonal reconstructions of the earth's surface at the last glacial maximum.* Boulder, Colo.: The Geological Society of America, Map Chart Series MC-36.

CLIMAP Project Members (1984) 'The last interglacial ocean', *Quaternary Research* **21**: 123–4.

Cline, R. M. and Hays, J. D. (eds) (1976) *Investigation of late Quaternary paleoceanography and paleoclimatology*, Memoir 145. Boulder, Colo.: The Geological Society of America.

Hays, J. D., Imbrie, J. and Shackleton N. J. (1976) 'Variations in the earth's orbit: pacemaker of the ice ages', *Science* **194**: 1121–32.

Imbrie, J. and Imbrie, K. P. (1979) *Ice ages: solving the mystery.* London: Macmillan.

Imbrie, J. and Kipp, N. G. (1971) 'A new micropalaeontological method for quantitative paleoclimatology: application to a late Pleistocene Caribbean core', in K. K. Turekian (ed.) *Late Cenozoic glacial ages.* New Haven: Yale University Press, pp. 71–181.

Shackleton, N. J. (1967) 'Oxygen isotope analyses and Pleistocene temperatures reassessed', *Nature* **215**: 15–17.

Shackleton, N. J. and Opdyke, N. D. (1973) 'Oxygen isotope and palaeomagnetic stratigraphy of equatorial Pacific core V28–238: oxygen isotope temperatures and ice volumes on a 10^5 and 10^6 year scale', *Quaternary Research* **3**: 39–55.

DSDP-ODP

Cullen, V. et al. (1994) '25 years of ocean drilling', *Oceanus* **36** (4).

Cumulative index to the initial reports of the Deep Sea Drilling Project, 1–96 (1991). Washington: National Science Foundation.

Initial Reports of the Deep Sea Drilling Project, I–XCVI (1969–86). Washington: National Science Foundation.

Open University Course Team (1989a) *The ocean basins: their structure and evolution.* Oxford: Pergamon Press.

Open University Course Team (1989b) *Ocean chemistry and deep-sea sediments.* Oxford: Pergamon Press.

Pisias, N. G., Mayer, L. A., Janecek, T. R., Palmer-Julson, A. and van Andel T. H. (eds) (1995) *Proceedings of the Ocean Drilling Program: scientific results*, vol. **138**. College Station, Tex.: Ocean Drilling Program.

Proceedings of the Ocean Drilling Program, vols **101/102** (*Initial Reports*) (1986–). College Station, Tex.: Ocean Drilling Program.

Proceedings of the Ocean Drilling Program, vols **101/102** (*Scientific Results*) (1988–). College Station, Tex.: Ocean Drilling Program.

Storms, M. A. (1990) 'Ocean Drilling Program (ODP) deep sea coring techniques', *Marine Geophysical Researches* **12**: 109–30.

Orbital chronology

Berger, A. and Loutre, M. F. (1991) 'Insolation values for the climate of the last 10 million years', *Quaternary Science Reviews* **10**: 297–317.

Hilgen, F. J. (1991a) ' Astronomical calibration of Gauss to Matuyama sapropels in the Mediterranean and implication for the geomagnetic polarity timescale', *Earth and Planetary Science Letters* **104**: 226–44.

Hilgen, F. J. (1991b) 'Extension of the astronomically calibrated (polarity) timescale to the Miocene–Pliocene boundary', *Earth and Planetary Science Letters* **107**: 349–68.

Pisias, N. G., Mayer, L. A. and Mix, A. C. (1995) 'Paleoceanography of the eastern equatorial Pacific during the Neogene: synthesis of Leg 138 drilling results', *Proceedings of the Ocean Drilling Program: scientific results*, vol. 138: 5–21. College Station, Tex.: Ocean Drilling Program.

Shackleton, N. J., Baldauf, J. G., Flores, J.-A., Iwai, M., Moore Jr, T. C., Raffi, I. and Vincent, E. (1995a) 'Biostratigraphic summary for Leg 138', *Proceedings of the Ocean Drilling Program: scientific results*, vol. 138: 517–36. College Station, Tex.: Ocean Drilling Program.

Shackleton, N. J., Berger, A. and Peltier, W. A. (1990) 'An alternative astronomical calibration of the lower Pleistocene timescale based on ODP Site 677', *Transactions of the Royal Society of Edinburgh: Earth Sciences* **81**: 251–61.

Shackleton, N. J., Crowhurst, S., Hagelberg, T., Pisias, N. G. and Schneider, D. A. (1995b) 'A new late Neogene timescale: application to Leg 138 sites', *Proceedings of the Ocean Drilling Program: scientific results*, vol. 138: 73–101. College Station, Tex.: Ocean Drilling Program.

Valet, J.-P. and Meynadier, L. (1993) 'Geomagnetic field intensity and reversals during the past four million years', *Nature* **366**: 234–8.

IMAGES

IGBP Secretariat (1990) *The International Geosphere–Biosphere Programme: a study of global change. The initial core projects.* Stockholm: IGBP.

Pisias, N., Jansen, E., Labeyrie, L., Mienert, J. and Shackleton N. J. (1994) *IMAGES International Marine Global Change Study: science and implementation plan.* Berne, Switzerland: PAGES Core Project Office.

Weaver, P. P. E. and Schultheiss, P. J. (1990) 'Current methods for obtaining, logging and splitting marine sediment cores', *Marine Geophysical Researches* **12**: 85–100.

8   Hydrothermal activity at mid-ocean ridges

Christopher R. German

Introduction

One of the most exciting developments since the beginning of the study of oceanography was the discovery in the late 1970s of submarine hydrothermal vents, associated with the volcanically active zone at the crest of the mid-ocean ridge system. Hydrothermal activity at mid-ocean ridges represents one of the fundamental processes that control the exchange of thermal energy and materials from the earth's interior to the oceans. Thus, hydrothermal interactions profoundly influence the composition of the ocean crust and sea water. In addition, hydrothermal vent areas support diverse and unique biological populations by means of microbiological communities, which link the transfer of thermal and chemical energy from the earth to the production of organic carbon.

A historical perspective

With the advent of advanced sea-floor mapping during the past few decades came the recognition of the existence of mid-ocean ridges – a vast, 50,000km long volcanic chain that encircles the Earth beneath the deep ocean and provides the driving force for plate tectonics (see Chapter 5). Since the earliest recognition of these features and their significance to plate tectonics, in the 1960s, geophysicists had argued that conductive heat flow measured through young ocean crust could not account for all the heat lost at these tectonic spreading centres. Instead, they predicted that some alternative *convective* heat transfer process must also occur near the crests of mid-ocean ridges. It was not until the late 1970s, however, that these predictions were proved correct.

Although the study of active hydrothermal venting is relatively young, evidence for unexplained emanations from the sea floor had been known for some time. During the *Challenger* Expedition, at Station 293 on the Tahiti–Valparaiso leg, a 'coherent, dark red-brown sediment' was recovered from the sea floor at a depth of 2025 fathoms (Murray and Renard 1891). This was the first documentation of ridge-crest metalliferous sediment, but it was some 60 years or more before such material received wider attention. From the 1960s onwards, various workers recognized that metal-rich sediments were associated with all the ocean's ridge crests and this was elegantly summarized by Boström and Peterson (1969), who depicted these enrichments by contouring the values of the ratio (Al+Fe+Mn)/Al for core top sediments on a global scale (Fig. 8.1). Boström, Peterson, Joensuu and Fisher (1969) invoked 'volcanic emanations' and deposition of metal-rich phases on the seabed to explain these distributions, and these explanations coincided with the first indications of hydrothermal circulation in the Red Sea (Charnock 1964, Degens and Ross 1969). With the

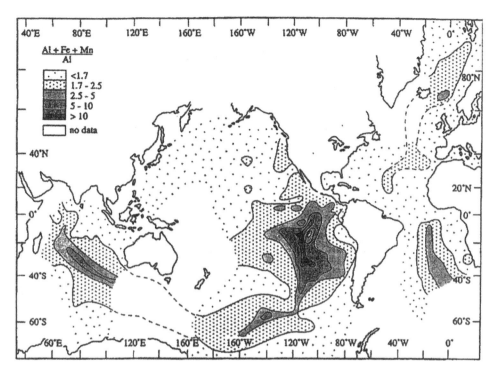

Figure 8.1 Global compilation of (Fe+Mn+Al)/Al ratios for surface sediments in the marine environment. High ratios delineate the mid-ocean ridges; low ratios represent background and pelagic sedimentation. (After Boström et al. 1969.)

advent of deep-diving submersibles such as the US DSV *Alvin*, it was not long before the mid-ocean ridges were coming under the close scrutiny of marine geologists. The first area to receive such attention was an area of the Mid-Atlantic Ridge between 36° and 37°N in the joint US–French expedition FAMOUS (Ballard and van Andel 1977, Bellaiche et al. 1974. See Chapter 5). Though that study identified many of the geological processes of volcanism and tectonism active at mid-ocean ridges, still no evidence for hydrothermal venting was discovered. Then came an expedition in 1977 to the Galapagos Spreading Centre, at which low-temperature (10–30°C) hydrothermal activity was discovered (Corliss et al. 1979) followed by a further French–American expedition in 1981 that discovered high-temperature (350°C) hydrothermal activity on the East Pacific Rise (Spiess and RISE Project Group 1980). The importance of this discovery went much further than simply proving correct the geophysicists' theories about convective heat-flow through oceanic crust. To marine geochemists, hydrothermal activity represented a new supply of chemicals to the oceans, comparable in importance to the influence of rivers flowing from the land. For marine biologists the discoveries were perhaps even more outstanding. Here, fed by the chemicals emanating from the vents, living in total darkness, and isolated from almost all other strands of evolution, were extraordinary new species. These animals were previously unsuspected and certainly not looked for, yet were soon recognized to be living in an entirely novel ecosystem whose origins could be traced back to the earliest life on Earth.

Hydrothermal vents: what are they and how do they work?

The base of the oceanic crust is extremely hot (above 1000°C), yet its upper boundary is in contact with sediments and sea water at temperatures that are only a few degrees above 0°C. The pattern of hydrothermal circulation is one in which sea water percolates downward through fractured ocean crust (the recharge zone) towards the base of the oceanic crust and, in some cases, close to molten magma (Fig. 8.2). In these hot rocks (the reaction zone) the sea water is first heated and then reacts chemically with the surrounding host basalt. As it is heated, the water expands and its viscosity reduces. If these processes occurred on land, at atmospheric pressure, catastrophic explosions would occur, as temperatures would rise above 100°C and the water would turn to steam. But, because mid-ocean ridges lie under 2000–4000m of sea water, at pressures 200–400 times greater than atmospheric pressure, the reacting sea water reaches temperatures up to 350–400°C without boiling. At these temperatures the altered fluids *do* become extremely buoyant, however, with densities only about two-thirds that of the downwelling sea water, so that they rise rapidly back to the surface (the discharge zone) as hydrothermal fluids. The movement of the fluid through the rock is such that, whereas the original downward flow proceeds by gradual percolation over a wide area, the consequent upflow is often much more rapid and tends to be focused into natural channels emerging at 'vents' on the sea floor.

Beneath the sea floor the reactions between sea water and fresh basalt remove the dissolved Mg^{2+} and SO_4^{2-} ions that are typically abundant in sea water and lead to precipitation of some sulphate and clay minerals. As the water seeps lower into the crust and the temperature rises, metals, silica and sulphide are all leached from the rock to replace the original Mg^{2+} and SO_4^{2-} ions. The hot and by now metal-rich (Fe, Mn, Cu, Zn, Pb) and sulphide-bearing fluids then ascend rapidly through the ocean crust to the sea floor (Von Damm 1995). When they begin to mix with the ambient cold, alkaline, well oxygenated deep-ocean waters, the result is instantaneous precipitation of a cloud of tiny metal-rich sulphide and

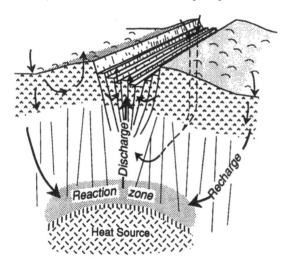

Figure 8.2 Schematic drawing illustrating the portions of submarine hydrothermal circulation discussed in the text. Sea water enters the crust in widespread 'recharge' zones and reacts at progressively increasing temperatures during penetration into the crust. High temperature (above 350°C) reactions occur in the 'reaction zone' immediately above the magmatic or hot-rock heat source, after which buoyant fluids rise rapidly along focused 'discharge' pathways. (After Alt 1995.)

oxide mineral grains. These rise within the ascending columns of hot water, giving the impression of smoke. Precipitation around the mouths of the vents over time builds chimneys through which the smoke pours, hence the term 'black smokers' (Fig. 8.3). Hot water gushes out of these tall chimney-like sulphide spires at temperatures of around 350°C and at velocities of 1–5m/sec. Upon eruption, the hydrothermal fluid continues to rise several hundred metres above the seabed, mixing with ordinary sea water all the time, in a buoyant turbulent plume (Lupton, Delaney, Johnson and Tivey 1985). In addition, more of the material initially transported in the fluid, subsurface, is precipitated in and around the vent orifices to produce large chimney-like constructs of Cu-, Zn-, Pb- and Fe-rich sulphide and Ca-sulphate minerals in highly concentrated ore-grade deposits (Hannington, Jonasson, Herzig and Petersen 1995). At slightly lower temperatures (below 330°C), the fluid may cool and mix with sea water sufficiently to deposit some metal-rich precipitates in the walls of the

Figure 8.3 Photograph of a cluster of black-smoker chimneys at the Broken Spur site (29°N), Mid-Atlantic Ridge. Acidic metal-rich fluids at temperatures of over 350°C are venting at tens of centimetres a second from chimneys that grow several metres high on the sea floor. The fauna in this area are dominated by shrimps, seen clustering around the chimney in the foreground. They are part of the ecosystem that is built up around the chemosynthetic bacteria that inhabit this extreme environment. The fluid within the chimneys is extremely clear but rich in dissolved metals and hydrogen sulphide. As soon as this fluid mixes with cold oxygen-rich sea water, at the very mouths of the vents, precipitation of a range of sulphide and oxide minerals occurs, giving rise to clouds of tiny black 'smoke' particles billowing upwards above the chimneys into the overlying sea water. (Reproduced courtesy Rachel Mills, Southampton Oceanography Centre.)

channels up which the fluids rise, before reaching the surface. In those cases, the particulate material formed when the ascending hot water finally emerges at the seabed is made up predominantly of amorphous silica and various sulphate and oxide minerals, yielding a white cloud of mineral precipitates and leading to the common name for these slightly cooler vents: 'white smokers' (Edmond et al. 1995, Tivey et al. 1995). Repeat study of the physical and chemical properties of hydrothermal vent fluids at both the East Pacific Rise (21°N) and the Mid-Atlantic Ridge (23°N) have demonstrated that the chemical composition of vent fluids at these sites can remain stable over time-scales of 5–10 years (Von Damm 1995). This need not always be the case, however. With the increase in experiments designed to identify and locate hydrothermal plumes and with increasing sophistication in the nature of our detection techniques, a spate of 'new' evolving vent sites have also been discovered in recent years, which show rapid temporal evolution, clearly linked to recent (days-, weeks-, months-old) volcanic eruptions (Fornari and Embley 1995).

Where do vents occur? How common are they?

Since their early discovery, hydrothermal vents have now been found at a range of spreading centres in a range of tectonic settings. The majority of these sites of venting are associated with tectonic plate boundaries: fast- and slow-spreading mid-ocean ridges, fracture zones and back-arc spreading centres (Fig. 8.4). Until 1984 it was widely predicted that hydrothermal activity might be restricted to fast-spreading ridges such as the East Pacific Rise and

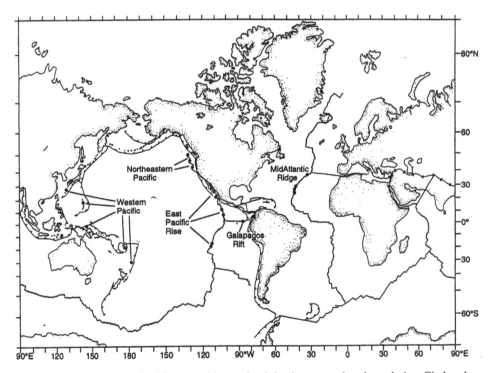

Figure 8.4 The locations of mid-ocean ridge and subduction-zone plate boundaries. Circles show locations of hydrothermal systems visited, thus far, by submersible. Some of these sites have been visited many times over periods of ten years or more. Others have enjoyed only cursory investigations. (After Hessler and Kaharl 1995.)

that at slow-spreading ridges such as the Mid-Atlantic Ridge, which exhibit relatively low magmatic/volcanic fluxes, heat-flow would be insufficient to support active high-temperature venting. It is now clear that such is not the case. Nevertheless, it was not until 1985 that detailed water-column surveys and deep-tow camera deployments identified the first-ever high-temperature vent field on the Mid-Atlantic Ridge, the TAG hydrothermal field at 26°N (Rona et al. 1986). Subsequent studies between the Kane and Atlantis fracture zones have confirmed the presence of at least two further high-temperature vent sites in the region, the Snakepit hydrothermal field at 23°N (Detrick et al. 1986) and the Broken Spur vent field at 29°N (Murton et al. 1994). Crudely, it remains the case that the abundance of hydrothermal activity along any section of mid-ocean ridge correlates directly with that ridge's spreading rate (Baker, German and Elderfield 1995). Nevertheless, with increased understanding of the nature of hydrothermal venting, and improved techniques for detecting and locating such activity, new sites are continuing to be discovered throughout the world's oceans – even at sites where it had previously been declared absent, including the first-ever submersible-investigated FAMOUS segment of the Mid-Atlantic Ridge (German et al. 1996).

Life at hydrothermal vents

Few events have come as such a surprise to oceanographers as the discovery of entire communities of bizarre animals thriving around hydrothermal vents. Life on the deep ocean floor is typically scarce because sources of food are also scarce (Gage and Tyler 1991). Further, hydrothermal fluids are enriched in dissolved hydrogen sulphide, a substance toxic to most forms of life and which is only found elsewhere in the oceans in lifeless and stagnant anoxic basins such as the Black Sea. It is quite remarkable, therefore, that sites of active hydrothermal venting are *not* barren wastelands where no life can exist. Instead, scientists diving at the very first vent site to be discovered, on the Galapagos rift in the eastern equatorial Pacific, were surprised to find dense concentrations of seabed-dwelling (benthic) large animals (megafauna) living within the vent-field area (Corliss et al. 1979). Since that time, new vent sites have been discovered around the world – and most have remarkably high concentrations of animals around them. The reason why these areas attract such abundant life to them is even more remarkable. A common observation at all the vent fields that have been studied to date has been that individual animals at any site often appear to be extremely large. This has raised the question: 'Where do the animals get food to grow at all, let alone enough to reach such a size?' The answer to this question is that hydrothermal vent animals derive their food from a food chain that is driven by geothermal (terrestrial) energy, unlike all other organisms that rely, directly or indirectly, upon sunlight for their survival. In hydrothermal vent communities, free-living bacteria that are anchored on the seabed or float free in the water column coexist with *symbiotic* sulphide-oxidizing bacteria that live within the larger vent-specific organisms, exploiting the free energy of reaction released when the hydrogen sulphide present in the vent fluids interacts with the dissolved carbon dioxide and oxygen in ordinary sea water to form organic matter:

$$CO_2 + H_2S + O_2 + H_2O = CH_2O + H_2SO_4.$$
$$(CARBOHYDRATE)$$

Because it is a *chemical*, hydrogen sulphide, which plays the role that sunlight plays in the more familiar process of photosynthesis in the warm surface waters of the oceans and on land, this unique deep-ocean, sunlight-starved process has been given the name *chemo*synthesis.

Approximately 95 per cent of all animals discovered at hydrothermal vent sites have been previously unknown species. So far, over 300 new species have been identified and for many of these the differences from previously known fauna are so great that new taxonomic families have had to be established in order for them to be classified satisfactorily (Tunnicliffe 1991). Some of the most exciting examples of hydrothermal vent species discovered include the spectacular tube-worms found along the East Pacific Rise vent sites, which can measure 2–3m long or more and which typically appear in thick clusters. Also common along the East Pacific Rise are giant clams and mussels, which can often reach the size of a large dinner plate. The total biomass at any one hydrothermal site is typically very high. Indeed, hydrothermal vent fields have been likened to submarine oases that punctuate the deserted barren plains of the deep-sea floor (Hessler and Kaharl 1995). In contrast, biodiversity at individual vent sites (i.e. the total range of different species present) is surprisingly low. Not only that, but the species present at vent sites in the different oceans show remarkably little similarity. For example, no giant tube-worms and giant clams have been found at any of the five known hydrothermal fields discovered so far in the North Atlantic Ocean. Instead, for example, the southernmost three of these sites are characterized by abundant small blind shrimp, which cluster in their thousands around the black-smoker chimneys (some can be seen in Fig. 8.3) (Van Dover 1995).

These completely isolated biological communities point to separate paths of evolution over many generations, perhaps stretching back millions of years. An important question, therefore – particularly given the globe-encircling nature of mid-ocean ridges – is: How do animals move from one hydrothermal field to colonize another on an intra-basinal scale, yet live within communities which, on an inter-basinal scale, are evidently quite isolated? One key area of research currently under way involves mapping the entire mid-ocean ridge system in sufficient detail to determine the total number of hydrothermal vent sites worldwide, their average spacing one from another, and how that spacing is controlled by the tectonic and volcanic nature of the mid-ocean ridges that host them. The second key issue to explain the biodiversity and separate evolution problems is to understand how animals reproduce and migrate along the lengths of these ridges. It is proposed that the processes of reproduction and migration must be intimately related to each other because a large majority of vent specific organisms exhibit quite sessile life-styles as adults. Therefore, only if these species can give rise to planktonic larval stages might there be any prospect for them to be able to migrate along mid-ocean ridge axes and colonize new vent fields as and when they occur (Mullineaux and France 1995). We know that individual vent fields and chimneys may remain active for periods of perhaps only 100–1000 years at a time before the chimneys themselves become choked with minerals and the flow of warm fluids becomes blocked. For any one species of hydrothermal organism to have survived down through the generations, therefore, we know that the ability to migrate from one vent site to another must have been vitally important.

Sea-floor hydrothermal deposits

Hydrothermal deposits on the sea floor range from single chimneys to large sprawling mounds topped by clusters of chimneys (Hannington et al. 1995). In general the size of hydrothermal deposits appears to be related to the length of time over which active venting has persisted at those sites. Pacific vent fields, with very localized chimney deposits, have life-spans of just tens of years to perhaps a hundred years or more before mineral precipitation cements their chimneys solid. By contrast, certain Atlantic sites are characterized by mounds stretching in excess of 200m across and reaching 30–50m high, which dating reveals have remained active for thousands and even tens of thousands of years (Lalou et al. 1993).

One such site, which has been studied extensively by the international community, is called TAG (named after a North Atlantic basin study in the 1970s – the Trans-Atlantic Geotraverse). The TAG site comprises one large active mound and several mounds that are now extinct. The deposit is one of the largest active sites known and is approximately 200m in diameter and 50m high. There is a vast range in the style of fluid venting and mineral deposits at different sites on the mound. Fluid flow through the mound is pervasive, fed by a complex network of channels through which high-temperature fluid flows, rising through the mound from the underlying basement (Fig. 8.5). Hot (360°C) acidic fluids gush from the summit at speeds of several metres per second and the whole mound apex is shrouded in black smoke, which is rapidly entrained upwards into a buoyant hydrothermal plume. Those chimneys that have been sampled are 1–10cm in diameter and consist predominantly of copper sulphides and iron sulphides (chalcopyrite, marcasite, bornite, pyrite) and calcium sulphate (anhydrite). Lower temperature 'white smokers' vent at rates of centimetres per second to the southeast of the main black smoker complex. Here the temperatures are up to 300°C and the chimneys are bulbous and zinc-sulphide (sphalerite) rich. The white 'smoke' consists of amorphous silica mixed with zinc sulphides and iron sulphides. Much of the rest of the mound surface is covered with red and orange iron oxides, through which diffuse, low-temperature fluids percolate. Areas of diffuse flow are delineated by clusters of white anemones (Fig. 8.6) and shimmering water (Thompson et al. 1988, Tivey et al. 1995). The range in fluid venting styles is evident from the distribution of fauna over the mound surface. Three different species of shrimp have been discovered at TAG; the most abundant is *Rimicaris exoculata*, which swarms over the black smoker edifice. The lower-temperature diffuse flow hosts a community of anemones and crabs. The distribution of organisms can give semi-quantitative information on the fluid flow regime (Van Dover 1995).

Hydrothermal activity is intermittent even at individual sites. Inactive chimneys have been discovered on the outer portions of the main mound and these have been oxidized to orange iron oxides, which are beginning to crumble and collapse. Eventually they will be completely weathered to the mound surface. The mound itself is unstable and subject to collapse and mass-wasting events in which submarine equivalents of landslides, perhaps triggered by minor earthquakes, sweep altered chimney material down off the slopes of the mound and out onto the surrounding sea-floor sediments. As a result, the sediments immediately adjacent to the flanks of the mound are also full of metal-rich sulphide and oxide minerals, just like the hydrothermal chimneys from which they are derived. The metal concentrations preserved within these sediments remain extraordinarily high – up to 45 per cent iron, 34 per cent copper, and as much as 10–15 per cent zinc (German et al. 1993, Mills, Elderfield and Thomson 1993).

Eventually, hydrothermal sediments are buried beneath subsequent lava flows and/or normal background pelagic sediments and are transported away from the ridge axis by plate-spreading processes. Deep-sea sediments have been drilled by the Ocean Drilling Program at over 700 sites. On those occasions where drilling has reached all the way to basement rock, a metal-rich layer has often been observed at the bottom of the sediment pile, representing ancient metalliferous sediments (Leinen 1981). Fossil records of hydrothermal activity have also been found on land. Ancient massive sulphides as old as 3.5 billion years have been identified in many regions hosting Archaean-age rocks or younger, for example in Western

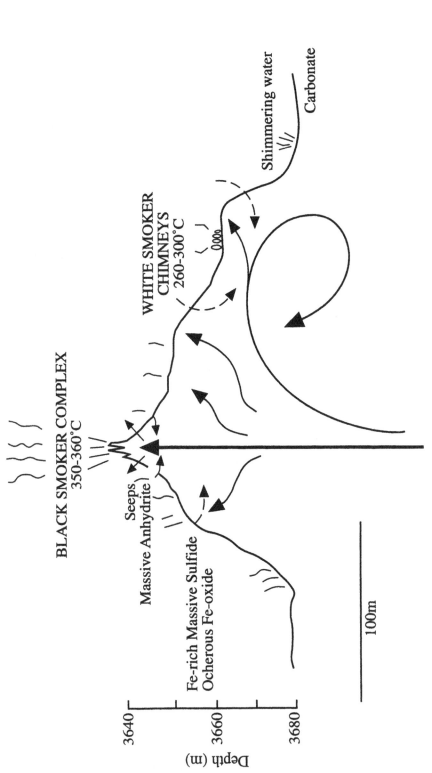

Figure 8.5 Schematic cross-section of the TAG active hydrothermal mound from the northwest to the southeast, showing suggested flow patterns, with entrainment of sea water into the upper conical edifice and into the mound as a whole. Flow patterns were derived from detailed studies of the mineralogy and chemistry of the deposits and the chemistry of the white-smoker and black-smoker fluids. (After Tivey et al. 1995.)

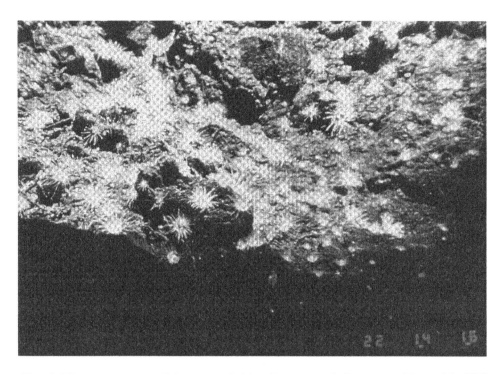

Figure 8.6 Deep-sea anemones living on metal-rich sediment towards the outer periphery of the TAG hydrothermal mound, 26°N, Mid-Atlantic Ridge. (Photograph reproduced by courtesy of Chris German, Southampton Oceanography Centre.)

Australia and South Africa (Barley 1992, de Ronde, de Wit and Spooner 1994). Detailed mineralogical and geochemical investigations of these deposits suggest that the nature of hydrothermal circulation in the ancient oceans may not have differed fundamentally from that in the modern world. This, in turn, lends further fuel to the argument that life at mid-ocean ridges may be of particular relevance to the origins of life on Earth (Holm et al. 1992).

Hydrothermal plumes

On the East Pacific Rise, the spreading plate boundary is sharply defined by a shallow axial summit caldera that sits on top of the axial volcanic ridge and is typically only a few hundreds of metres wide and a few tens of metres deep. Hydrothermally active sources are typically confined to this narrow band along the trend of the ridge axis, but their effects can often be detected much further afield. When fluids erupt from a black smoker they are hot and buoyant, and they rise, mixing turbulently with local sea water, carrying their mix of particles and fluid with them. Eventually, this rising plume reaches a stage where it is no longer more buoyant than the overlying water column and a level of neutral buoyancy is attained (Lupton et al. 1985). This particle-rich fluid is then dispersed as an approximately horizontal layer flowing along constant-density (isopycnal) surfaces. The exact height of rise of a hydrothermal plume is a function of both the strength of the hydrothermal source itself and the intensity of the stratification of the water column into which it is emitted (Speer and Rona 1989). For hydrothermal vents along the East Pacific Rise, the height of rise is typically of the order of 100m – that is, sufficient to rise clear of the surrounding, constraining topography – such that the resulting plume effluent can be dispersed very widely across the Pacific Ocean.

Because hydrothermal fluids are extremely enriched in certain key tracers (e.g. dissolved Mn, CH_4, ^3He), relative to typical oceanic deep waters, chemical anomalies associated with hydrothermal plumes can often be detected at significant distances away from hydrothermal vent sites. One particularly good example is the hydrothermal plume at around 15°S on the East Pacific Rise. Lupton and Craig (1981) demonstrated that ^3He enrichments emitted from this portion of the EPR could be traced over distances up to 2000km away from the ridge axis (Fig. 8.7). Interestingly, the oceanic circulation patterns linked to this dispersion of dissolved hydrothermal ^3He coincided almost exactly with the contours of anomalous metal enrichments that had been mapped previously in the underlying surface sediments from the flanks of this section of the East Pacific Rise (Boström et al. 1969). Dissolved manganese (Mn) and methane (CH_4) are enriched approximately 10^6-fold over ordinary sea water in high-temperature vent fluids and so, even though these fluids undergo approximately 10^4-fold dilution in buoyant hydrothermal plumes, neutrally buoyant hydrothermal plumes directly above hydrothermal vent sites exhibit dissolved Mn and CH_4 concentrations that are approximately 100-fold enriched relative to typical oceanic deep water, in addition to their high suspended particulate matter concentrations (Lilley, Feely and Trefry 1995). Because neutrally buoyant plumes overlie a much greater area of the mid-ocean ridge crest than is occupied by active hydrothermal chimneys and mounds, these water-column enrichments afford geochemists an important and valuable tool with which to prospect for and predict the occurrence of new hydrothermal vent sites (Baker et al. 1995).

An important role that hydrothermal plumes play in the marine geochemistry of submarine vents is that they modify the raw flux of dissolved chemicals to the oceans. In this regard, they can be seen as analogous to the estuaries that play their part in modifying the raw fluxes of dissolved material from rivers to the oceans. In addition to the initial precipita-

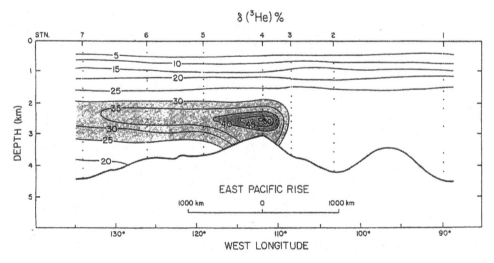

Figure 8.7 Contours of δ (^3He) in a zonal section spanning the East Pacific Rise at 15°S, showing a huge plume of ^3He-rich water emanating from the ridge crest and extending over 2000km to the west. The samples for this section were collected during South Tow expedition in 1972. (After Lupton 1995.)

tion of dissolved metals to form particulate oxide and sulphide phases, the hydrothermal particles, once formed, may also scavenge further dissolved species from the ambient water column, such that – in the extreme case – hydrothermal circulation may act as a net sink for some elements in the dissolved chemical budget of the oceans (Kadko, Baross and Alt 1995).

Recent developments in the study of hydrothermal plumes have included the application of plume theory to help understand the physical dispersion of hydrothermal products through the oceans. One of the most intriguing theories has been the proposal that buoyant hydrothermal plumes should force a baroclinic vortex, which is unstable and would be expected to give rise to isolated 'hydrothermal eddies' that propagate away from their source (Speer and Helfrich 1995). A natural development of this theory is the proposal that oceanic hydrothermal plumes might shed isolated eddies, which are capable of transporting chemically and physically anomalous characteristics long distances away from individual vent sources. No such features have been positively identified in the real oceans yet, but the possibility that they might exist raises some very important questions.

If hydrothermal eddies are formed regularly above vent sites, could they represent 'incubating' parcels of chemically enriched sea water in which young larval stages of vent fauna might survive for hundreds of days as they are carried tens of kilometres up and down the world's mid-ocean ridge axes? Such a mechanism might go a long way towards resolving the key problem of how vent fauna colonize newly formed hydrothermal vent sites. In the extreme case, on a fast-spreading section of mid-ocean ridge crest where abundant hydrothermal activity occurs at very close spacing, could the cumulative heat flux and consequent action of many hydrothermal eddies acting in concert be of sufficient magnitude to begin to exert some control on the general physical circulation of the deep-water column over the length scale of entire ocean basins?

Diffuse flow versus black smoker vents

In recent work, Stein and Stein (1994) have re-evaluated the magnitude and age-variation of hydrothermal heat-flow at mid-ocean ridges. Their study, based on an updated global heat-flow model, predicts that no more than 30 per cent of hydrothermal heat-flow occurs in crust younger than 1Ma, whereas the majority occurs off-axis, away from active spreading centres, in crust up to 70Ma. Further, for the approximately 30 per cent of hydrothermal heat-flow that *is* concentrated on young oceanic crust, on-axis, not all may be linked to focused black smoker venting. Increasing attention has been paid, in recent years, to the importance of diffuse venting at axial hydrothermal fields (Rona and Trivett 1992, Schultz, Delaney and McDuff 1992, Baker, Massoth, Walker and Embley 1993). Diffuse-flow fluids may follow tortuous routes along networks of fractures through fresh or altered crust or hydrothermal sulphide structures. Consequently, the resultant fluids emerging from the sea floor might be expected to exhibit a lower temperature than the primary high-temperature fluid, to be diluted with local (un)altered sea water, and to exhibit a chemistry which is not a direct linear mixture of vent fluid and sea water because non-conservative mixing would occur for many of the dissolved species originally present in the high-temperature fluid. Current best estimates indicate not only that on-axis hydrothermal circulation may account for only about 30 per cent of the total hydrothermal heat-flux that escapes through 0–65Ma oceanic crust but, further, that of this 30 per cent, only some 10 per cent (i.e. just 3 per cent of the global total) may be concentrated through focused black-smoker type chimneys – the remainder being dispersed through lower-temperature, diffuse, on-axis venting. To date, such studies have been restricted almost exclusively to one particular sub-section of the Juan de Fuca Ridge, northeast Pacific. However, if the result obtained there proves to be the general case, then there will be serious implications for, for example, predicted global fluxes of many non-conservative elements that have been calculated previously, assuming that all on-axis heat-flow *is* derived from high-temperature black smoker venting. Also, because it is this lower-temperature axial hydrothermal flow that gives rise to those environmental niches most commonly associated with vent-specific organisms, the occurrence of axial diffuse-flow venting may play an important role in the colonization of hydrothermal fields by vent-specific organisms, which must migrate along-axis from one hydrothermal 'oasis' to another.

Hydrothermal activity as a natural analogue for anthropogenic pollution

Recently, German and Angel (1995) have argued that the net flux of hydrothermal metals to the oceans is insignificant when compared to that from global anthropogenic pollution. Though raw fluxes from rivers and from the mouths of high-temperature black-smoker vents may agree to within as close as, perhaps, an order of magnitude for some elements, the very efficiency of hydrothermal plumes ensures that very little of this material escapes as a true dissolved metal flux to the surrounding oceans. Instead, the vast majority of hydrothermal material is deposited to the immediately adjacent seabed producing the high incidences of metal-rich sulphide and oxide-rich sediments (Mills and Elderfield 1995) that host the abundant life at hydrothermal vents (discussed earlier). Controversially, Nisbet and Fowler (1995) have subsequently adopted this argument to propose that one could safely use such environments to dispose of anthropogenic waste products such as the redundant Brent Spar oil storage platform. Though this may be seen as a rather extreme interpretation, it is undoubtedly the case that hydrothermal vent fields host a range of exotic animals that evidently thrive in what, to prior human understanding, has been considered as an extremely hostile, toxic and environmentally unfriendly habitat. Because the study of hydrothermal systems is so young, full identification of the species present is incomplete and even for those 350–400 new species discovered to date, little of their lifestyles, physiology, and so on is well understood. It is not unreasonable to expect, therefore, that many secrets as to how these organisms cope with and, indeed, thrive upon this environment, remain to be discovered. The opportunities for remedial application in our own near-shore coastal polluted waters are without doubt, therefore, both great and very real. It is to be hoped that these issues will benefit from continuing mid-ocean ridge research studies significantly in the years and decades to come.

References

Alt, J. C. (1995) 'Subseafloor processes in mid-ocean ridge hydrothermal systems', *American Geophysical Union Geophysical Monograph* **91**: 85–114.

Baker, E. T., German, C. R. and Elderfield, H. (1995) 'Hydrothermal plumes over spreading-center axes: global distributions and geological inferences', *American Geophysical Union Geophysical Monograph* **91**: 47–71.

Baker, E. T., Massoth, G. J., Walker, S. L. and Embley, R. W. (1993) 'A method for quantitatively estimating diffuse and discrete hydrothermal discharge', *Earth and Planetary Science Letters* **118**: 235–49.

Ballard, R. D. and van Andel, T. H. (1977) 'Project FAMOUS: operational techniques and American submersible operations', *Geological Society of America Bulletin* **88**: 495–506.

Barley, M. E. (1992) 'A review of Archean volcanic-hosted massive sulfide and sulfate mineralization in Western Australia', *Economic Geology* **87**: 855–72.

Bellaiche, G., Cheminee, J. L., Francheteau, J., Hekinian, R., LePichon, X., Needham, H. D. and Ballard, R. D. (1974) 'Inner floor of the Rift Valley: first submersible study', *Nature* **250**: 558–60.

Boström, K. and Peterson, M. N. A. (1969) 'The origin of aluminum-poor ferromanganoan sediments in areas of high heat flow on the East Pacific Rise', *Marine Geology* **7**: 427–47.

Boström, K., Peterson, M. N. A., Joensuu, O. and Fisher, D. E. (1969) 'Aluminum-poor ferromanganoan sediments on active oceanic ridges', *Journal of Geophysical Research* **74**: 3261–70.

Charnock, H. (1964) 'Anomalous bottom water in the Red Sea', *Nature* **203**: 591.

Corliss, J. B., Dymond, J., Gordon, L. I., Edmond, J. M., Von Herzen, R. P., Ballard, R. D., Green, K., Williams, D., Bainbridge, A., Crane, K. and van Andel, T. H. (1979) 'Submarine thermal springs on the Galapagos Rift', *Science* **203**: 1073–83.

de Ronde, C. E. J., de Wit, M. J. and Spooner, E. T. C. (1994) 'Early Archean (>3.2Ga) Fe-oxide-rich, hydrothermal discharge vents in the Barberton greenstone belt, South Africa', *Geological Society of America Bulletin* **106**: 86–104.

Degens, E. T. and Ross, D. A. (1969) *Hot brines and recent heavy metal deposits in the Red Sea*. New York: Springer Verlag.

Detrick, R. S., Honnorez, J. and ODP Leg 106 Scientific Party (1986) 'Drilling the Snake Pit hydrothermal sulphide deposit on the Mid-Atlantic Ridge, Lat. 23°22' N', *Geology* **14**: 1004–7.

Edmond, J. M., Campbell, A. C., Palmer, M. R., Klinkhammer, G. P., German, C. R., Edmonds, H. N., Elderfield, H., Thompson, G. and Rona, P. (1995) 'Time series studies of vent fluids from the TAG and MARK sites (1986, 1990) Mid-Atlantic Ridge: a new solution chemistry model and a mechanism for Cu/Zn zonation in massive sulphide ore bodies', in L. M. Parson, C. L. Walker and D. R. Dixon (eds) *Hydrothermal vents and processes. Geological Society Special Publication* **87**: 77–86.

Fornari, D. J. and Embley, R. W. (1995) 'Tectonic and volcanic controls on hydrothermal processes at the mid-ocean ridge: an overview based on near-bottom and submersible studies', *American Geophysical Union Geophysical Monograph* **91**: 1–46.

Gage, J. D. and Tyler, P. A. (1991) *Deep-sea biology: a natural history of organisms at the deep-sea floor*. Cambridge: Cambridge University Press.

German, C. R. and Angel, M. V. (1995) 'Hydrothermal fluxes of metals to the oceans: a comparison with anthropogenic discharge', in L. M. Parson, C. L. Walker and D. R. Dixon (eds) *Hydrothermal vents and processes. Geological Society Special Publications* **87**: 365–72.

German, C. R., Higgs, N. C., Thomson, J., Mills, R. A., Elderfield, H., Blusztajn, J., Fleer, A. P. and Bacon, M. P. (1993) 'A geochemical study of metalliferous sediments from the TAG hydrothermal field, 26°08'N, Mid-Atlantic Ridge', *Journal of Geophysical Research* **98**: 9683–92.

German, C. R., Parson, L. M. and HEAT Scientific Team (1996) 'Hydrothermal exploration near the Azores Triple-Junction: tectonic control of venting at slow-spreading ridges?', *Earth and Planetary Science Letters*, **138**: 93–104.

Hannington, M. D., Jonasson, I. R., Herzig, P. M. and Petersen, S. (1995) 'Physical and chemical processes of seafloor mineralization at mid-ocean ridges', *American Geophysical Union Geophysical Monograph* **91**: 115–57.

Hessler, R. R. and Kaharl, V. A. (1995) 'The deep-sea hydrothermal vent community: an overview', *American Geophysical Union Geophysical Monograph* **91**: 72–84.

Holm, N. G., Cairns-Smith, A. G., Daniel, R. M., Ferris, J. P., Hennet, R. J.-C., Shock, E. L., Simoneit, B. R. T. and Yamagawa, H. (1992) 'Future research', in N. G. Holm (ed.) *Marine hydrothermal systems and the origin of life*. Reprinted from *Origins of Life and Evolution of the Biosphere* **22**: 181–90. Dordrecht: Kluwer Academic.

Kadko, D., Baross, J. and Alt, J. (1995) 'The magnitude and global implications of hydrothermal flux', *American Geophysical Union Geophysical Monograph* **91**: 446–66.

Lalou, C., Reyss, J.-L., Brichet, E., Arnold, M., Thompson, G., Fouquet, Y. and Rona, P. A. (1993) 'New age data for Mid-Atlantic Ridge hydrothermal sites: TAG and Snakepit chronology revisited', *Journal of Geophysical Research* **98**: 9705–13.

Leinen, M. (1981) 'Metal-rich sediments from northeastern Pacific Deep Sea Drilling Project sites', *Initial Reports of the Deep Sea Drilling Project* **63**: 667–76.

Lilley, M. D., Feely, R. A. and Trefry, J. H. (1995) 'Chemical and biochemical transformations in hydrothermal plumes', *American Geophysical Union Geophysical Monograph* **91**: 369–91.

Lupton, J. E. (1995) 'Hydrothermal plumes: near and far field', *American Geophysical Union Geophysical Monograph* **91**: 317–46.

Lupton, J. E. and Craig, H. (1981) 'A major helium-3 source at 15°S on the East Pacific Rise', *Science* **214**: 13–18.

Lupton, J. E., Delaney, J. R., Johnson, H. P. and Tivey, M. K. (1985) 'Entrainment and vertical transport of deep ocean water by buoyant hydrothermal plumes', *Nature* **316**: 621–3.

Lutz, R. A. and Kennish, M. J. (1993) 'Ecology of deep-sea hydrothermal vent communities: a review', *Reviews of Geophysics* **31**: 211–42.

Mills, R. A. and Elderfield, H. (1995) 'Hydrothermal activity and the geochemistry of metalliferous sediment', *American Geophysical Union Geophysical Monograph* **91**: 392–407.

Mills, R., Elderfield, H. and Thomson, J. (1993) 'A dual origin for the hydrothermal component in a metalliferous sediment core from the Mid-Atlantic Ridge', *Journal of Geophysical Research* **98**: 9671–81.

Mullineaux, L. S. and France, S. C. (1995) 'Dispersal mechanisms of deep-sea hydrothermal vent fauna', *American Geophysical Union Geophysical Monograph* **91**: 408–24.

Murray, J. and Renard, A. F. (1891) 'Deep-sea deposits', *Report on the scientific results of the voyage of HMS Challenger during the years 1873–1876*. London: HMSO.

Murton, B. J., Klinkhammer, G., Becker, K., Briais, A., Edge, D., Hayward, N., Millard, N., Mitchell, I., Rouse, I., Rudnicki, M., Sayanagi, K., Sloan, H. and Parson, L. (1994) 'Direct evidence for the distribution and occurrence of hydrothermal activity between 27–30°N on the Mid-Atlantic Ridge', *Earth and Planetary Science Letters* **125**: 119–28.

Nisbet, E. G. and Fowler, C. M. R. (1995) 'Is metal disposal toxic to deep oceans?' *Nature* **375**: 715.

Rona, P. A. and Trivett, D. A. (1992) 'Discrete and diffuse heat transfer at ASHES vent field, Axial Volcano, Juan de Fuca Ridge', *Earth and Planetary Science Letters* **109**: 57–71.

Rona, P. A., Klinkhammer, G., Nelsen, T. A., Trefry, J. H. and Elderfield, H. (1986) 'Black smokers, massive sulphides and vent biota at the Mid-Atlantic Ridge', *Nature* **321**: 33–7.

Schultz, A., Delaney, J. R. and McDuff, R. E. (1992) 'On the partitioning of heat flux between diffuse and point-source seafloor venting', *Journal of Geophysical Research* **97**: 12,299–314.

Speer, K. G. and Helfrich, K. R. (1995) 'Physics of hydrothermal plumes', in L. M. Parson, C. L. Walker and D. R. Dixon (eds) *Hydrothermal vents and processes. Geological Society Special Publication* **87**: 373–85.

Speer, K. G. and Rona, P. A. (1989) 'A model of an Atlantic and Pacific hydrothermal plume', *Journal of Geophysical Research* **94**: 6213–20.

Spiess, F. N. and RISE Project Group (1980) 'East Pacific Rise: hot springs and geophysical experiments', *Science* **207**: 1421–33.

Stein, C. A. and Stein, S. (1994) 'Constraints on hydrothermal heat flux through the oceanic lithosphere from global heat flow', *Journal of Geophysical Research* **99**: 3081–95.

Thompson, G., Humphris, S. E., Schroeder, B., Sulanowska, M. and Rona, P. A. (1988) 'Active vents and massive sulfides at 26°N (TAG) and 23°N (Snakepit) on the Mid-Atlantic Ridge', *Canadian Mineralogist* **26**: 697–711.

Tivey, M. K., Humphris, S. E., Thompson, G., Hannington, M. D. and Rona, P. A. (1995) 'Deducing patterns of fluid flow and mixing within the TAG active hydrothermal mound using mineralogical and geochemical data', *Journal of Geophysical Research* **100**: 12527–55.

Tunnicliffe, V. (1991) 'The biology of hydrothermal vents: ecology and evolution', *Oceanography and Marine Biology Annual Review* **29**: 319–407.

Van Dover, C. L. (1995) 'Ecology of Mid-Atlantic Ridge hydrothermal vents', in L. M. Parson, C. L. Walker and D. R. Dixon (eds) *Hydrothermal vents and processes. Geological Society Special Publication* **87**: 257–94.

Von Damm, K. L. (1995) 'Controls on the chemistry and temporal variability of seafloor hydrothermal fluids', *American Geophysical Union Geophysical Monograph* **91**: 222–47.

Part 3

Ocean circulation

Introduction
Ocean circulation

Henry Charnock and Margaret Deacon

'It is one of the chief points of a seaman's duty to know where to find a fair wind, and where to fall in with a favourable current.' That remark is attributed by Admiral Robert FitzRoy, who in his younger days as Captain of HMS *Beagle* had invited Charles Darwin to sail as naturalist on a surveying voyage to South America, to Captain Basil Hall (1788–1844), but the importance of winds to navigation had been appreciated much earlier. We have inherited evocative names for the winds of the world: the Roaring Forties, the Horse Latitudes, the Trade Winds and the Doldrums.

Over most of the ocean the importance of the surface currents was less immediately recognized, apart from the narrow swift flows near the western boundaries of the ocean, which were soon apparent. Its situation on the lucrative trade route to the east led to early awareness of the Agulhas Current, by Bartholomew Diaz and by Vasco da Gama at the end of the fifteenth century. By 1513 the Gulf Stream (specifically the Florida Current) had been described by Ponce de Leon.

Before the availability of chronometers, the weaker currents, away from the western boundaries, were almost impossible to measure – the delineation of the general circulation of the ocean awaited later workers, such as James Rennell (1832). Even today, the surface currents shown in climatological atlases are based on the difference between the course steered and the course made good by merchant vessels. They show the major contrast between the western boundary currents and those of the ocean interior, but it must be remembered that the boundary currents vary in space and time, so that the climatological average shows a current like the Gulf Stream, for example, as much broader and steadier than it would be found on any particular occasion.

Rennell was one of the first to attempt systematic collection and analysis of information on surface currents and winds – something made possible by the general introduction of the marine chronometer around 1800. However, there was already speculation that density differences, caused by variations in either the salinity or temperature of the sea water, or both, might lead to deep circulation within the ocean (Deacon 1985, 1997) – what ocean-ographers now call the thermohaline circulation. Controversy over which of these two mechanisms, winds or density-driven circulation, was responsible for major ocean currents such as the Gulf Stream, played a decisive role in the dispatch of the *Challenger* Expedition. During the voyages of the *Lightning* and *Porcupine*, between 1868 and 1870 (see General Introduction and Chapter 1), one of the scientists involved, W. B. Carpenter, though primarily a zoologist, became increasingly fascinated by deep-sea temperature data collected by the ship's officers, and how the distribution of marine organisms was clearly related to it (Deacon 1997).

Carpenter's version of density theory was challenged by James Croll, who thought, mistakenly, that admitting such ideas would undermine the importance of changing wind patterns, especially trade winds, and their effect on surface water movements, which in his view had played a crucial role in past climate change, leading to the onset of the Ice Age (Hamlin 1982). Carpenter's enthusiasm for collecting new data to support his position in the debate led him to conceive the idea of a circumnavigation (see Chapter 2) and to champion it with great vigour. However, as Tony Rice points out in Chapter 1, for a variety of reasons the study of ocean circulation was the one aspect of the main areas of investigation on board the *Challenger* that failed to give significant results. It was not until long after that the expedition's data were used, together with later material, to produce meridional sections showing the distribution of Atlantic water masses, by members of the German *Meteor* expedition in the 1920s.

Both the theoretical understanding and detailed observation of ocean circulation have been twentieth-century developments. Effective work began with Scandinavian scientists in the early 1900s (Schlee 1973), and was to a large extent inspired by Arctic exploration (see Chapter 9). The winds and currents of the Arctic Ocean were of crucial importance to Fridtjof Nansen in his attempt to sail to the North Pole in the *Fram*. A friend and colleague of John Murray, well acquainted with the *Challenger*'s work, Nansen passed on his scientific observations to scientists at home, resulting in the crucial discovery by V. W. Ekman (1905) that the effect of the rotation of the Earth leads to a mass transport of water at right angles to the surface wind direction. Developments in Scandinavia and Germany during the next few decades led to identification of the geostrophic and other forces that set in motion and influence ocean currents, and how to calculate their effect – the science of dynamical oceanography. This was the scientific tradition in which Harald Ulrik Sverdrup trained and worked, though not without difficulty as shown in Chapter 9, and which he helped to transport to the New World.

The foundations for modern ideas about ocean circulation were laid in European centres during the first half of the twentieth century, but the group of papers that have defined the course of more recent research were published in North America during the immediate postwar period. In 1947 Sverdrup found that the variation of wind from place to place would produce currents in the meridional (north–south) direction. In 1948 Henry Stommel, working independently of Sverdrup on the east coast of the USA, demonstrated in a classic paper that the strong western boundary currents were a consequence of the variation with latitude of the force due to the rotation of the Earth. A development of Stommel's model by Walter Munk (1950), formerly one of Sverdrup's pupils, at the Scripps Institution of Oceanography, related the climatological distribution of the wind to the major currents of the Northern Hemisphere oceans in a generally satisfying way. At that time Stommel believed that not only the Gulf Stream, but the entire horizontal circulation of the world's oceans, might be attributed quantitatively to the wind stress. But he added that a quantitative study of the thermohaline circulation had not been made, so there remained a possibility that in the deep ocean, and locally, the thermohaline circulation might dominate.

At that time there were virtually no direct measurements of currents below the sea surface. Observations of temperature, from the *Challenger* onwards, had confirmed that the deep ocean, at all latitudes, was so cold that the water must have originated at the surface in polar regions: the general view, as elaborated in the results of the *Meteor* Expedition, was that there was a broad meridional convective cell bringing sinking polar water equatorwards. It was Stommel's own extension of his 1948 theory that demonstrated that this was incorrect. He predicted that polar water travels equatorwards in a narrow western boundary current at depth, returning polewards over most of the ocean.

This was in the mid-1950s, and it happened that John Swallow had found a way of measuring deep currents by tracking the motion of neutrally buoyant floats drifting at known depths (see Chapter 10). Soon Swallow and Val Worthington were able to confirm Stommel's prediction of a deep southward current beneath a generally northward flowing Gulf Stream. Attempts to measure the slow northward return flow were, however, less successful. Instead it was found that the ocean is populated with what were named mesoscale eddies – quasi-circular circulations of water with a space-scale in the order of a hundred kilometres and a time-scale of around a hundred days. Speeds within the eddies were about 10cm per second, making it impossible to measure the much slower predicted northwards drift.

Developments in the forty years since then have been determined by improvements in technology. There have been great advances in the use of deep drifting floats and there is a highly developed technology of mooring vertical arrays of current meters in the deep sea to record for months or years before being recovered. Earth orbiting satellites are providing images of the surface structure of the ocean as well as measuring the mean slope of the sea surface, which provides information on surface currents (see Chapter 11). Much improved methods for measuring temperature, salinity, oxygen and other tracers, both natural and artificial, from ships (see Chapter 12) contribute to improved understanding of the long-term movements of water masses in the body of the ocean.

Yet in spite of these technological developments the ocean remains very sparsely observed, with many significant processes that are still not sufficiently clearly defined or well understood. Rapid advances in computing power have made it possible to construct detailed numerical models that simulate the structure and motion of the ocean, though not yet in sufficient detail to represent the mesoscale eddies completely. The mutually beneficial interaction between observers and modellers is leading to increased understanding of the ocean as an important component, and arguably the most important yet least well known component, of the Earth's climate system.

References

Deacon, M. B. (1985), 'An early theory of ocean circulation: J. S. von Waitz and his explanation of the currents in the Strait of Gibraltar', *Progress in Oceanography* **14**: 89–101.

Deacon, M. B. (1997). *Scientists and the sea, 1650–1900*. Aldershot: Ashgate.

Ekman, V. W. (1905) 'On the influence of the earth's rotation on ocean-currents', *Archiv för Mathematik, Astronomi och Fysik* **2**(11): 1–52.

Hamlin, C. (1982) 'James Geikie, James Croll and the eventful Ice Age', *Annals of Science* **39**: 565–83.

Munk, W. H. (1950) 'On the wind-driven ocean circulation', *Journal of Meteorology* **7**: 79–93.

Rennell, J. (1832) *An investigation of the currents of the Atlantic Ocean*. London: Rivington.

Schlee, S. (1973) *The edge of an unfamiliar world. A history of oceanography*. New York: Dutton.

Stommel, H. (1948) 'The westward intensification of wind-driven ocean currents', *Transactions of the American Geophysical Union* **29**: 202–6.

Sverdrup, H. U. (1947) 'Wind-driven currents in a baroclinic ocean; with application to the equatorial currents of the eastern Pacific', *Proceedings of the National Academy of Sciences* **33**: 318–26.

9 Polar dreams and California sardines

Harald Ulrik Sverdrup and the study of ocean circulation prior to World War II

Robert Marc Friedman[1]

Introduction

History, it may seem, comes in many different sizes and shapes. What we choose to remember is not merely a question for academic historians. Professional and social identities are created in part through historical awareness; the stories we tell of the past provide a resource for creating a culture for the future. When confronted with the topic of the history of ocean circulation, we might be tempted to try piecing together a history of ideas on this subject. We might also try juxtaposing a history of ingenious instruments that enabled assembling observations with which patterns and theories could be developed. We may also add considerations of funding: why individuals, foundations, or governments have been willing to provide money, sometimes even relatively large sums, towards studying the oceans. And of course the broader question of oceanography as some form of discipline or speciality or collection of specialities offers perspectives on professional identity, institutional growth, and ability to train as well as employ new practitioners. All of these themes are naturally essential for a history of oceanography, but, in addition, part of our historical endeavour should try to illuminate the changing practice of oceanography: what type of skills – intellectual, technical, social, entrepreneurial, and so forth – enabled individuals to shape and enrich this scientific endeavour.

We all too often take for granted that young women and men today can consider studying oceanography at a university and following one of many career paths. Professional organizations and journals flourish on national and international levels, conferences are regularly held, academic programmes in ocean science ensure a supply of practitioners, while a relatively stable network of patronage and publics sustains research and pedagogical activity. As we enter the post-Cold War era, with its greater uncertainty as to funding and rationale for research, we might gain some inspiration from the recent past when what we recognize as professional oceanography did not exist. Neither the present contours of science nor its relations with society are inevitable or natural. Changing political, economic and cultural circumstances provided opportunities and constraints for scientists to define and pursue oceanographic research programmes. By telling a tale of one such scientist, Harald Ulrik Sverdrup, I would like to open a window onto some of the social processes, human drama and moral values that entered into constructing the professional study of ocean circulation. His story – and those of many others – will help us appreciate what it meant and still means to be an ocean scientist.

Sverdrup and the *Maud* Expedition

Sverdrup was born in 1888 in the west Norwegian town of Sogndal (Revelle and Munk 1958). His father was a teacher, but more significantly he was also a minister of considerable Lutheran piety. Although Harald eventually decided not to follow the Sverdrup family tradition into the clergy, he nevertheless derived from his religious exposure a moral imperative of duty, obligation and hard work. As a student in Oslo hoping to study astronomy, he found himself following, in 1908, Vilhelm Bjerknes's engaging lectures describing a visionary project: a research programme to establish an exact physics of the atmosphere and oceans. Sverdrup abandoned his first love and decided instead on geophysical science. In 1912 he began his first expedition, of sorts – that towards becoming a researcher and – more geographically – moving to the University of Leipzig, to which Bjerknes had been called to direct a new geophysical institute. Sverdrup followed as a research assistant and doctoral student. He joined Bjerknes's massive project of precalculating changes in atmospheric and oceanic conditions by graphical means (Friedman 1989, 1994). In particular Sverdrup sought to factor in the effects of friction, turbulence and energy balance. He authored or co-authored twenty articles during his five years in Leipzig. His doctoral dissertation, completed in 1917, on the structure, dynamics and thermodynamics of the North Atlantic trade wind became a classic.

Sverdrup was off to a bright start. He was also astute enough to be assessing the question

Figure 9.1 Fridtjof Nansen on his return from the *Fram* Expedition. (Nansen 1897, vol. 2, facing page 466.)

that most doctoral students ask: What on earth am I going to do when I'm finished? When approached by Roald Amundsen to join his much delayed expedition to drift with the ice over the polar sea, Sverdrup responded affirmatively. Why would a most promising young scientist agree to spend what was projected to be four or more years drifting in the Arctic ice? True, no job was waiting for him, nor did his skills and expertise fit any clear academic category. To family friends, who opposed his plans, he wrote the following:

> For me this is to be presented with a large, interesting, and also honourable job, which if all goes well, can provide the basis for my future . . . And not the least, if I am now able to make a small scientific contribution, then it will be a contribution to *Norwegian* science. If I get home safely, then I will be connected to Norway, which is the grandest fatherland anyone can find.[2]

To colleagues he noted that he had worked almost exclusively with theoretical problems; the time had now come to obtain practical experience and direct contact with Nature. Undoubtedly adventure attracted him as well. Sverdrup sailed on the *Maud* in July 1918 towards the Arctic, thus beginning an expedition that proved crucial for much of his subsequent career (Sverdrup 1933).

The *Maud* Expedition should be recognized as part of a development beginning with Fridtjof Nansen's pioneering polar activities, which proved crucial for Norwegian polar and broader geophysical research as well as for oceanography internationally. In 1890 this young zoologist proposed building a ship with a reinforced and rounded bottom to resist the pressure of ice (Nansen 1897). He sailed north of Siberia in 1893 and intentionally locked the ship in the pack-ice to allow it to drift with the ice across the central Arctic and – hopefully, eventually – emerge north of Spitsbergen. Around the world explorers and scientists had scorned Nansen's plan. Few believed that any ship could withstand the pressure of the churning ice field; moreover, many experts believed that the central Arctic contained a large, as yet undiscovered, land mass. When, after having disappeared without a trace for three years, Nansen and his ship returned in 1896, having attained the farthest northerly human presence to date, he became an international celebrity. He also became an ocean-ographer, as he began analysing the many data collected and, subsequently, devising new instruments to improve the precision of this fledgeling science. Nansen asserted that no land existed in the central Arctic, that the polar sea was extremely deep and that the currents entering and leaving the Arctic play a fundamental role in hemispheric ocean circulation. Again, few researchers accepted these claims. He therefore welcomed Amundsen's plan, first announced in 1908, to repeat the drift, but now with vastly improved instrumentation.

Amundsen planned to sail the *Maud* through the northeast passage and enter the pack-ice north of the Bering Strait before winter. During the ensuing three years the *Maud* was to drift with the ice, coming, he hoped, as close to the North Pole as possible. All the while the ship was to serve as a floating laboratory, collecting measurements of terrestrial magnetism, atmospheric electricity, weather conditions near the surface and aloft, oceanic conditions, northern lights, and virtually anything else that might be of scientific and cultural interest.

The *Maud* Expedition had many adventures. Unusual ice conditions locked the ship, winter after winter, close to the coast, delaying the drift for four years. Sverdrup spent the winters with native tribes in northern Siberia living exclusively on reindeer while adopting other local customs. The team experienced tragedies and mishaps ranging from deaths to Amundsen having part of his buttocks removed by a swipe from a polar bear, which was in hot pursuit of the rest of him. When the expedition came to Alaska in 1920 to resupply after

Figure 9.2 Cartoon of Sverdrup as a young man – dreaming polar dreams? (Reproduced by kind permission of the Norsk Polarinstitutt, Tromsø.)

the initial two years of fiasco, Amundsen was willing to allow members of the crew to abandon ship; Sverdrup declined. Although frustrated and despondent, Sverdrup had given his word: as long as he was physically capable, he would not abandon his position as chief science officer. Although he had collected a vast array of data during the first years of the voyage, the real work began when the *Maud* finally entered the pack-ice in August 1922 and began drifting west-northwest. And now he would be aided by two new members of the crew: Odd Dahl, who had been recruited to maintain and pilot a small plane carried on ship, proved to be a brilliant instrument maker; and Finn Malmgren, a young Swedish meteorologist who was to serve as Sverdrup's scientific assistant.

Based on a strict regime of discipline and innovative instrumental practices, Sverdrup set up a comprehensive geophysical observatory on the drifting ice. He overcame the seemingly insurmountable difficulties of making reliable precision measurements in a most hostile environment: frost instantly formed on the eyepieces of measuring instruments and needles and dials had an unfortunate tendency to freeze solid; sounding balloons had to be made visible in the weeks of perpetual darkness; means had to be devised for launching instrument-carrying kites and then bringing them down from several thousand feet in stiff Arctic breezes; holes in the ice for oceanographic instruments had to be re-cut every several hours. He also appreciated the advantages of working on a ship locked into an ice platform that is not available to ships at sea. The adventures were many: the ice suddenly opening on

Figure 9.3 The *Maud*. (From H. U. Sverdrup (ed.) (1939) *The Norwegian North Polar Expedition with the* Maud, *1918–25: scientific results* 1b, frontispiece. Bergen: A. S. John Griegs Boktrykerri.)

the side of the ship, carrying off instruments and dogs on a voyage of their own; or giant mounds of ice being forced upwards by the shifting currents and threatening to capsize the ship. They witnessed extraordinary blizzards, skies aflame with northern lights and the relentless monotony of the summer fog above which the sun never set. All the while, three sledges with provisions for forty days always stood ready for quick escape. At any time churning waves of ice could capsize the ship.

But what of their drift across the polar sea? In September 1923 they were buoyed by the prospect of crossing on a path north of Nansen's, but then strong winds began to blow from the north; day after day they helplessly watched as the wind blew the ice and the entrapped *Maud* farther and farther to the south. Now their path would be likely to be longer and less interesting scientifically. They avoided descending into extreme depression and personal antagonism through strict work discipline and keeping contact to a minimum. No longer did they finish the Saturday evening toddy drinking with a gramophone sing-along of 'It's a long way to Tipperary'. Finally in February 1924 they abandoned the attempt to cross the Arctic and decided to try returning home. They were locked in the ice; if during the summer the ice did not open sufficiently towards the south to allow an escape, they could remain adrift for three or four more years. As it happened, they did escape, but only to be blocked once again from proceeding to the Bering Strait by coastal ice. Another long winter was spent waiting. Sverdrup began analysing data and writing what was later published as 'Dynamics of

Figure 9.4 Scientific practice at 82 degrees north: Sverdrup titrating sea water in the *Nautilus*. (Reproduced by kind permission of the Norsk Polarinstitutt, Tromsø.)

tides on the north Siberian shelf' (Sverdrup 1927), in which he elegantly proves, among other important findings, that no major land mass exists in the central Arctic. Finally, in October 1925 the *Maud* arrived in Seattle and the expedition came to an end. Sverdrup noted on many occasions that the greatest achievement of the voyage was that the men departed as friends. He thanked Amundsen 'not just because you provided me with a wonderful opportunity to work with things that interest me, but even more because you helped make a man of me'.[3]

Polar dreams

To the general public Amundsen's expedition was nothing less than a complete fiasco. One opportunity did exist to save its honour and for Sverdrup to justify the many years on the ice: to convert the enormous amount of observation material into scientific reports. But how? This task would require many years of intensive work as well as considerable publishing subsidies. Those huge volumes of expedition reports, such as those from the *Challenger* or Nansen's north polar sea expeditions, don't just 'happen'. The processes of analysing and publishing often prove to be an adventurous expedition in themselves. Bjerknes wrote to Sverdrup that the economic situation in Norway could scarcely be worse, but he and oceanographer Bjørn Helland-Hansen would do everything possible to assist him. They and others felt that Sverdrup had sacrificed much for the glory of Norway and the advancement of geophysical science. The country should reward him appropriately. Fortunately, during the years of the expedition Helland-Hansen and Bjerknes had established a major geophysical institution, connected with the Bergen Museum, which was to be the cornerstone for a new university for western Norway. They appealed successfully to the government to fund a professorship in theoretical meteorology for Sverdrup which would largely be devoted for the foreseeable future to analysing *Maud*'s observations and editing the expedition reports (Sverdrup 1933). The few Norwegian endowments and foundations for research pledged funds for publishing the massive volumes; their publication was considered a matter of national pride.

Although Sverdrup assumed the role of a professor, he was not finished with polar dreams. The *Maud* had never left the continental shelf; the deep waters of the polar sea remained as yet largely unknown. Sverdrup appreciated that over 70 per cent of the Arctic remained unexplored; more importantly for oceanography, he grasped the importance of the polar sea for understanding the hemispheric system of ocean currents. But little prospect existed for any comparable expensive, time-consuming expedition. Appealing to national honour and the need for Norway to contribute to civilization no longer opened the coffers for polar activities. In most nations, raising money for polar expeditions increasingly demanded the promise of sensational extravaganzas, as media moguls often held the purse strings. And although scientific research could endow a sense of respectability to sheer adventurousness and nationalist fervour, by the mid-1920s the needs of science were becoming more difficult to reconcile with polar exploration, as dramatic and rapid flights by plane and small dirigibles began capturing the public's attention (see also Chapter 4). Some possibilities did nevertheless arise through which polar oceanography could, in principle, be developed. First, Sverdrup joined forces with Nansen in 1926 in an international project to use a giant Zeppelin airship for systematic polar exploration. Amundsen and Ellsworth had already taken a small airship across the North Pole. Now, Nansen and several German polar researchers considered the possibility of using a huge Zeppelin airship which could transport large teams of scientists and tons of supplies to otherwise inaccessible parts of the Arctic. Moreover, echo-soundings to determine the polar sea's depth could be made from the airship while it hovered over the multitude of openings in the ice during the summer.

When in 1928 the *Graf Zeppelin* successfully flew around the world, scientists realized that here was an airship capable of serious polar work. Nansen, Sverdrup and others in this international endeavour began drawing up plans for a voyage during the summer of 1929 or 1930. But delay followed delay. Nansen died in 1930; the British R101 airship tragedy dampened enthusiasm for airship travel. Sverdrup was asked to lead the entire costly enterprise, but declined. When finally a trial voyage was secured for the summer of 1931, based on newspaper sponsorship, Sverdrup again had to decline. He had already made plans to take part in another expedition, one that seemed to promise an alternative to the airship for polar research: Hubert Wilkins's attempt to use a submarine to cross the Polar Sea under the ice.

Wilkins had already shown a flair for sensationalism when he had flown across the Arctic; now he caught the public's eye and the Hearst newspaper empire's purse by planning to reach the North Pole by submarine. Wilkins also wanted the expedition to support scientific research; he asked Sverdrup to be scientific leader and promised ample support for equipment. Sverdrup enthusiastically accepted.

Many of the public, including most of Sverdrup's colleagues, considered Wilkins's plan to be nothing less than scatter-brained and immensely dangerous. Still, Sverdrup could not resist the opportunity to obtain data, in a few summer weeks, which had eluded him in the course of several years with the *Maud*.

How would this adventure turn out? Mechanical problems resulted in delays and worries from the start. Finally, Wilkins and the *Nautilus* arrived in Bergen. After installing the scientific equipment, they headed north to Spitsbergen, to the Arctic pack-ice beyond – and to adventure. Sverdrup's excellent physical condition and self-discipline helped immensely in this effort to advance oceanography. The quarters were extraordinarily cramped, the submarine rolled and rocked when cruising on the surface, the smell of diesel oil pervaded everything, smoke frequently found its way into the narrow corridor, the drinking water was discoloured while also tasting and smelling wretched, water for washing was oily and perhaps worst of all, far in the rear, accessible only by climbing though endless crates of supplies as the ship mercilessly rocked and swayed, was the ship's only toilet. Obviously, the practice and culture of oceanography posed challenges that differed immensely from laboratory sciences. Try imagining Max Planck in the *Nautilus*!

Once the *Nautilus* entered the waters north of Spitsbergen, Sverdrup began a highly successful programme of making observations from a pressurized diving chamber. They reached 82 degrees north and climbed onto the ice for fresh water, air and magnetic observations. Then, at last, Wilkins gave the signal to submerge under the ice. But the diving rudder didn't respond. Donning a diving suit and diving into the freezing water, a crew member investigated. What he found did not make Wilkins happy: the diving rudder had disappeared. Either it had fallen off or had been sabotaged by one of the crew. They turned back.

Again, another major expedition ended in public ridicule. Sverdrup still believed that submarines and giant airships held the key for systematic study of the Arctic; the *Nautilus* was just the wrong ship. Using the extensive observations made during the short cruise, Sverdrup made several significant findings, which included a convincing depiction of the Gulf Stream's branching into the polar sea. Who might support further exploration? What

Figure 9.5 The submarine *Nautilus*. (Reproduced by kind permission of the Norsk Polarinstitutt, Tromsø.)

rationale could be offered for massive support? Large-scale oceanographic exploration in general was scarcely institutionalized at the time; polar oceanography fared no better.

Back in Bergen Sverdrup was finally finishing the last of the *Maud* observations. He became restless; he thirsted for new data to analyse. Work from the *Maud* focused his attention on problems of heat and energy transfer between the atmosphere and the ocean. Laboratory and theoretical studies provided some clues; Sverdrup wanted direct measurement. Although he began devising instruments for making measurements at the ocean surface, he accepted that he could begin by first analysing the heat budget above and below a layer of smooth snow. His insight was informed in part by his understanding of the funding situation at the time: 1933 could not have been a worse year to plan even a minor expedition, or for obtaining funding for any scientific research not promising direct economic impact. But Sverdrup and his Swedish friend, the glaciologist Hans W:son Ahlmann, both knew that a political reason could be found for funding such an expedition.

Having been given sovereignty over Spitsbergen and surrounding islands in the peace treaties following World War I, Norway extended her territory to the new Arctic colony now named Svalbard. Norway had also hoped to keep rights to eastern Greenland, which otherwise was claimed by Denmark. In 1933 the increasingly bitter dispute was settled in the international court against Norway. In the meantime Soviet activity was increasing on Svalbard. Legitimation of territorial claims came through scientific work, surveying and other forms of on-site activity; territorial claims controlled access to fishing grounds, seal and bear hunting, coal deposits, as well as other resources that might as yet be undiscovered. Sverdrup and Ahlmann proposed a small expedition to the top of a glacier in the Spitsbergen mountains. Sometimes oceanographers must settle for something less than a voyage out to sea. With the assistance of the Norwegian Prime Minister, whom they contacted directly, they received a relatively substantial grant through the Ministry of Commerce, which had jurisdiction over Svalbard affairs.

Sverdrup arranged transport to the desolate northwest coast where the sea meets the glaciers. Then with the help of seventeen Greenland huskies, they trekked with over two tonnes of equipment and provisions up 3000ft and 20 miles inland to the Isachsen Plateau. Although a relatively minor expedition, all contingencies had to be covered. Sverdrup again refined yet further his extraordinary organizational abilities. On a list of provisions that were needed, Sverdrup drew up neat columns of numbers and supplies, such as cans of fish balls, crackers and dried reindeer meat, but also scrawled across the top of the page, in what appears to be a sudden last-minute revelation – circled to ensure that it not be forgotten again – the all-important commodity that even oceanographers occasionally have use for – toilet paper! Here again the need for skills irrelevant for many other sciences, in this case that of organizing and leading expeditions, proved critical for the advance of geophysical research.

And once again extraordinary discipline permitted them to obtain exceedingly valuable data. Although the sun never set, the temperature often hovered at freezing while dense fog swirled around them for days on end. Every hour was accounted for; the scientists collected over 20,000 observations. Sverdrup was thrilled with the data. Among other findings, he extended the theory of geophysical turbulence. He began to consider further studies on the transport of heat and water vapour at the ocean surface, foreshadowing the tricky but vital subject of air–sea interaction, which has become an important topic in late twentieth-century oceanography. Then, in 1936, he was asked whether he would accept becoming director of the Scripps Institution of Oceanography at La Jolla, California.

Sverdrup and Scripps

Sverdrup had previously received offers to work in America, where physical oceanography was scarcely developed. He had by the late 1920s acquired a reputation for being exceptionally hard-working and talented. In 1930, for example, he received an intriguing letter from Henry Bigelow (Graham 1968), doyen of American oceanography. He related that money was now in place to incorporate the new Woods Hole Oceanographic Institution. As had been the case much earlier in Scandinavia, important American constituencies on the east coast (Burstyn 1980) recognized that fisheries needed to be rationalized through extensive physical and biological oceanographic research. Bigelow expected to act as director for the first several years; his reason for writing to Sverdrup was clear: 'We are wondering if you would still be at all interested in a position with the new institution, or whether you have definitely decided to remain in Norway? . . . All I can do at present is to ask, in a perfectly informal way, whether you would care at all to consider the position as Chief Oceanographer . . . if our plans mature we can at least promise that there will be a comfortable lab, well equipped; a ship (120 feet long or so) of a type fit to go anywhere, and perhaps personal freedom for research.'[4] This was very tempting, but Sverdrup had already searched his soul for principles by which he might live in peace with himself. He would prefer to work where he was most needed. And as long as he felt needed in Norway he felt compelled to remain there. He was proud that Norway was regarded as a leading international centre for oceanography as well as meteorology; he hoped that this situation could be maintained. But a few years later, California beckoned.

What awaited Sverdrup when he arrived in La Jolla in late summer, 1936? What was this institution that was first established as a marine biological station in 1903 (Raitt and Moulton 1967) that joined the University of California in 1912, and that added the word oceanography to its name in 1925 (Shor 1980)? Sverdrup's impressions during the first months were decidedly negative.

The Scripps Institution was oceanographic in name only. Sverdrup observed, in a series of letters to colleagues back in Scandinavia, that land-based laboratories dominated it and no possibilities existed for systematic work at sea. The one vessel owned by the institution, the 64-foot *Scripps*, was not capable of venturing beyond the immediate coastal waters: Sverdrup considered it 'A filthy cramped washtub'. He recognized that to comprehend even the immediate coastal waters required greater insight into the currents further out in the ocean. Without a clear oceanographic mission, the institution merely developed as an umbrella for independent specialities housed in separate laboratories. He also found relations with the university poor, a pedagogical programme that was deficient, and personnel suffering low morale. But there was another side to Scripps that Sverdrup also quickly appreciated:

> I like the men here very well. They are easy to work with, willing and helpful . . . most are enthusiastic about expanding the scope of the institution to the ocean. . . . The group here . . . is largely a collection of unusually pleasant people, who have done everything possible to make us feel at home. Therefore it would be especially satisfying to do something for them: get them better housing, increase contact with the university, bring in bigger and more interesting problems for discussion. Whether I can accomplish this I don't know.[5]

Sverdrup believed that by showing goodwill he might make greater headway than his predecessor. He threw himself fully into the task.

Sverdrup received help in unexpected ways. First, not long after he arrived, an explosion ripped through the institution's ship, which caught fire and sank. Now, he might more easily raise money both to procure a seaworthy vessel and to sustain an active-at-sea research programme. By early 1937 he had achieved excellent relations with the president of the University of California and with Robert Scripps, who managed the Scripps estate in relation to the institution; they agreed to assist him. In the end, Sverdrup obtained Scripps family support for buying and renovating a 104-foot schooner, as well as for matching funds to those promised by the university, for maintaining a crew. In addition, insurance cover on the sunken ship yielded funds for scientific equipment. Sverdrup was elated.

The next unexpected good fortune involved sardines. Sverdrup arranged to use the California Fish and Game Commission's ship, the *Bluefin*, to attempt a first systematic study of the ocean off the coast in order to understand the conditions at the time that sardines spawn, which in turn had significant consequences for the economic livelihood of many Californians. He and Scripps personnel made three cruises during the spring of 1937 at the same locations. Almost immediately they realized the importance of this investigation, which provided clear insight into the mechanism of upwelling and even allowed them to follow the mixing of different waters in detail. Consequently, Sverdrup recognized how he might transform the institution. First, as he had suspected all along, the ocean right outside their door posed a vast wealth of interesting problems. The great differences in surface layers and the life histories of these currents must certainly be reflected in biological conditions. Sverdrup decided that when the institution's newly acquired ship was ready, it should be used not for long expeditions, but rather to concentrate repeatedly on a limited area. Moreover, the study of this limited area could be used as a research programme to unify the institution. The whole institution could be mobilized 'to work according to a *plan* in which individual problems mutually reinforce one another' and consequently establish an overall picture. He even decided to immerse himself in biological oceanography in order to better coordinate the different investigations, and 'not the least to make physical and chemical oceanography useful to the biologists'.[6] And possibly also to California's important fishery concerns.

We should remember that, as Sverdrup attempted to create a viable institution and discipline, he could see but few patrons or markets for oceanography. The massive wartime and postwar involvement of the military and other national security agencies in supporting ocean science could scarcely then be imagined. Indeed, Sverdrup commented at this time in disgust, 'The Navy, when you get right down to it, doesn't care about oceanographic research.'[7]

Within only two years of arriving, Sverdrup was making significant progress towards transforming the Scripps. Everything was fine – except for a terrible decision that had to be confronted. Sverdrup, as well as his wife Gudrun, had become painfully homesick. He could not find an acceptable successor to take over after his self-imposed three-year term.

Sverdrup shared his dilemma with friends back in Scandinavia. He regretted in part that he was not the type of person who avoids getting involved. Had he simply come and worked on his own projects, he could now pack up, say thank you, and return home:

> But I guess I'm not made that way. Instead I've worked away – and you don't do that without getting punished. You give something of yourself. . . . In all my work here – in all my plans – I can't constantly think that this is an intermezzo. But often I long so damned to be home – to the land and friends; I have a feeling of living in exile. . . . Oh no, this is not at all easy. 'An expedition,' Bjørn [Helland-Hansen] said. No, it's not that

at all, because the institute here is something permanent – with a group of people whose trust I think I've won – and who gladly will hold on to me.[8]

Complicating matters further, Sverdrup admitted that his working conditions at home were actually far from favourable. In Bergen he had experienced the irony of having a position at an institution proclaiming to be devoted to 'pure research', where the scholar, unfettered by teaching and administration, was 'free' to follow his curiosity. Such rhetoric might serve well as cultural ideology, but it certainly does not serve an oceanographer well. Sverdrup needed access to a research vessel – a costly affair at best – or the data collected regularly on one, if he was to follow *his* curiosity. He confessed – perhaps over-modestly:

> At home I was a 'free lance' . . . I am best suited to work with systematic observations and to make some sense of them. I'm not sufficiently a theoretician to be able to work solely theoretically . . . my strength lies in analysing data and using the tiny bit of theory I master. At home I tried several times to get my fingers on the massive oceanographic material that Bjørn has collected through the years. But Bjørn would not let me get hold of it. . . . To put it bluntly: to be a geophysicist without being able to get hold of the data one wants results in leading an uncertain and doubtful existence.[9]

In contrast at Scripps, he was the leader: 'here I can plan and have an influence on others' work . . . and can draw upon all my earlier experiences'. Finally, he agreed to stay an additional two years at Scripps, until the end of 1941.

Wartime and postwar developments

When the European war began, Sverdrup understood that he would have to remain in La Jolla even longer. After the Germans occupied Norway in April 1940, Sverdrup accepted the director's position on a permanent basis. Fortunately he had a project that could keep his thoughts from straying to the depressing world situation and to his own forced exile. As part of Sverdrup's desire to bring the many specialities concerned with the oceans into a greater scientific and institutional unity, with the help of his colleagues, Richard H. Fleming and Martin W. Johnson, in 1939 he began to write a comprehensive textbook for oceanography. The task of preparing this volume was enormous. After much difficult and time-consuming work, the book appeared in December 1942: *The oceans: their physics, chemistry, and general biology*. Sverdrup noted privately, 'I hope [it] will be a useful reference book for some years to come.'[10] His hopes were more than realized. Just as the *Challenger Reports* had formed the basis for oceanographic research at the start of the twentieth century, *The oceans* synthesized existing knowledge and, more than any other work, provided a unifying foundation, and springboard, for the efforts of oceanographers as the science entered a period of rapid development during the second half of the century.

At Scripps, as elsewhere, the war stimulated the mobilization for the war effort of first-rate research teams, the training of many new oceanographers and the embarking on innovative research programmes with the help of generous government funding to assist on strategic problems. Ocean science benefited greatly from these activities, most notably Sverdrup and Walther Munk's pioneering work on the prediction of swell and breakers, which proved very significant for several crucial Allied military operations.

Once peace was restored, Sverdrup again had to confront his and his wife's longing to return home. In his letters he reveals a continued inner struggle to find a just and satisfactory solution. A man of principle, Sverdrup was inclined to remain in California because he did not want to be a burden on Norway during its recovery from the German occupation. He would consider returning only if Norway had a specific use for him and if his return did not compromise any younger person's professional advancement. Moreover, he knew full well that the time had arrived for oceanography to emerge as a truly vigorous discipline, especially in America. Wartime experience showed the importance of oceanography for the nation's security. And Sverdrup was a central figure in a country ready to invest heavily in studying the oceans: academic departments were being planned and government funding was expected to remain at levels inconceivable before the war. All this certainly made the prospect of staying longer more palatable. An even greater consideration for staying was, again, a sense of duty to individuals who were counting on his aid. And yet shortly thereafter, he indicated a willingness to leave, but not until he secured all of his personal obligations. What changed his mind? Why, as a Norwegian journalist asked him, did he leave paradise in California?

We must change scenes. Immediately following victory in Europe, Ahlmann attended a celebration at the Russian Academy of Sciences and visited many installations. He became alarmed over the extraordinary measures being taken by the Soviet Union to step up polar research, exploration and colonization. He also knew that during the war the Americans had dramatically increased their engagement in the Arctic. Ahlmann feared for the Nordic nations; he especially feared for Norway's vital interests in the Arctic and Antarctic. Territorial claims were still disputed; and only vigorous activity could legitimate these claims. In discussions with the Norwegian Prime Minister and cabinet ministers, Ahlmann urged establishing a major institution to oversee greatly expanded research, surveying and commercial activity in the polar regions. And such an institution would have to be led by an internationally respected scientist – Harald Ulrik Sverdrup.

In 1948 Sverdrup came home to lead the Norwegian Polar Institute and to take over the planning of the strategically urgent Norwegian–Swedish–British Antarctic expedition. Although very much missing the excitement at Scripps, he also expressed joy over ending his twelve-year longing to experience the Norwegian spring again. His exceptional capacity to work continued; he found it difficult to say no to tasks for which he felt qualified to help, even when these took him far from his research and administration. But he felt increasingly suffocated by administrative obligations that reduced his research time to virtually nothing. Then, on 28 July 1957 he died suddenly from a heart attack at the age of 69.

History is of course as much about the present, if not the future, as it is about the past. As we attempt to define a cultural heritage for ocean science, as well as to comprehend how and why its practices and structure developed as they have, we might also want to keep in mind issues that normally do not arise in our studies. Yes, we should continue to comprehend the manner by which conceptual and methodological innovation has occurred in ocean science. And, yes, we should understand the political economies of national and international oceanographic communities. But we might also include in our construction of oceanography's self-understanding questions of values that informed and sustained the growth of this science. How did individuals interact as colleagues? What sorts of behaviour and obligations were expected and accepted? Who acquired access to resources and rewards? How will oceanographers' identity change if and when expeditions are replaced by remote sensing devices? And how have changing cultures of financing oceanographic research impacted on the style and quality of both research and life as a researcher? Questions of purpose and questions about the internal culture of professional science have increasingly been in the thoughts and discussions of researchers and students. I suspect Sverdrup would find some

Figure 9.6 Harald Ulrik Sverdrup, leader in postwar oceanography. (Reproduced courtesy of Scripps Institution of Oceanography Archives, University of California, San Diego.)

features of more recent science distasteful. Sverdrup believed in duty and obligation to institutions and to students. He put service and duty ahead of careerist self-interest and ego. Sverdrup believed in service; he was a devoted teacher. He also recognized and welcomed the need for science to serve a broader society than merely a group of specialists. On his life-long expedition through vastly different contexts, Sverdrup developed and perfected instruments to guide his actions: modesty, duty, kindness and personal sacrifice. These are values that I hope will also be part of the history of oceanography currently being made and about to be made.

Notes

1 Major portions of this text have been published in *The expeditions of Harald Ulrik Sverdrup: contexts for shaping an ocean science*, William E. and Mary B. Ritter Memorial Lecture, 29 October 1992 (La Jolla, CA: Scripps Institution of Oceanography, 1994), 49pp. Full references to source materials can be found in the original publication.

2 Letter, Harald Ulrik Sverdrup to provsten Eilert Patrick Juul, 6 October 1916, Sverdrup biographical files, Norsk Polarinstitutt, Tromsø.

3 Letter, Sverdrup to Amundsen, 18 August 1926, Roald Amundsen papers, 480B, University of Oslo Library.

4 Letter, Henry Bigelow to Sverdrup, 22 January 1930, Harald Ulrik Sverdrup papers, 634A, University of Oslo Library.

5 Letter, Sverdrup to Helland-Hansen, 17 November 1936, B. Helland-Hansen papers, Geophysical Institute, University of Bergen.
6 Letters, Sverdrup to Helland-Hansen, 10 March, 3 June and 27 August 1937; Sverdrup to Ahlmann, 28 December 1937, H. W. Ahlmann papers, Royal Academy of Sciences, Stockholm; Sverdrup to Devik, n.d. [June 1937], Devik papers, Riksarkivet [National Archive], Oslo; Sverdrup to Bjerknes, 20 February 1937, Bjerknes family papers, University of Oslo Library.
7 Letter, Sverdrup to Helland-Hansen, 2/3 January 1938.
8 Letter, Sverdrup to Ahlmann, 25/28 December 1937; see also Sverdrup to Helland-Hansen, 2 January 1938.
9 Ibid.
10 Letter, Sverdrup to Ahlmann, 19 January 1943, Ahlmann papers.

References

Burstyn, H. L. (1980) 'Reviving American oceanography: Frank Lillie, Wickliff Rose and the founding of the Woods Hole Institution', in M. Sears and D. Merriman (eds) *Oceanography: the past*. New York: Springer-Verlag, pp. 57–66.

Friedman, R. M. (1989) *Appropriating the weather: Vilhelm Bjerknes and the construction of a modern meteorology*. Ithaca: Cornell University Press.

Friedman, R. M. (1994) *The expeditions of Harald Ulrik Sverdrup: contexts for shaping an ocean science*. San Diego: Scripps Institution of Oceanography.

Graham, M. (1968) 'Henry Bryant Bigelow', *Deep-Sea Research* **15**: 125–32.

Nansen, F. (1897) *Farthest north*. London: Constable.

Raitt, H. and Moulton B. (1967) *Scripps Institution of Oceanography: the first fifty years*. San Diego: Ward Ritchie Press.

Revelle, R. and Munk, W. (1958) 'Harald Ulrik Sverdrup – an appreciation', *Journal of Marine Research* **7**: 127–38.

Shor, E. N. (1980) 'The role of T. Wayland Vaughan in American oceanography', in M. Sears and D. Merriman (eds) *Oceanography: the past*. New York: Springer-Verlag, pp. 127–37.

Sverdrup, H. U. (1927) 'Dynamics of tides on the North Siberian shelf', *Geofysiske Publikasjoner* **4**: 1–75.

Sverdrup, H. U. (1933) 'General report of the expedition', in H. U. Sverdrup (ed.) *The Norwegian North Polar expedition with the* Maud, *1918–25: scientific results* **1**, 3–22. Bergen: Geofysisk Institutt.

Sverdrup, H. U., Johnson, M. W. and Fleming R. H. (1942) *The oceans*. New York: Prentice-Hall.

10 Direct measurement of subsurface ocean currents

A success story

W. John Gould

Introduction

There are two fundamental types of measurement of the physical state of the open ocean: the distributions of important properties (e.g. temperature and salinity) and the movement of the water itself. The two are closely related, but require very different observational techniques.

From early last century there was a method to derive the large-scale ocean circulation from the temperature and salinity (and hence density) fields by means of the so-called dynamical method (Helland-Hansen and Nansen 1909, Sandström and Helland-Hansen 1903). However, this method left unsolved the problem of determining the arbitrary constant in the geostrophic calculation. The determination of the reference velocity seemed to demand that currents in the open ocean should be measured directly. In what follows I will look at the rapid advances that were made in the direct measurement of ocean currents primarily since 1950 and will set these in the context of the techniques available to the *Challenger* scientists.

Current measurements on the *Challenger* Expedition and before 1950

Accuracy of navigation has been and remains a key factor in our ability to measure currents, since, in the open ocean, there are no readily available reference points. The *Challenger* scientists had two means of determining currents. Across broad areas of the ocean basins, surface currents were estimated from the mismatch between dead reckoning positions and between positions determined by astronomical fixes using sextant and chronometer (Fig. 10.1). This was a technique that was basically no different from that used by Rennell (1832) almost 100 years earlier.

Locally, surface currents could be measured by tracking buoys that had drogues ('current drags') attached to them to 'lock' them to the surface water masses (Fig. 10.2). However, the determination of the drift track of a drogued buoy was still dependent on knowing the observing ship's position and was subject to many uncertainties (Tizard, Moseley, Buchanan and Murray 1885: 82):

> These results were assumed as giving the rate and direction of the current at different depths with sufficient accuracy to ascertain any marked movements, but it is evident that they are not strictly accurate, as no allowance was made for the retarding or accelerating influence of the surface water on the watch buoy, or of the intermediate water on the line.

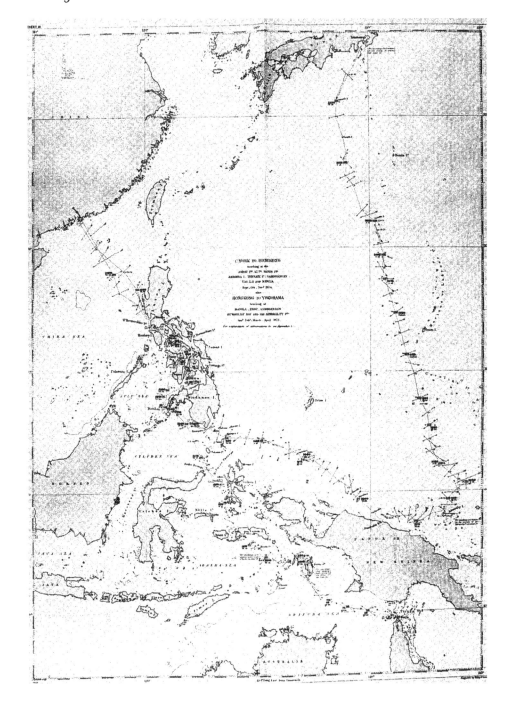

Figure 10.1 Estimates of the surface currents in the western Pacific from the navigation of HMS *Challenger*. Wind direction is shown by straight arrows, and currents by wavy ones. (Tizard et al. 1885.)

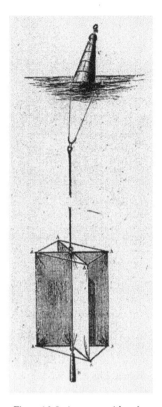

Figure 10.2 A current 'drag' as used on HMS *Challenger*. (Tizard et al. 1885.)

Mechanical inventions and the advent of electronics

During the early part of the twentieth century a number of devices were used to try to expand the techniques available to scientists. These consisted of drifters of various types and ingenious, primarily mechanical, recording devices that could be lowered from ships. The most successful of this latter group and the most widely used were the Ekman (1905) or the improved Merz–Ekman (Merz 1921) current meters. Both were used on the 1925–7 *Meteor* Atlantic Expedition (Spiess 1985) (Fig. 10.3).

These instruments used a propeller to measure current speed, the revolutions being geared down and registered on a mechanical counter and a vane to align the instrument and its propeller into the flow. A magnetic compass sensed current direction. The current direction was recorded by the ingenious device of allowing six small lead balls to be released for every 100 propeller rotations. The balls fell down a chute and were captured in thirty-two boxes, each corresponding to an 11.25 degree segment of the compass card. The period of recording could be started and stopped by dropping a messenger down the non-magnetic aluminium-bronze line from which the instrument was suspended. This ensured that the record was not contaminated from the flow past the meter as it was lowered to depth and then retrieved. A 'repeat' version of the Ekman meter was used on the *Meteor* expedition that allowed up to forty-eight individual recordings to be made, and hence a short time series could be recorded without having to recover the instrument. However ingenious an instrument of this design might have been, it was still of limited use. It had to be suspended

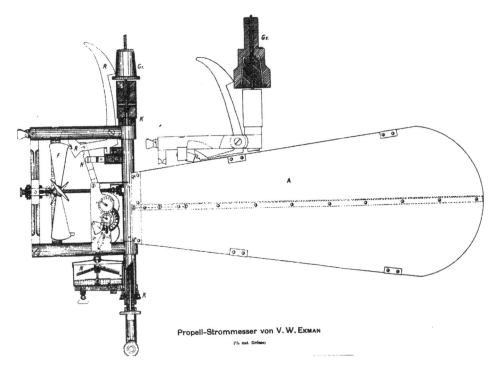

Propell-Strommesser von V. W. EKMAN
(¹/₁ nat. Grösse)

Figure 10.3 The Ekman current meter (1905), deployed by lowering from a ship.

from a ship which, in deep water, could not be anchored and so currents were always measured relative to the ship. Ship drift was often as large as or larger than the currents at depth and was difficult to determine with any accuracy.

World War II forced the development of electronics in general, but of prime importance to marine scientists was the advent of radio navigational aids, radar and sonar. The marine science community, many of whom had been involved with the use of these techniques in wartime, were quick to see the potential of these new methods to solve the outstanding problems of oceanography – among them the measurement of open ocean currents. An excellent overview of the general impact of this development was given by Snodgrass (1968).

A theoretical perspective and practical considerations

Von Arx (1953: 13) stated that:

> Since physical intuition is acknowledged as an unsafe basis for the theoretical description of the oceans, it is necessary at first to describe transient conditions from observations. This description must be based mainly on direct observations taken simultaneously over a sufficiently large area to have synoptic meaning, and for a long enough time to distinguish the longer periodic variations from secular trends . . . this is a more ambitious program of observations than can be managed at present.

Clear evidence of the magnitude of the problem is given in the fact that Bowden (1954) published a summary of all the reported direct current measurements that had been made in the open ocean up to that date. It is reproduced here as Table 10.1

From the perspective of the late 1990s it is amazing that these measurements could be contained in a single table. Little progress had been made since the *Challenger* and *Meteor* expeditions – at least little in any quantitative sense. The longest record in Bowden's table was only 166 hours. All of the measurements were made from instruments lowered from ships that were anchored and so were subject to the uncertainties caused by ship motions. The measurements were typically restricted to the top 1000m and were only of any significance where the currents were strong. At best they could give only very uncertain evidence of tidal currents.

Seminal papers by Stommel (1948), Munk (1950) and by Stommel and Arons (1960a, 1960b) provided a theoretical background for a view of the open ocean circulation consisting of wind-driven, western-intensified gyres, deep boundary currents as the lower limbs of the thermohaline circulation and weak interior flows in ocean basins. However, there were no means of directly testing these hypotheses (Stommel 1954).

If oceanographers were really to understand deep ocean currents from direct measurements, a significant technology breakthrough was required. As we have already seen, there are two basic approaches to measuring currents: tracking tracers or indicators of water movement (such as the drogued buoys used by *Challenger*), known as Lagrangian techniques, and repeatedly measuring the flow vector at a fixed point (such as the Ekman current meter technique), known as Eulerian techniques. Both approaches were and still are used. The development of both techniques went on in parallel and largely independent of one another, so they will be dealt with separately here.

Eulerian techniques

If major advances from the status of the Bowden summary were to be made, problems had to be overcome:

1   A means had to be found of deploying current meters that was independent of having an attendant ship throughout the recording period, i.e. moorings had to be designed and built.
2   The current meters had to be capable of measuring the current speed and direction (and speeds might vary from zero up to several hundred centimetres per second) and of recording these data.
3   Finally, the mooring had to be relocated and recovered.

Moorings

The simplest moorings were those with an anchor on the seabed, a mooring line and a surface buoy. The current meters could be attached anywhere on the mooring line and the surface buoy could be as large as required, subject to the deploying ship being able to handle it. The initial obvious disadvantage was that the mooring had to be rugged enough to withstand the surface winds and waves and the mooring line had to have enough 'scope' (ratio of mooring line length to water depth) to prevent the buoy from being towed under in strong currents. Radar reflectors, radio beacons and lights would serve as relocation devices – provided that the mooring did not break loose and drift away. Toroids and discus shapes became the preferred buoy design (Fig. 10.4a).

Table 10.1 Table from Bowden (1954: 36) summarizing subsurface current observations made prior to 1940. (Reproduced with kind permission of Elsevier Science Ltd, The Boulevard, Langford Lane, Kidlington, OX5 1GB, UK.)

Ship	Date	Area	Reference	Ship's length and tonnage	No. of stations	Duration of stations	Depth of water	Length of cable / Depth of water	Type of current meter	Depths of current measurements
Blake	1888–9	Gulf Stream, West Indian waters	PILLSBURY (1891) WÜST (1924) v. SCHUBERT (1932)	147 ft. 218 tons	39	Up to 166 hr.	Mostly to 1000m once 4000m	1.5 to 2 sometimes 3	Pillsbury	6 to 238m (a few deeper)
Michael Sars	1910	(a) Straits of Gibraltar (b) S. of Azores	HELLAND-HANSEN (1930)	125 ft. 226 tons	1 / 1	15 hr. / 14 hr.	400m / 1000m	1.5	Ekman / Ekman	To 250m / To 730m
Meteor	1925–7	Atlantic between 24°N & 28°S	DEFANT (1932)	220 ft. 1180 tons	9 (+ 3 test stations)	29 to 51 hr.	2150 to 5365m	About 1⅓	(i) Ekman–Merz (ii) Ekman repeating	Various depths to 2500m (few below 1000m)
Willebrord Snellius	1929–30	Dutch East Indian waters	LEK (1938)	204 ft. 1055 tons	7	24 to 89 hr.	1150 to 4850m	1⅓ to 1½	(i) Ekman–Merz (ii) Ekman repeating	To 3000m (few below 1000m)
Armauer Hansen	1930	NE Atlantic	EKMAN & HELLAND-HANSEN (1931) HEKMAN (1953)	76 ft. 59 tons	6	49 to 141 hr.	1805 to 3970m	1.05 to 1.33	(i) Ekman (ii) Ekman repeating (iii) Sverdrup–Dahl	5 to 1000m (a few at 1400 and 2000m)
Meteor	(a) 1937 (b) 1938	N. Atlantic N. Atlantic	DEFANT (1937) BÖNECKE (1937) v. SCHUBERT (1939)	as above	1 / 2	52 hr. / 60 hr.	2140m / 2950 and 2210m	1.7 to 1.8	Böhnecke / Böhnecke	8 depths to 800m
Altair	1938	N. Atlantic	DEFANT (1940)	4000 tons	1	90 hr.	1100 to 1330m	About 4	Böhnecke	5 to 800m

(a)

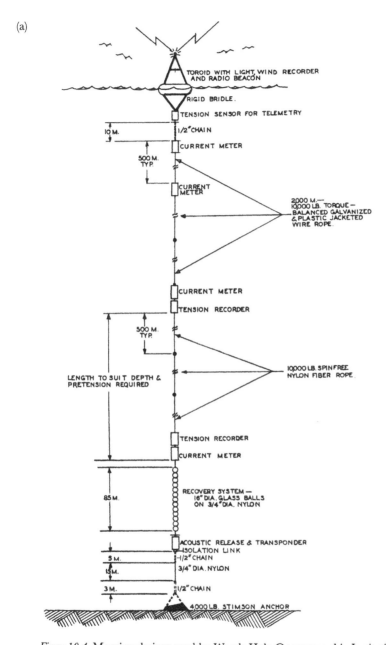

TOROID WITH LIGHT, WIND RECORDER AND RADIO BEACON

RIGID BRIDLE.

TENSION SENSOR FOR TELEMETRY

10 M.

1/2" CHAIN

CURRENT METER

500 M. TYP.

CURRENT METER

2000 M.— 10,000 LB. TORQUE — BALANCED GALVANIZED & PLASTIC JACKETED WIRE ROPE.

CURRENT METER

TENSION RECORDER

500 M. TYP.

LENGTH TO SUIT DEPTH & PRETENSION REQUIRED

10,000 LB. SPIN FREE NYLON FIBER ROPE.

TENSION RECORDER

CURRENT METER

85 M.

RECOVERY SYSTEM — 16" DIA. GLASS BALLS ON 3/4" DIA. NYLON

ACOUSTIC RELEASE & TRANSPONDER

ISOLATION LINK

5 M.

1/2" CHAIN

15 M.

3/4" DIA. NYLON

3 M.

1/2" CHAIN

4,000 LB. STIMSON ANCHOR

Figure 10.4 Mooring designs used by Woods Hole Oceanographic Institution in the 1970s: (a) a mooring with surface buoyance (a toroidal float) and backup buoyance recovery system, and (b, overleaf) a mooring with subsurface buoyancy provided by glass floats. (Heinmiller and Walden 1973: 13, 16.) (Reproduced courtesy of the Woods Hole Oceanographic Institution.)

(b)

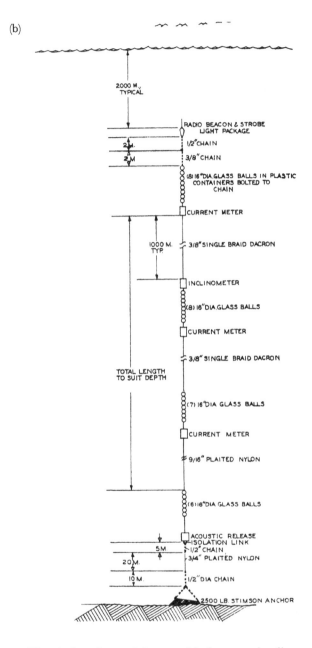

2000 M.,
TYPICAL

RADIO BEACON & STROBE
LIGHT PACKAGE

1/2"CHAIN

2 M.

3/8" CHAIN

2 M

(8) 16"DIA.GLASS BALLS IN PLASTIC
CONTAINERS BOLTED TO
CHAIN

CURRENT METER

1000 M.
TYP.

3/8" SINGLE BRAID DACRON

INCLINOMETER

(8) 16"DIA GLASS BALLS

CURRENT METER

3/8" SINGLE BRAID DACRON

TOTAL LENGTH
TO SUIT DEPTH

(7) 16"DIA GLASS BALLS

CURRENT METER

9/16" PLAITED NYLON

(6)16"DIA GLASS BALLS

ACOUSTIC RELEASE
ISOLATION LINK

5 M.

1/2" CHAIN

20 M.

3/4" PLAITED NYLON

10 M

1/2"DIA CHAIN

2500 LB. STIMSON ANCHOR

The choice of materials was critical; a mooring line could be entirely of steel wire rope, but this meant that for a mooring in 6000m of water, the weight of the wire (typically 1 tonne) was high and this increased the required buoyancy and hence the size of the surface buoy. Wires themselves were subject to corrosion and all elements of the mooring system had to be electrochemically similar or insulated from each other to prevent corrosion taking place in contact with sea water. Stainless steel wires, which were expected to solve the corrosion problem, proved to be disastrously subject to crevice corrosion, and galvanized or plastic-coated wires became the norm.

Weight could be reduced only by replacing the wire with materials such as polypropylene and nylon, which had lower density. Then, rope stretch under tension became a problem, as did cutting of the ropes by fish bite (Stimson 1965). It is only within the past five to ten years that satisfactory solutions to the problems of mooring materials have been found by using plastic-coated wire rope in the top 1000–2000m to prevent fishbite and modern high-strength, low-density materials with low stretch characteristics (e.g. Kevlar) at the deepest levels.

Moorings with surface buoyancy were, as has been said, vulnerable to damage by wind and waves and by interference from passing ships (particularly by inquisitive fishermen). A mooring with subsurface buoyancy was a solution to this problem but again this posed problems of finding suitable buoyancy capable of providing a high buoyancy-to-weight ratio and of being able to withstand pressures.

Solutions were found with spun steel and aluminium spheres, syntactic foam moulded to a variety of low-drag shapes and, perhaps surprisingly, glass spheres. Glass spheres in plastic protective cases ('hard hats') rapidly became the preferred buoyancy element for moorings near the ocean bed and as a means of providing backup buoyancy distributed along moorings covering most of the water column. Recovery of failed moorings using the backup buoyancy was the means by which the modes of failure of early moorings became known, even if the upper part of the mooring, with its main buoyancy element, was lost.

Finally, subsurface moorings needed a release to separate them from the anchor. Timed releases were clearly the simplest, but could result in the mooring surfacing with no ship present or in very poor weather, which precluded recovery (Fig. 10.4b).

Such releases were not a viable option and instead acoustic releases were developed that allowed the mooring to be separated from its anchor when the attendant ship was present and when weather conditions were suitable. Problems to be solved centred on discrimination of coded acoustic signals from background noise, using the acoustic beacon to locate the mooring and developing reliable electromechanical systems that would allow release of a tension of over a tonne at a pressure of 500 atmospheres. A variety of devices were used, including explosive bolts (used for separating missile stages). With hindsight it was amazing that there were no serious injuries in these early development phases.

Reliability and mooring recovery steadily improved with major development programmes at Woods Hole Oceanographic Institution, the National Institute of Oceanography in the UK and at the Institut für Meereskunde in Kiel, Germany.

Current meters

But what of the instruments to be deployed? The development of the transistor and hence of solid state electronics was the key to real progress.

The development of internally recording current meters capable of fulfilling the needs outlined by Von Arx had US, European and Russian foci. Richardson (1961, 1962) developed a current meter that used a Savonius rotor and small vane to sense current speed and direction. The sensors were at either end of a pressure case some 1.3m long and the data were recorded on a film in a series of dots and lines. The data then had to be read either manually or by a rather complex decoding machine. The Richardson current meter was eventually put into commercial production by the Geodyne company and formed the backbone of US development of current-measuring techniques.

In Europe the NATO-sponsored development of a recording current meter was carried out at the Chr. Michelsen Institute in Bergen, Norway (Aanderaa 1963, 1964). As the titles of these two works by Aanderaa imply, the plan was for more than just a current meter; other sensors for temperature and pressure were included and acoustic data telemetry was a feature. The NATO instrument was very advanced compared with the Richardson device. It used digital (10 bit) recording on magnetic tape (the encoder was an electromechanical one that owed more to the watchmaker's skill than to mass-produced electronics), the electronics were 'potted' in resin so that reliability was ensured and the instrument would not be destroyed if it leaked. The NATO current meter was later put into production by Aanderaa himself and its derivatives are still in production to this day (Fig. 10.5).

In the Soviet Union the Alexeev BPV—2 current meter, which used a paper-tape recording medium and was powered by clockwork, was manufactured in large numbers. Its current sensor was a half-shielded paddle wheel. Its maximum number of observations per deployment was 1440 (Fig. 10.6).

Early current meter deployments were aimed at improving the reliability of both the current meters and the mooring systems, rather than addressing particular scientific problems. Woods Hole Oceanographic Institution was at the forefront of this effort through the 1960s and early

Figure 10.5 A prototype of the Aanderaa current meter being deployed from RRS *Discovery* on the International Indian Ocean Expedition in 1962. Below it is a prototype conductivity, temperature depth (CTD) probe. (Southampton Oceanography Centre.)

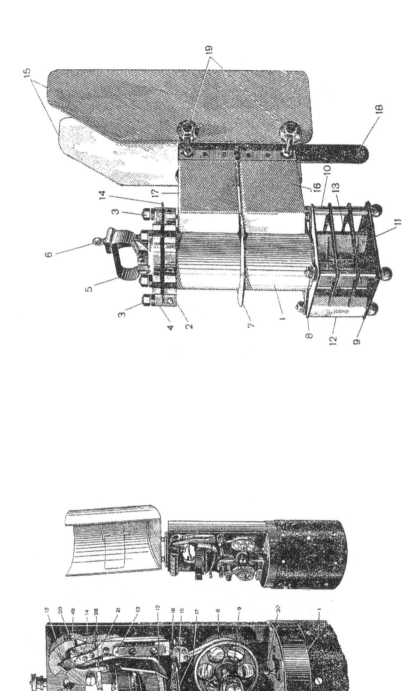

Figure 10.6 The Soviet Alexeev BPV-2 current meter used in the various Polygon experiments of the 1960s and 1970s. A semi-shielded paddle wheel measures current speed. The data are printed onto ticker tape and later transcribed for further analysis. (Reproduced from *Handbook of current meter Alexeev types BPV-2 and BPV-2R.* Leningrad: Isdatelstvo Morskoi Transport 1963.)

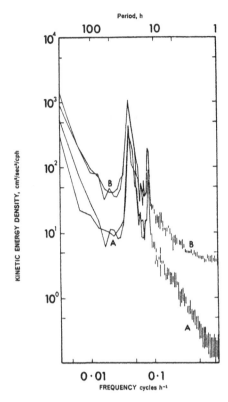

Figure 10.7 Horizontal kinetic energy spectra of currents at the same location from surface and subsurface moorings. Wave-induced mooring motion, rectified by the current meter rotor and undersampled by the vane, contaminates the surface mooring spectra throughout the frequency range. (Gould et al. 1974: 922.) (Reproduced with kind permission of Elsevier Science Ltd, The Boulevard, Langford Lane, Kidlington, OX5 1GB, UK.)

1970s. These developments are very well documented in many Woods Hole technical and data reports and the early stages are summarized by Fofonoff and Webster (1971).

Current meters were initially deployed on both surface and subsurface moorings but it was not until the analysis of Woods Hole site D data by Gould and Sambuco (1975) that it became clear that the two data sets were very different and that this was explained by the inadequate response of the current meters to the high-frequency motions induced by wave action of surface moorings (Fig. 10.7).

It is interesting to note that, in the first flush of enthusiasm for current measurement and before all the problems with the development had been appreciated, grand plans were laid for arrays of buoys to populate the ocean and to provide real-time data – a global ocean-observing system (Garth 1964, IOC 1969, 1971). It was only in the mid-1990s that such a system was seen as a practical possibility (IOC 1993).

Through the 1960s, current meter types proliferated and it would be impossible to document them all. Each had its own strengths and weaknesses and many comparisons between instruments were carried out under a wide range of conditions. Some were initiated by individual scientists, but some sponsored by a SCOR working group chaired by John Swallow (SCOR WG 21 1975).

The comparisons left many questions unanswered, but almost all highlighted the fact that most instruments did a poor job of measuring near-surface currents or currents in the presence of high frequency mooring motion.

Recent developments

By the early 1970s the techniques of both mooring deployment and instrument reliability had improved to the point at which a major attempt could be made to document the mesoscale currents that had been hinted at by float measurements near Bermuda. The Mid-Ocean Dynamics Experiment (MODE) (MODE Group 1978) was a joint US/UK venture that was actually predated by an experiment (Polygon-70) (Kort and Samoilenko 1983) mounted by Soviet scientists in the tropical eastern Atlantic. These experiments explored the variability of currents on spatial scales of many tens of kilometres and over periods of months. The Russian data were subject to surface wave contamination and in MODE a vector-averaging derivative of the Geodyne meter attempted to avoid this problem.

An experiment as large as MODE stretched resources and ingenuity to the full, and WHOI explored methods to deploy the moorings (some twenty-four of them) as quickly as possible. One experiment, the so-called 'faking box' trial, had an entire mooring, current meters, buoyancy, 5000m of wire and rope all connected together and laid out in a large, subdivided box on the stern of the ship. In the appointed position the anchor was released and as it dropped towards the seabed it took with it all the mooring components. It was eventually successful, but the risk of a snag, with the attendant dangers to both instruments and personnel, precluded its use.

From that point onwards the story of deep-sea current measurement from moored buoys has been one of continuing success, as more reliable electronics for current meters and releases and, most importantly, improved materials such as Kevlar and titanium have been used. Records of many years' duration have now been collected; some, from a site started in 1980 (Zenk and Müller 1988), continue to this day (Zenk, personal communication). Statistics of ocean currents have been compiled. Dickson (1989) documents the statistics from some 2400 current meter years of data from the deep ocean, all from records longer than seven months. Deployments of two years or more are becoming routine. Records of current profiles are also now being made using acoustic Doppler and correlation sonars from ships, buoys and units mounted on the sea floor. Clearly, we have come a long way since the Bowden table was published.

Lagrangian techniques

At the same time as current meters were being developed, a radically different approach was being pursued in the UK. John Swallow, who had been developing marine seismic techniques at the University of Cambridge under Bullard, and had used these techniques on the 1950–2 world circumnavigation by HMS *Challenger* (Ritchie 1957), was employed by George Deacon at the National Institute of Oceanography to develop a technique to measure deep-ocean currents.

The Swallow float

Swallow's expertise in marine acoustics was applied initially to the concept of an oceanic meteorological balloon. (A slowly ascending acoustic source would be tracked and its horizontal displacement converted into a current profile.) Swallow found that this approach was not feasible using the hardware available at the time, and instead started to develop a drifter that would stay at one level and move with the currents and be tracked acoustically. The prerequisites were:

1   a pressure case that could be ballasted to a set level,
2   a sound source, and
3   a means of locating the source from an attendant ship.

The underlying principle was that a pressure vessel that was less compressible than sea water could be ballasted to sink at the ocean surface and would gain buoyancy relative to the surrounding water as it sank. If ballasted correctly it would find its neutral density level and then drift with the currents. The first versions were in the true British 'sealing wax and string' tradition. Aluminium scaffold poles formed the pressure case. Two were needed for sufficient buoyancy and the wall thickness had to be adjusted in a bath of caustic soda. The transducers were 10kHz nickel scrolls, supplied by the Admiralty. The floats had then to be assembled with their batteries, primitive electronics and nickel scroll transducers, and the complete assembly of several tens of kilogrammes weighed in salt water to within 1 gramme in order for them to attain the planned pressure level (Fig. 10.8).

The initial Swallow float trials in May–June 1955 over the Iberian Abyssal Plain (Swallow 1955) gave the first direct and reliable measurements of deep currents in the open ocean. The problems of navigation were immense. The ship's position was measured by radar fixes

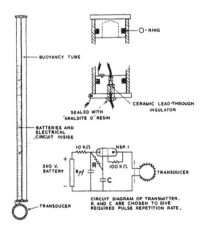

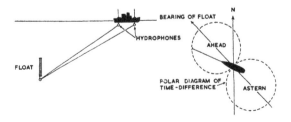

Figure 10.8 A schematic of the prototype Swallow float. (Swallow 1955: 76, 77.) (Reproduced with kind permission of Elsevier Science Ltd, The Boulevard, Langford Lane, Kidlington OX5 1GB, UK.)

on a dhan buoy moored in over 5000m of water and checked on the echo-sounder relative to the position of a small abyssal hill. The float's bearing from the ship was determined by using an oscilloscope to measure changes in the difference in signal arrival times at two hydrophones on the ship's hull as the ship's heading changed. It was only through Swallow's dogged persistence and attention to detail that useful results were obtained. Detection of the rather weak ocean currents over the abyssal plan was a severe test of the technique.

Following the method's success, neutrally buoyant floats were used in 1956 to observe the stronger subsurface Mediterranean outflow at Gibraltar (Swallow 1969) and the flow of Arctic water past the Faroes (Crease 1965).

Exploring the ocean

Henry Stommel of Woods Hole Oceanographic Institution in the USA had predicted that there would be a southward-going undercurrent beneath the Gulf Stream and in 1957 a joint US/UK experiment on RRS *Discovery II* and RV *Atlantis* used Swallow floats to observe this flow off North Carolina (Swallow and Worthington 1961).

The success of the Gulf Stream experiment led to an attempt in 1959/60 to measure the predicted slower interior poleward recirculation of the oceans using Swallow floats. The expectation of very sluggish (order a few millimetres per second) flows were this time found wanting. Currents were an order of magnitude greater and were discovered to vary on time

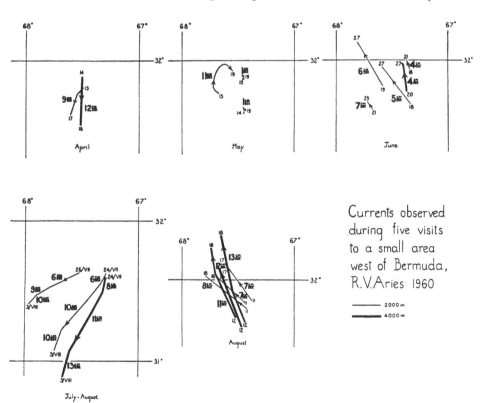

Figure 10.9 Results from the 1960 *Aries* experiment off Bermuda. (Crease 1963: 3174.) (Published by the American Geophysical Union.)

and space scales of tens of days and kilometres. These results from the *Aries* experiment (Fig. 10.9), named after the 100ft ketch on which the work was carried out south of Bermuda, gave the first glimpses of an ocean now known to be populated with energetic eddies – the analogues of atmospheric weather systems. This was arguably the most significant discovery about the nature of the ocean made in the twentieth century (Crease 1963, Swallow 1971).

Extending the range

As mentioned in the previous sections, both current meter and float technologies reached maturity in the late 1960s and by 1970 plans were made for a Mid-Ocean Dynamics Experiment that would employ on surface and subsurface moorings the latest Vector Averaging Current Meters.

These methods were complemented by two developments of neutrally buoyant floats. In the United Kingdom, Swallow, McCartney and Millard (1974) developed a transponding version of the Swallow float, the Minimode float, that could achieve longer ranges (up to 100km) by using lower tracking frequencies. On a much larger scale Rossby and Webb (1970, 1971) had developed floats that could be tracked over basin-scale distances using much lower frequencies (around 200Hz) and receiving these signals at hydrophone sites maintained by the US Navy. These were used in MODE (MODE Group 1978) and in the subsequent US–USSR Polymode experiments (Kamenkovich, Monin and Voorhis 1986, Collins and Heinmiller 1989).

The initial restriction of SOFAR floats to the western North Atlantic, within range of fixed hydrophones, was relaxed when in the late 1970s Autonomous (moored) Listening Stations (ALS) were developed (Bradley 1978). From then on the use of SOFAR floats expanded rapidly and enabled current statistics to be accumulated (Owens 1991, Ollitrault 1994).

Further developments enhanced the floats' capabilities. Rossby, Dorson and Fontaine (1986) developed RAFOS (SOFAR backwards) floats that used moored sound sources, the signals from which were received by the float and transmitted back to an overflying satellite when the float surfaced at the end of its life. Floats had previously been isobaric (their density dependent on the pressure level at which they settled). The addition of a compressee (Rossby, Levine and Connors 1985) enabled the floats to follow isopycnals and hence be better markers of the true movement of water masses. In the mid-1990s the state of the art in acoustically tracked floats was the ALFOS float. This is a RAFOS float that surfaces at regular intervals throughout its life and transmits its navigation via satellite (Ollitrault 1993).

Acoustic tracking implies a geographical restriction to areas of the ocean that are insonified. For a truly global experiment such as the World Ocean Circulation Experiment (WOCE) this restriction was not acceptable and therefore neutral buoyancy floats were developed that surfaced regularly and relayed their position to the ARGOS satellite (Davis, Webb, Regier and Dufour 1992). WOCE will have deployed over a thousand of these ALACE floats before the end of the experiment in 1997 (Fig. 10.10).

Conclusion

This chapter documents the progress made in solving a major challenge to the oceanographic community. The progress made in the past forty years has been huge and now

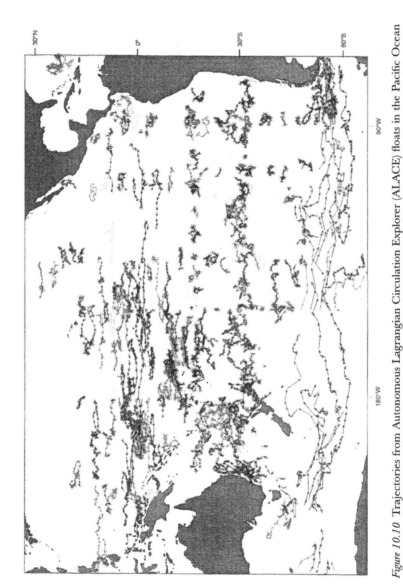

Figure 10.10 Trajectories from Autonomous Lagrangian Circulation Explorer (ALACE) floats in the Pacific Ocean from the WOCE project. (Courtesy Professor Russ Davis, Scripps Institution of Oceanography.)

Figure 10.11 John Swallow in the early 1980s, preparing to weight a SOFAR float. The tube on the right side is the 'organ pipe' sound source. The spheres on the left side contain electronics and batteries. When deployed, the two units are in line. (Southampton Oceanography Centre.)

enables the currents of the world's oceans to be measured in an almost routine fashion. The application of these techniques in operational monitoring of the state of the ocean is upon us.

The progress was a cooperative effort between scientists and engineers but was driven by scientific requirements. John Swallow (Fig. 10.11) played a major role in both strands (Eulerian and Lagrangian) of the unravelling of the puzzle. John's death in 1994 makes it appropriate to dedicate this brief summary to his memory. Without his insight and dogged persistence one might well argue that the subject of this chapter might not have been such a success story.

References

Aanderaa, I. (1963) *A recording and telemetering instrument for collecting data from the sea. As reported to the Symposium on Anchored Buoys held in Kiel, 6–7 June 1963.* Bergen: Chr. Michelsens Institutt.

Aanderaa, I. (1964) *A recording and telemetering instrument.* NATO Subcommittee on Oceanographic Research Technical Report **16**.

Bowden, K. F. (1954) 'The direct measurement of subsurface currents in the oceans', *Deep-Sea Research* **2**: 33–47.

Bradley, A. (1978) 'Autonomous listening stations', *Polymode News* **4** (unpublished manuscript).

Collins, C. A. and Heinmiller, R. H. (1989) 'The POLYMODE program', *Ocean Development and International Law* **20**: 391–408.

Crease, J. (1963) 'Velocity measurements in the deep water of the western North Atlantic', *Journal of Geophysical Research* **67**: 3173–6.

Crease, J. (1965) 'The flow of Norwegian Sea water through the Faroe Bank Channel', *Deep-Sea Research* **12**: 143–50.

Davis, R. E., Webb, D. C., Regier, L. A. and Dufour J. (1992) 'The Autonomous Lagrangian Circulation Explorer (ALACE)', *Journal of Atmospheric and Oceanic Technology* **9**: 264–85.

Dickson, R. R. (1989) *Flow statistics from long-term current meter moorings. The global data-set in January 1989.* WCRP-30, WMO/TD–No. 337. World Climate Research Programme, World Meteorological Organisation.

Ekman, V. W. (1905) 'Kurze Beschreibung eines Propell-Strommessers', *Conseil International pour l'Exploration de la Mer, Publications de Circonstance* **24**.

Fofonoff, N. P. and Webster, F. (1971) 'Current measurements in the western Atlantic', *Philosophical Transactions of the Royal Society of London*, **A270**: 423–36.

Garth, F. M. (1964) 'An initial look at oceanographic buoy system requirements', in *Buoy Technology – Supplement*, 13–30. Washington: Marine Technology Society.

Gould, W. J. and Sambuco, E. (1975) 'The effect of mooring type on measured values of ocean currents', *Deep-Sea Research* **22**: 55–62.

Gould, W. J., Schmitz Jr, W. J. and Wunsch, C. (1974) 'Preliminary field results for a Mid-Ocean Dynamics Experiment' (MODE–0)', *Deep-Sea Research* **21**: 911–31.

Heinmiller, R. H. and Walden, R. G. (1973) 'Details of Woods Hole moorings', Woods Hole Oceanographic Institution Technical Report **73–71** (unpublished manuscript).

Helland-Hansen, B. and Nansen, F. (1909) *The Norwegian Sea: its physical oceanography based upon the Norwegian researches 1900–1904.* Report on Norwegian Fishery and Marine Investigations **2**(2).

IOC (1969) *Perspectives in oceanography, 1968.* Intergovernmental Oceanographic Commission Technical Series **6**. Paris: UNESCO.

IOC (1971) *IGOSS (Integrated Ocean Station System). General plan and implementation proposal for Phase 1.* Intergovernmental Oceanographic Commission Technical Series **8**. Paris: UNESCO.

IOC (1993) *Global Ocean Observing System. Status report on existing ocean elements and related systems.* Intergovernmental Oceanographic Commission Information Bulletin IOC **INF-958**.

Kamenkovich, V. M., Monin, A. S. and Voorhis, A. D. (1986) *The POLYMODE atlas.* Woods Hole, Mass.: Woods Hole Oceanographic Institution.

Kort, V. G. and Samoilenko, V. S. (1983) *Atlantic hydrophysical Polygon-70: meteorological and hydrophysical investigations.* New Delhi: Amerind Publishing Co. (for US National Science Foundation).

Merz, A. (1921) 'Stark- und Schwachstrommesser', *Veröffentlichungen des Instituts für Meereskunde, Berlin*, N.S. **A7**.

MODE Group (1978) 'The Mid-Ocean Dynamics Experiment', *Deep-Sea Research* **25**: 859–910.

Munk, W. H. (1950) 'On the wind-driven ocean circulation', *Journal of Meteorology* **7**: 79–93.

Ollitrault, M. (1993) 'The SAMBA0 experiment: a comparison of subsurface floats using the RAFOS technique', *WOCE Newsletter* **14**: 28–32 (unpublished manuscript).

Ollitrault, M. (1994) 'The TOPOGULF experiment Lagrangian data', *Repères Océan* **7**. Brest: IFREMER.

Owens, W. B. (1991) 'A statistical description of the mean circulation and eddy variability in the northwestern Atlantic using SOFAR floats', *Progress in Oceanography* **28**: 257–303.

Rennell, J. (1832) *An investigation of the currents of the Atlantic Ocean, and of those which prevail between the Indian Ocean and the Atlantic.* London: J. G. and F. Rivington.

Richardson, W. S. (1961) 'Current measurement from moored buoys', *Oceanus* **8**: 14–19.

Richardson, W. S. (1962) 'Instruction manual for recording current meter', Woods Hole Oceanographic Institution **62–6** (unpublished manuscript).

Richardson, W. S., Stimson, P. B. and Wilkins, C. H. (1963) 'Current measurements from moored buoys', *Deep-Sea Research* **10**: 369–88.

Ritchie, G. S. (1957) Challenger: *the life of a survey ship*. London: Hollis and Carter.

Rossby, T. and Webb, D. 1970. 'Observing abyssal motions by tracking Swallow floats in the SOFAR Channel', *Deep-Sea Research* **17**: 359–65.

Rossby, T. and Webb, D. 1971. 'The four month drift of a Swallow float', *Deep-Sea Research* **18**: 1035–9.

Rossby, T., Dorson, D. and Fontaine, J. (1986) 'The RAFOS system', *Journal of Atmospheric and Oceanic Technology* **3**: 672–9.

Rossby, H. T., Levine, E. R. and Connors, D. N. (1985) 'The isopycnal Swallow float – a simple device for tracking water parcels in the ocean', in J. Crease, W. J. Gould and P. M. Saunders (eds), *Essays on oceanography: a tribute to John Swallow. Progress in Oceanography* **14**: 511–25.

Sandström, J. W. and Helland-Hansen, B. (1903) *Über die Berechnung von Meeresströmungen'*. Report on Norwegian Fishery and Marine Investigations 2(4).

SCOR Working Group 21 (1975) *An intercomparison of some current meters III*. Technical Papers in Marine Science **23**. Paris: UNESCO.

Snodgrass, J. M. (1968) 'Instrumentation and communications', in J. F. Brahtz (ed.) *Ocean engineering: goals, environment, technology*. New York: John Wiley, pp. 393–477.

Spiess, F. (1985) *The* Meteor *expedition: scientific results of the German Atlantic expedition 1925–1927*, trans. W. J. Emery. New Delhi: Amerind Publishing Co. (for US National Science Foundation).

Stimson, P. B. (1965) 'Synthetic-fiber deep-sea mooring cables: their life expectancy and susceptibility to biological attack', *Deep-Sea Research* **12**: 1–8.

Stommel, H. (1948) 'The westward intensification of wind-driven ocean currents', *Transactions of the American Geophysical Union* **29**: 202–6.

Stommel, H. (1954) 'Why do our ideas about the ocean circulation have such a peculiarly dream-like quality?' Woods Hole Oceanographic Institution (unpublished manuscript).

Stommel, H. and Arons, A. B. (1960a) 'On the abyssal circulation of the world ocean – I. Stationary planetary flow patterns on a sphere', *Deep-Sea Research* **6**: 140–54.

Stommel, H. and Arons, A. B. (1960b) 'On the abyssal circulation of the world ocean, II: An idealized model of the circulation pattern and amplitude in oceanic basins', *Deep-Sea Research* **6**: 217–33.

Swallow, J. C. (1955) 'A neutral-buoyancy float for measuring deep currents', *Deep-Sea Research* **3**: 74–81.

Swallow, J. C. (1969) 'A deep eddy off Cape St Vincent', *Deep-Sea Research* **16** (Supplement): 285–95.

Swallow, J. C. (1971) 'The *Aries* current measurements in the western North Atlantic', *Philosophical Transactions of the Royal Society of London* **A270**: 451–60.

Swallow, J. C. and Worthington, L. V. (1961) 'An observation of a deep countercurrent in the western North Atlantic', *Deep-Sea Research* **8**: 1–19.

Swallow, J. C., McCartney, B. S. and Millard, N. W. (1974) 'The Minimode float tracking system', *Deep-Sea Research* **21**: 573–95.

Tizard, T. H., Moseley, H. N., Buchanan, J. Y. and Murray, J. (1885) 'Narrative of the cruise of HMS *Challenger*, with a general account of the scientific results of the expedition', in C. W. Thomson and J. Murray (eds) *Report on the scientific results of the voyage of HMS* Challenger *during the years 1872–76* (1880–95) **1**. London: HMSO.

Von Arx, W. S. (1953) 'Measurements of the oceanic circulation in temperate and tropical latitudes', in J. D. Isaacs and C. O'D. Iselin (eds) *Symposium on Oceanographic Instrumentation*, Rancho Santa Fe, California, 21–23 June 1952. Washington, D.C.: National Academy of Sciences, National Research Council, pp. 13–35.

Zenk, W. and Müller, T. J. (1988) 'Seven-year current meter record in the eastern North Atlantic', *Deep-Sea Research* **35A**: 1259–68.

11 Oceanography from space
Past success, future challenge

*T. H. Guymer, P. G. Challenor and
M. A. Srokosz*

Introduction

Figure 11.1 is a photograph of the Mediterranean taken by an oceanographer on board
Challenger – the space shuttle *Challenger*, that is! From a height of two hundred miles or so it
shows complex eddy patterns, visible in the glitter pattern produced by the Sun, and gives a
completely new perspective on the world ocean. Such an image prompts questions as to how
oceanic features are observed and how we might use the new knowledge to best effect. In this
chapter, we review developments in the subject, emphasizing particularly the deep-sea
aspects and the contribution made by UK scientists. The review is not intended to be
exhaustive but to highlight some key milestones and ways in which the data are being used;
for detailed explanation of sensor design and data processing the reader is referred to
Stewart (1985) and Robinson (1985). A more general description of present-day satellite
data, the sampling obtained and the uses to which the data can be put is given by Robinson
and Guymer (1996). Before discussing the progress made in different application areas, we
make some general observations.

Key milestones

The history of space oceanography is comparatively short (Table 11.1), the first artificial
satellite having been launched by the USSR in 1957. However, in the following forty years
spectacular advances have been made, often exceeding the dreams of the early pioneers.
Much of this resulted from the race to put a man on the Moon. (So great was the effort
devoted to this task that the goal was achieved before detailed exploration of mid-ocean
ridges.) The spin-off from the Apollo programme was considerable; looking back at the
Earth, astronauts took thousands of photographs and documented what they saw. Many of
them showed features in the ocean. In 1964, several years before Armstrong walked on the
lunar surface, NASA gathered about 150 oceanographers together to assess the potential of
using space-borne instruments for studying the sea. The report produced by the symposium
'Oceanography from Space' (Ewing 1965) laid the foundation for much of what followed,
with speculation, for example, about the possibility of measuring sea level to 10cm by radar
altimetry and hence deriving information on surface currents. The fundamental variables
measurable from space and the main oceanographic quantities that can be inferred are listed
in Table 11.2.

Subsequently, several series of meteorological and research satellites were launched.
NASA's Skylab and GEOS–3 in the 1970s paved the way for wind, wave and sea-level
measurements. However, 1978 provided a leap forward in widespread appreciation of this

Figure 11.1 Eddies in the Mediterranean revealed by sun-glitter, photographed from space shuttle *Challenger* in 1984.

new technology, for three major satellites were launched, carrying between them a significant combination of oceanographic sensors. Not only did these provide a wealth of data providing new insights on oceanographic phenomena, but they spurred other agencies to develop their own programmes. Even by the launch of Seasat, the European Space Agency had drawn up specifications for a Coastal Oceans Monitoring Satellite System (COMSS) consisting of an optical imaging instrument (a successor to NASA's Coastal Zone Color Scanner (CZCS), a synthetic aperture radar and an imaging microwave radiometer. UK scientists were among those pressing for a radar altimeter to be included and ESA commissioned a study on the usefulness of such a sensor for oceanography, glaciology and climatology (Duchossois et al. 1980). Largely as a result of that study, an altimeter was incorporated on COMSS but unfortunately it was at the expense of the ocean colour sensor. The satellite was renamed as ERS-1 and was eventually launched in 1991. Given that there has been an interval of at least eighteen years between the launch of CZCS and its successor,

Table 11.1 Key milestones

1957	First artificial satellite
1964	Oceanography from Space Symposium
1960s	TIROS and NIMBUS satellites, attempts to measure SST
1973	Skylab (scatterometer and altimeter)
1975	GEOS–3 altimeter for geodesy
1978	Seasat, AVHRR, CZCS launched, ERS-1 conceived (originally COMSS)
1980	Altimeter study by RAL and IOS
1981, 1984	Shuttle-imaging radars
1985	Geosat altimeter launched
1991	ERS-1 launched
1992	Topex/Poseidon launched
1995	ERS-2 launched
1996	ADEOS launched (failed June 1997)
1997	SeaWiFS launched
1999	Terra (EOS-AM1) launches
2000	JASON-1 due for launch
2000	CHAMP due for launch (first direct measurement of the geoid)
2000/1	Envisat due for launch
2001	GRACE due for launch
2003	GOCE due for launch
2004/5?	SMOS due for launch (salinity from space)

Table 11.2 Variables measurable from space

Basic quantity	*Sensor type*	*Derived information*
Radiance	infra-red, passive MW visible	SST, eddies, chlorophyll, sediments
Topography	radar altimeter	sea-level change, currents
Reflectivity	visible, radar	winds, waves, eddies, internal waves, slicks

the marine biology community may consider itself very unfortunate compared with the physical oceanographers, who have had several imaging radar and altimeter missions during this time. This serves to illustrate both the compromises that are sometimes necessary and the very long lead times between initial design decisions and the flow of useful data.

Sea-surface temperature

The first successful measurements of sea-surface temperature (SST) from space were acquired by an infrared radiometer on NIMBUS-II in 1966 (Warnecke, McMillin and Allison 1969). Prior to that, some SST data had been obtained by similar instruments on ships and aircraft. Lower resolution instruments had also been flown on the TIROS meteorological satellites, but appreciable absorption of radiation by atmospheric water vapour prevented reliable extraction of surface temperatures.

The thermal radiation measured at satellite altitude is a combination of that emitted by the underlying surface – sea, ice, land, clouds – and absorption and re-emission by the intervening atmosphere. The emissivity of water is known, so the temperature can be calculated from the radiance, provided the effect of the water vapour can be corrected for. By choosing an atmospheric 'window', e.g. 11/12 micron, where absorption is at a minimum, this problem can be reduced. This means using instruments that measure in very narrow wavelength bands. Unfortunately, the early meteorological satellites could not achieve this, hence their lack of success in measuring SST. Interestingly, for NIMBUS-II the sensor operated in another region (3.5–4.1 micron, the near infra-red). Modern sensors have reverted to 11/12 micron. Figure 11.2 shows an image of the Southwest Atlantic obtained with this sensor. The quality is very poor compared with what we are now used to, but we have to remind ourselves that the data are from the first quarter of the space era so far.

In 1978, one of the key milestone years, the first Advanced Very High Resolution Radiometer (AVHRR) was launched and, for those areas within range of suitable receiving stations, 1km resolution images have been available on a daily basis ever since. These data have been used for examining a very wide variety of oceanic features, including fronts and eddies, coastal upwelling and tropical waves, and there is now a sufficiently long and consistent data set from which to study the seasonal and interannual variations of basin-scale SST. Improved atmospheric correction is achieved through the 'split window' technique, whereby measurements are made at two wavelengths lying close together in the 11/12 micron window.

The most accurate sensor for measuring SST is the UK-built Along-Track Scanning Radiometer (ATSR) flying on ERS-1 and ERS-2. Figure 11.3 shows rich mesoscale activity west of Greenland and illustrates ATSR's very good radiometric resolution. However, the main advantage over other sensors is its high absolute accuracy (better than 0.5K) which makes it more able to meet the stringent accuracy criteria for climate purposes than other IR sensors. This is achieved by more stable instrument calibration and by adding a dual viewing-angle approach to the split window method. ATSR scans in a cone and looks at the same spot on the ocean, first at an incidence angle of 50° or so, then some time later at nadir. This gives different optical path lengths and hence an additional means of correcting for atmospheric effects.

Ocean topography and currents from radar altimeter measurements

In the 1964 symposium it had been speculated that it might be possible to measure sea level using a radar altimeter. The concepts involved are very simple (see Fig. 11.4). If the altimeter sends a radar pulse down to the sea surface and the time taken for the pulse to be reflected back to the antenna is measured very accurately, then, knowing the velocity of the signal (which requires knowledge of the intervening atmosphere), the distance between the altimeter and the surface can be determined. This can be done very precisely – in fact, to an accuracy of considerably better than 10cm from a satellite height of 1000km (1 part in 10^7). If the height of the satellite, with respect to an Earth coordinate system, is known, then the along-track variations of the topography of the sea surface are obtained. To obtain currents you need to know the difference between the sea-surface topography and the geoid – the geopotential surface that the sea surface would adopt if at rest. As the geoid does not change over the time-scales of interest to oceanographers, this difference is known as the dynamic topography, from which the surface geostrophic current may be calculated (Stewart 1985). The variations in the geoid are of order ±100m (relative to the reference ellipsoid; see Marsh and Martin 1982) and reflect local variations in the earth's gravity field. In contrast, the variations in dynamic topography associated with the strongest currents, such as the Gulf Stream, are of the order of 1m, with much smaller values in 'quieter' regions of the ocean. Clearly, to obtain currents from the altimeter data, it is necessary to know the geoid accurately in order to calculate the absolute dynamic topography. (See also Chapter 5.)

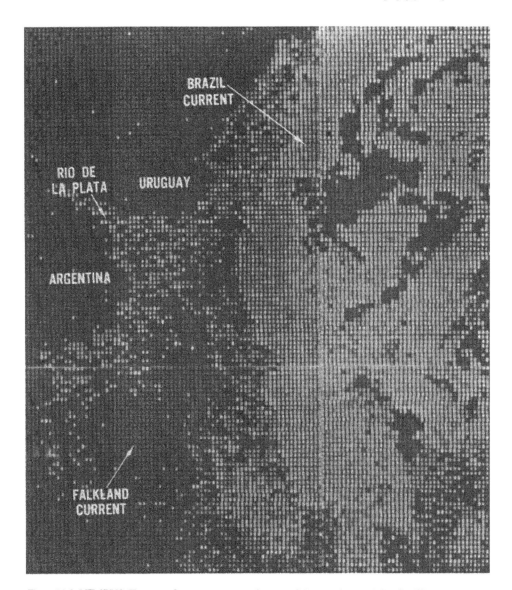

Figure 11.2 NIMBUS-II sea-surface temperature image of the southwest Atlantic. The temperature difference between the warm Brazil current and the cool Falkland current is 8K. (From Warnecke et al. 1969.)

In fact, the altimeter data themselves provide good information on the geoid which has proved of interest to geodesists and geophysicists (Marsh and Martin 1982, Rapp 1993), but this topic will not be pursued here. It is sufficient to note that the time-invariant part of the altimeter height measurements includes both the geoid and the mean circulation, so that an independent measurement of the geoid would enable the mean circulation to be determined by altimetry. Geoid (strictly, mean sea surface) measurements from altimetry have also provided information on the bathymetry of the oceans (Smith and Sandwell 1994). The bathymetry is poorly known in many places, and these data have proved useful in identifying

Figure 11.3 Along-Track Scanning Radiometer image of eddies west of Greenland. (Courtesy Rutherford Appleton Laboratory.)

features such as sea-mounts (Craig and Sandwell 1988), whose presence can affect the currents. Another problem with the topography data is that they include the effects of the tides, which have to be removed if the ocean circulation is to be studied. This can be done using tidal models, and the altimeter data themselves can be used to study tides in their own right (Ray 1993). Fu et al. (1994) discuss these and other corrections that need to be made to the data before ocean currents can be measured (see also Stewart 1985).

A prototype altimeter was flown on Skylab in 1973, but its measurement precision was low (around 1m; Stewart 1985) so it was not that useful for ocean circulation studies. The GEOS–3 altimeter could measure to higher precision (around 20cm; Stanley 1979), but only regional data were obtained (Stanley 1979). The first truly global picture, of the sea-level

Altimetry Schematic

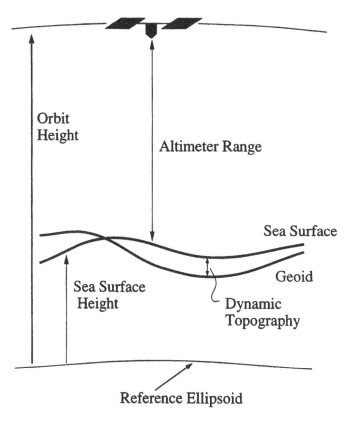

Figure 11.4 Schematic of altimetry measurement showing the relationship between the instant-aneous sea surface and the geoid. (After Cromwell et al. 1996.)

variability, was obtained by Cheney, Marsh and Beckley (1983) from only twenty-four days of Seasat altimeter data (see Fig. 11.5). At the time of Seasat the geoid was not well enough known to be able to separate it from the mean circulation signal, so Cheney et al. (1983) used a succession of repeat measurements over the same area to calculate the vari-ations in sea-surface topography relative to the mean, the so-called variability. The results (Fig. 11.5) show the high variability regions associated with the western boundary currents (such as the Gulf Stream and the Kuroshio) and the Antarctic Circumpolar Current (ACC), and the lower values in the 'quieter' (less energetic currents) areas of the oceans. Similar calculations can be made for the variability in the currents, which are derived from the gradient of the topography (see Shum et al. 1990, for an example using Geosat data).

Subsequent altimeters, Geosat, ERS-1 and Topex/Poseidon have made increasingly precise measurements of the sea-surface topography (Topex precision is around 3cm; Fu et al. 1994). For a review of recent work see Fu and Cheney (1995). Figure 11.6 shows results for variability in the South Atlantic from Topex/Poseidon, with somewhat better resolution than those from Seasat, but with similar features. The resolution is related to the different space–time sampling pattern explained by the ten-day repeat orbit of Topex/Poseidon

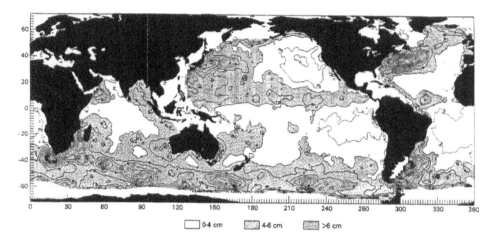

Figure 11.5 Global mesoscale sea-height variability measured by the Seasat altimeter. High variability regions are associated with the Gulf Stream, the Kuroshio Current, the Agulhas Retroflection, the Brazil–Falklands Confluence and the Antarctic Circumpolar Current. (Courtesy R. E. Cheney; see Cheney et al. 1983.)

(which gives better spatial and poorer temporal resolution than the Seasat three-day repeat data used by Cheney et al. 1983). The variability is related to the passage, growth and decay of individual eddies and to meandering of the current systems. By processing the data in a slightly different way one can obtain these height anomalies. A height anomaly map is obtained from a single ten-day cycle of Topex data, from which the long-term mean has been subtracted, and highlights the individual eddy-scale features. A sequence of anomaly maps can be generated for the period of Topex operation. Such sequences show the eddies spinning off from the Agulhas, south of Africa, and making their way across the Atlantic and likewise being squeezed out from the Brazil–Falklands confluence off Argentina. Altimetry is proving useful in tracing the history of eddies which might be observed only once or twice by a ship (Smythe-Wright, Gordon, Chapman and Jones 1996).

Topex has provided a significant advance over earlier altimeters not only because it has been able to improve the precision of the range measurement, but also because the orbit of the satellite has been determined more accurately (Fu et al. 1994). In conjunction with this, there has been commensurate improvement in our knowledge of the geoid (Nerem et al. 1994a). Therefore with Topex it is possible to measure the large-scale ocean circulation (Nerem et al. 1994b). Figure 11.7 shows one year of sea surface height data, from the South Atlantic, from which the best available geoid has been subtracted. The ACC, the subtropical gyre and the Agulhas Retroflection are clearly visible. Due to the continuing lack of knowledge of the small-scale geoid (Nerem et al. 1994a), this picture has to be viewed with caution. Some of the smaller-scale features are probably artefacts attributable to this uncertainty.

Apart from making independent measurements of the geoid at small scale, it is possible to remove its effects by combining the altimeter data with *in situ* measurements of the hydrography and currents, made by a ship travelling along the satellite ground track. By referencing all the altimeter data taken along the ground track to that obtained at the time of the ship traverse, and using the ship-measured surface currents, it is possible to obtain absolute

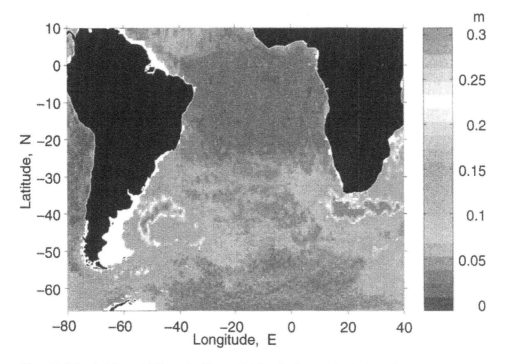

Figure 11.6 Sea-height variability as for Figure 11.5, but for South Atlantic from Topex data. (Courtesy M. S. Jones.)

surface currents from the altimeter data every time the altimeter passes along that ground track. This technique has been developed by Challenor, Read, Pollard and Tokmakian (1996) and successfully applied to ERS-1 altimeter data across the Drake Passage (Challenor et al. 1996) and in the region of the Azores Current (Cromwell, Challenor, New and Pingree 1996). Clearly it is not possible to make such measurements along all the ground tracks of all the altimeters, but the technique allows the monitoring of specific regions. The need to improve our knowledge of the geoid at small scales remains the outstanding problem if full use is to be made of altimeter data for ocean circulation studies.

Waves

The history of the measurement of waves by remote sensing goes back to the late 1940s when the National Institute of Oceanography in Britain conducted experiments with a radar altimeter on an aircraft. Although they obtained good results, interest moved to the measurement of waves *in situ*, leading to the development of the ship-borne wave recorder, and the method was not pursued. However, the approach was taken up in the US, initially at Woods Hole and then at the Naval Research Laboratory. These studies showed that it was possible to measure wave properties from aircraft using radar, and were the forerunners of the airborne systems used today and hence of all satellite systems. The 1964 symposium considered sea state as one of the parameters that could be measured from space. Although the details have changed, the two techniques identified at that meeting, a nadir-pointing

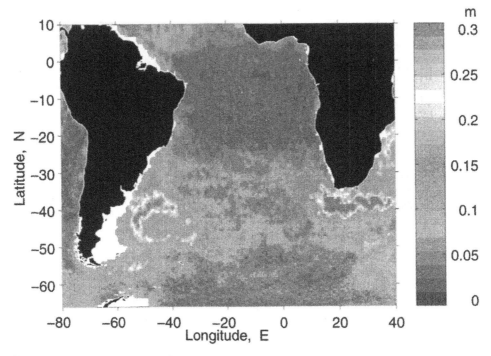

Figure 11.7 South Atlantic sea-surface height field from one year of Topex data, after removal of the geoid. (Courtesy M. S. Jones.)

radar altimeter and a side-looking imaging radar, have proved successful for the measurement of surface waves.

Radar altimeters

The basic theory for the measurement of significant wave height by radar altimeters was developed by Brown (1977) and Barrick (1968). Significant wave height is defined to be four times the standard deviation of the surface elevation and is thus a measure of surface roughness. It is also the mean height of the highest third of the waves and is roughly equivalent to a visual estimate of wave height. In essence an altimeter measures significant wave height by measuring the degree to which a narrow pulse of radar energy is smeared when it is reflected from the sea surface. This smearing is caused by the pulse being reflected first from the crests of the waves and later from the troughs. A higher sea state results in a greater degree of smearing.

GEOS–3 was the first satellite to carry a radar altimeter that could measure waves to a useful accuracy (an altimeter on Skylab demonstrated the possibility but the results were too crude to be useful). A special issue of the *Journal of Geophysical Research* was dedicated to GEOS–3 results and this contains a number of papers on wave measurement, mainly on calibration (Fedor et al. 1979, Mognard and Lago 1979, Parsons 1979, Priester and Miller 1979, Walsh 1979) but with one on applications (Pierson and Salfi 1979). Unfortunately GEOS–3 did not possess any on-board storage so measurements could be taken only when the satellite was visible from a ground station. This led to very patchy data recovery, with good coverage around the US coast and progressively poorer coverage elsewhere. However, the coverage was good enough to allow the first measured global wave climatology to be compiled (Sandwell and Agreen 1984).

Seasat was launched (and failed) while GEOS–3 was still operating. However, because of the true global coverage and the additional sensors, it quickly took the limelight. Chelton, Hussey and Parke (1981) is an important milestone paper because it put the measured global wave climate in colour on the cover of *Nature*. This was an important step in gaining wider acceptance of the unique view of the oceans provided by satellite observations.

After the failure of Seasat, and the end of the GEOS–3 mission, both in 1978, there was a gap of six years until the launch of Geosat. Initially the first eighteeen months' data from Geosat, including the wave data, were classified so all scientific work had to be carried out on the later part of the mission when the satellite was in a seventeen-day repeat period. The seasonal structure of the wave climate could now be considered (Challenor, Foale and Webb 1990, Carter, Foale and Webb 1991). The wave data from the classified mission were quite quickly declassified and it started to be possible to look at the interannual changes in the wave climate.

Geosat produced data for about five years in total, although towards the end of its life the quality deteriorated somewhat. A short gap ensued before ERS-1 started to produce good data. This was followed a year later by Topex/Poseidon. For one year while ERS-2 was being calibrated we had the luxury of three altimeters operating simultaneously. The data stream from ERS-1 stopped on 6 June 1996, although the satellite could be reactivated if necessary.

Although, following the work of Brown (1977), there is a theoretical model of how altimeters measure waves, there is still a need for calibration. This is particularly true if we wish to look for small interannual changes using more than one satellite. If two satellites are flying simultaneously, such as Topex/Poseidon and ERS-1 it is possible to look at the cross-overs where both altimeters measure at the same time and in the same place and calibrate one satellite relative to another. However, in some cases this is not possible. For instance the Topex and Poseidon altimeters cannot both be on at the same time and, more importantly for looking for climate change, Geosat was not active at the same time as any other satellite altimeter. We therefore need another data source that can be used as a calibration standard. NOAA has for some years maintained a network of data buoys around the US coast and these have been used extensively to calibrate altimetry (e.g. Carter et al. 1992). There is a slight problem in that these buoys are all in the northern hemisphere and to allow calibrations to be extended to the highest waves, we need many more buoys so that we can capture the rarer extreme events.

The great advantage of satellite observations of wave height compared with more conventional sources is that they enable us to look at the global wave climate. Measurements from buoys or ships equipped with ship-borne wave recorders are few and in the main are restricted to coastal areas, with the notable exception of a few weather ships, while visual observations are both of uncertain quality and are concentrated along the major shipping routes. For areas such as the South Pacific it is the first time that we have been able to have any measurements, let alone ones that are reliable. Figure 11.8 shows the global wave climate averaged over two months, January 1995 and July 1995. The difference in seasonality between the northern and southern hemispheres is immediately apparent. In the North Atlantic, for example, the average wave height goes from 5m in January to 2m in July; in the Southern Ocean, on the other hand, the difference is only a metre or two.

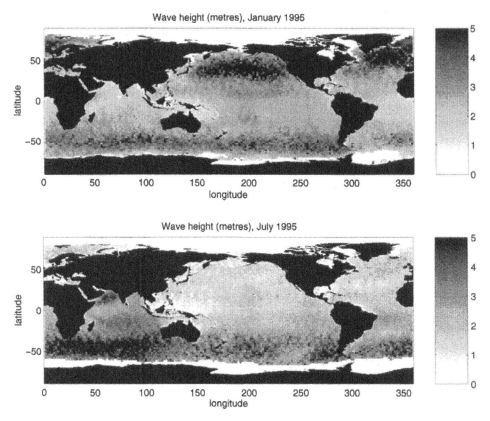

Figure 11.8 Monthly mean significant wave height (metres) from ERS-1 altimeter data in 1995. Top: January; bottom: July. (Courtesy P. D. Cotton.)

Over the past twenty years or so the modelling of ocean waves has made great progress. Relatively accurate wave forecasts are now routinely produced by forecasting centres such as the UK Meteorological Office and the European Centre for Medium Range Weather Forecasts. Like all such models these contain errors, both in the physics described by the model and in the wind field used to force it. One way to improve the forecast is to assimilate measured data into the models, that is, to combine the current model forecast and the measured data in some way to create a better starting value for the next forecast.

Imaging radars

The altimeter is not the only satellite-borne instrument capable of measuring sea-state information. Imaging radars can also 'see' waves and are capable of giving information on both wave length and direction. In space the imaging systems have used synthetic apertures (although some Russian missions have flown real-aperture radars). Synthetic aperture radars (SAR) use Doppler information from radar pulses along the direction of travel to 'synthesize' a much larger aperture than is actually being flown. This gives unprecedented resolution from relatively small antennas. However, the waves on the sea surface are moving themselves so the technique of aperture synthesis confounds the movement of the waves with the

Doppler information from the satellite motion. There is thus great controversy over how the wave spectrum has been modified by the aperture synthesis. It turns out that the crucial parameter is the ratio of the range from the instrument to the sea surface to the velocity of the instrument platform (R:V). For aircraft SAR this is small and waves are unambiguously imaged, but for satellite-borne systems R:V is large and the wave spectrum has been modified by a modulation transfer function.

Computing the inverse of this modulation transfer function to derive the wave spectrum from the image spectrum is very difficult if not impossible. However, it is relatively easy to compute the forward function, or in other words to compute what the SAR would see, given the wave spectrum. Using this method, Hasselmann and co-workers have come up with an elegant method of assimilating SAR data into a wave model (Brüning et al. 1993). They change the wave model output until the transformed spectrum matches the image spectrum from the SAR. Thus they can assimilate SAR data without having to solve the more difficult inverse problem.

Radar determination of wind

It had been known from World War II that the background noise on ship's radar displays (clutter) increased with sea state. Although this was a problem in tracking targets, the observation also suggested the possibility of being useful if one could quantify the effect of sea state on the radar signal. A series of airborne radar measurements were made in the 1960s and 1970s showing that there was a sensitivity of the amount of backscattered energy to wind speed. For altimeters, which view at nadir, there is an inverse relationship; scatterometers on the other hand view at between 20° and 50° from nadir, and higher winds are associated with increased backscatter. The reason for this is that in calm conditions the incident beam is reflected from the surface away from the antenna so no return signal is measured. When the wind blows, small-scale roughness is generated on the sea surface (the 'cat's paws' familiar to sailors as gusts approach are an example). These ripples interact with the radar pulse to scatter energy back along the beam. Moreover, the backscatter is found to vary with viewing direction. When the radar looks into the wind or downwind the returns are stronger than when viewing across-wind. Therefore, by viewing the same area of sea from different angles, information on wind direction can be obtained. This is achieved by having more than one antenna.

Scatterometers have been flown on Skylab, Seasat and ERS-1/2. Seasat demonstrated that accuracies of 2 m/s in speed and 20° direction were achievable (Jones et al. 1981). The data have been used for large-scale, open-ocean studies (Guymer 1983) and in semi-enclosed seas (Guymer and Zecchetto 1993). Figure 11.9 shows a comparison of global monthly mean winds from ERS-1 and a numerical weather forecast model. Major features are well reproduced. Scatterometer data sets will be increasingly used for forcing ocean circulation models and, as for ERS-1 and ERS-2, their availability within three hours of acquisition makes them suitable for assimilation into weather forecast models. In both cases they offer significantly improved winds in data-sparse regions such as the Southern Ocean.

Ocean biology from ocean colour measurements

Because of the presence of chlorophyll in phytoplankton, it is possible to see, by eye, changes in the colour of the ocean when biological activity is occurring. Indeed, the first observations from space of biological activity (plankton blooms) in the ocean were made by

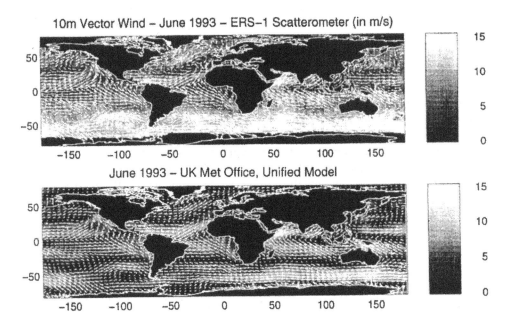

Figure 11.9 Monthly mean surface wind vectors corrected to 10m height. Top: ERS-1 scatterometer (courtesy ESA); bottom: UK Meteorological Office Unified Model.

astronauts using simple hand-held cameras (Stewart 1985). The first instrument to produce measurements of ocean colour from space was the multi-spectral scanner (MSS) flown on the Landsat series of satellites (Stewart 1985). This instrument was designed for making measurements over land and had limitations for oceanic applications (Robinson 1985). The major step forward in the use of satellite observations of ocean colour came with the launch of CZCS, in 1978, on the Nimbus–7 satellite. CZCS was specifically designed for ocean use and produced a wealth of data on ocean biology and on coastal processes, such as sediment transport. The basic principle of operation is to measure water leaving radiance in spectral bands (six for CZCS) in the visible and near infra-red. These measurements can then be interpreted in terms of in-water constituents, such as chlorophyll. The main difficulty in making such measurements from space is the presence of the intervening atmosphere, as atmospheric scattering of light contributes around 80–90 per cent of the signal measured by the satellite sensor. Correcting for the effects of the atmosphere is absolutely vital if oceanographic measurements are to be made. Details of how this is done and how chlorophyll and phytoplankton information is retrieved are discussed by Robinson (1985) and Stewart (1985). The availability of data from CZCS is summarized by Feldman et al. (1989) and a review of applications is given by Abbott and Chelton (1991). Here an example of one area of application is given.

CZCS provided the first global perspective on biological activity in the surface waters of the ocean and this global picture has been confirmed by data from the recently launched Sea WiFS ocean colour sensor. Figure 11.10 shows the seasonal cycle of phytoplankton concentrations, using data from January, April, July and October 1998. The spring 'bloom' in the North Atlantic and the presence of the nutrient-rich upwelling region along the coast of West Africa are clearly seen. Given the potential importance of the ocean biology for the global carbon cycle (Siegenthaler and Sarmiento 1993), data such as these will help in understanding the processes involved. They can be used to study the differences in seasonal cycles between different oceans (Yoder, McClain, Feldman and Esaias 1993) and to test and validate coupled biophysical models of the ocean (Sarmiento et al. 1993). Sarmiento et al. (1993) give a comparison between the satellite-derived annual mean chlorophyll concentration and that obtained from a model. This shows that the model is capturing the main features of the biology, but that deficiencies exist. For example, the plankton levels at the equator are too high, due to too much upwelling in the model physics bringing up nutrients and increasing the biological activity there. Seasonal comparisons (Sarmiento et al. 1993) show that the model reproduces the North Atlantic spring bloom seen in Figure 11.10.

CZCS operated from 1978 until 1986, but the sensor deteriorated significantly after 1982. Thus the data available are somewhat limited (Feldman et al. 1989). In 1996 the Japanese ADEOS mission was launched, carrying the new ocean colour sensor OCTS (unfortunately the satellite failed due to a problem with the solar array in June 1997). Another ocean colour sensor, SeaWiFS, was launched in 1997, and two more are due for launch shortly (MODIS on Terra and MERIS on Envisat; see Table 11.1.) So after a ten year gap, these provide new opportunities for further studies of the global ocean biology.

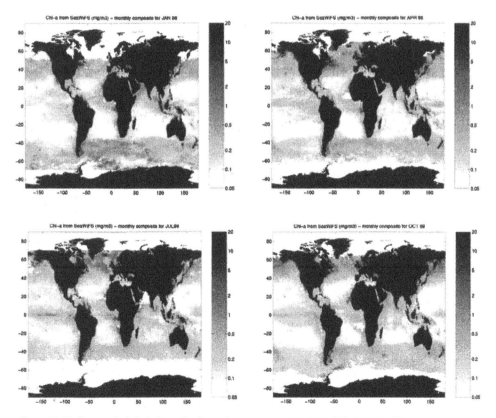

Figure 11.10 Seasonal global phytoplankton changes as seen by SeaWiFS in 1998. The spring bloom in the North Atlantic can be clearly seen (the scale is in mg Chl m⁻³), as can the effects of the upwelling zone off West Africa. Gaps due to cloud cover have been filled with a Kalman filter. (Courtesy NASA.)

Perspective on the future

In a sense, marine remote sensing has come of age and over the next few years it is unlikely that we will witness any major new advances in the technology of the sensors or platforms. What we can look forward to is an expansion in the number of data sets and in the length of these series. In order to calibrate correctly the sensors to allow us to look at small changes in the ocean over time it is important that there is overlap between sensors. Although it is possible to cross-calibrate new and old sensors using *in situ* data series, it is difficult to prove that there has been no drift in the *in situ* data.

There are three areas where we may see technological advances that could lead to new instruments, which could have a major impact. These are salinity sensors, hyperspectral scanners and gravity measurements.

Changes in the salinity of sea water affect the dielectric constant and thus it should be possible to measure salinity using microwave techniques. So far, work has concentrated on passive microwave radiometer systems, where the frequencies used are comparatively low and very large antennas are needed to obtain reasonable spatial resolution. Several possible missions have been proposed, and ESA are funding the Soil Moisture and Ocean Salinity (SMOS) mission, with a probable launch of 2004/5.

Hyperspectral scanners work in the visible and infra-red. They differ from the instruments being flown today in that they have many more (order 50+) channels. By increasing the spectral resolution it is possible to extract much more information. Similar sensors are currently flying on aircraft and new techniques are being developed. For example, it may prove possible to extract the relative abundance of phytoplankton species in addition to the amount of chlorophyll from the current generation of ocean colour sensors.

The importance of the gravity field for interpreting altimeter measurements has been explained earlier. The measurement of the geoid on the scales and to the accuracy required is now feasible and within the next ten years we will see a number of dedicated missions to carry out this task. The first will be Catastrophes and Hazards Monitoring and Prediction (CHAMP) due for launch in 2000. This will be followed by Gravity Recovery and Climate Experiment (GRACE) in 2001 which will measure not only the time-invariant geoid but also that part of the gravity signal which changes with time. This will enable us to calculate the pressure at the *bottom* of the ocean. Combined with the measure of the sea surface height from altimetry, this opens up many new applications. The final gravity mission is Gravity and Ocean Circulation Experiment (GOCE) which will measure the geoid to an accuracy of 1–2cm on scales down to 100km.

For the most part, though, the measurements we will have in ten years' time will be very similar to the ones we have now. However, our methods of analysing them will probably have changed radically. One area of great interest is the combination of data from a satellite sensor with other data sources, including *in situ* data and model output, as well as other satellite sensors. We believe that this will increase in the future. Figure 11.11 is an example in which altimeter-derived currents have been superimposed on infra-red SSTs off South Africa. The extent to which features can be seen in both data types tells us something of the vertical extent of the structure. Both satellites and models give a synoptic, global picture of the ocean. However, both have disadvantages. From space it is not possible to measure the interior of the ocean directly and model output is subject to errors in the forcing fields and in the model physics. Satellite data will impact on ocean modelling in three ways: (1) through the measurement of winds and sea-surface temperature we can look forward to much better fluxes and hence better forcing fields to drive the models; (2) once we have performed a

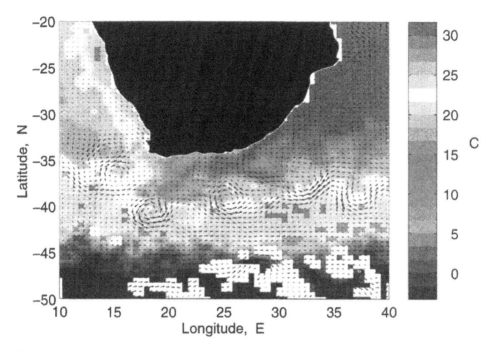

Figure 11.11 Surface geostrophic current vectors off South Africa derived from Topex altimeter data superimposed on ATSR sea-surface temperatures. (Courtesy M. S. Jones.)

model simulation, independent satellite data, e.g. altimetry, can be used to verify the models; (3) by assimilating satellite data into models it should be possible to use them for ocean forecasting in much the same way as for forecasting the weather. However, it should be noted that oceanography does not yet possess the equivalent of the World Weather Watch. There is a difference between making measurements of limited duration for research purposes (whether from space or at sea) and ensuring that the same quantities are available on a routine basis. Planning for a Global Ocean Observing System is under way; although the details of such a system are still being defined, it is clear that remote sensing satellites will have a major role to play. (See Introduction to this volume.)

Finally, it is interesting thirty years on to remind ourselves of part of Gifford Ewing's preface to the report of the 'Oceanography from Space' symposium (Ewing 1965):

> Intuition tells us that future generations of investigators scanning the oceans from a new vantage point will have the imagination to ask new questions of it and the ingenuity to devise new ways of answering them. It is unthinkable that oceanographers will not find ways to exploit this burgeoning technology for the advancement of their science.

Without doubt, this process is well under way.

References

Abbott, M. R. and Chelton, D. B. (1991) 'Advances in passive remote sensing of the ocean', *Reviews of Geophysics* **29** (suppl.): 571–89.

Barrick, D. E. (1968) 'Rough surface scattering based on the specular point theory', *IEEE Transactions. Antennas and Propagation* **AP-16**: 449–54.

Brown, G. S. (1977) 'The average impulse response of a rough surface and its applications', *IEEE Transactions. Antennas and Propagation* **AP–25**: 67–74.

Brüning, C., Hasselmann, S., Hasselmann, K., Lehner, S. and Gerling, T. (1993) 'On the extraction of ocean wave spectra from ERS-1 SAR wave mode image spectra', *Proceedings of the First ERS-1 Symposium – Space at the Service of our Environment, Cannes, France, 4–6 November 1992*, ESA **SP–359**, ed. B. Kaldeich, vol. 2: 747–52. Noordwijk: European Space Agency.

Carter, D. J. T., Challenor, P. G. and Srokosz, M. A. (1992) 'An assessment of Geosat wave height and wind speed measurements', *Journal of Geophysical Research* **97**: 11383–92.

Carter, D. J. T., Foale, S. and Webb, D. J. (1991) 'Variations in global wave climate throughout the year', *International Journal of Remote Sensing* **12**: 1687–97.

Challenor, P. G., Foale, S. and Webb, D. J. (1990). 'Seasonal changes in the global wave climate measured by the Geosat altimeter', *International Journal of Remote Sensing* **11**: 2205–13.

Challenor, P. G., Read, J. F., Pollard, R. T. and Tokmakian, R. T. (1996) 'Measuring surface currents in the Drake Passage from altimetry and hydrography', *Journal of Physical Oceanography* **26**: 2748–59

Chelton, D. B., Hussey, K. J. and Parke, M. E. (1981) 'Global satellite measurements of water vapour, wind speed and wave height', *Nature* **294**: 529–32.

Cheney, R. E., Marsh J. G. and Beckley B. D. (1983) 'Global mesoscale variability from collinear tracks of SEASAT altimeter data', *Journal of Geophysical Research* **88**: 4343–54.

Craig, C. H. and Sandwell, D. T. (1988) 'Global distribution of seamounts from Seasat profiles', *Journal of Geophysical Research* **93**: 10408–20

Cromwell, D., Challenor, P. G., New, A. L. and Pingree, R. D. (1996) 'Persistent westward flow in the Azores Current as seen from altimetry and hydrography', *Journal of Geophysical Research* **101**: 11923–33.

Duchossois, G., Foster, J. E., Thomas, L., Guymer, T. H. and Crease, J. (1980) 'Study on satellite radar altimetry in climatological and oceanographic research' (Final report prepared by Science Research Council and Natural Environment Research Council). European Space Agency, Contract Report, under ESA Contract No. 4355/80/F/FC (SC).

Ewing, G. C. (ed.) (1965) *Oceanography from space, Proceedings of conference on the feasibility of conducting oceanographic explorations from aircraft, manned orbital and lunar laboratories. Woods Hole, Massachusetts, 24–28 August 1964*. Woods Hole Oceanographic Institution Technical Report. **65–10**.

Fedor, L. S., Godbey, T. W., Gower, J. F. R., Guptil, R., Haynes, G. S., Rufenach, C. L. and Walsh, E. J. (1979) 'Satellite altimeter measurements of sea state – an algorithm comparison', *Journal of Geophysical Research* **84**: 3991–4001.

Feldman, G. C., Kuring, N., Ng, C., Esaias, W., McClain, C., Elrod, J., Maynard, N., Endres, D. and others (1989) 'Ocean color – availability of the global data set', *EOS, Transactions, American Geophysical Union* **70**: 634–5, 640–1.

Fu, L.-L. and Cheney, R. E. (1995) 'Application of satellite altimetry to ocean circulation studies: 1987–1994', *Reviews of Geophysics* **33** (Suppl.): 213–23.

Fu, L.-L., Christensen, E. J., Yamarone, C. A., Lefebvre, M., Ménard, Y., Dorrer, M. and Escudier, P. (1994) 'TOPEX/POSEIDON mission overview', *Journal of Geophysical Research* **99**: 24369–81.

Guymer, T. H. (1983) 'A review of Seasat scatterometer data', *Philosophical Transactions of the Royal Society of London* **A 309**: 399–414.

Guymer, T. H. and Zecchetto, S. (1993) 'Applications of scatterometer winds in coastal areas', *International Journal of Remote Sensing* **14**: 1787–1812.

Jones, W. L., Boggs, D. H., Bracalente, E. M., Brown, R. A., Guymer, T. H., Shelton, D. and Schroeder, L. C. (1981) 'Evaluation of the Seasat wind scatterometer', *Nature* **294**: 704–7.

Marsh, J. G. and Martin, T. V. (1982) 'The SEASAT altimeter mean sea surface model', *Journal of Geophysical Research* **87**: 3269–80.

Mognard, N. and Lago, B. (1979) 'The computation of wind speed and wave heights from Geos 3 data', *Journal of Geophysical Research* **84**: 3979–86.

Nerem, R. S. et al. (1994a) 'Gravity model development for TOPEX/POSEIDON: Joint Gravity Models 1 and 2', *Journal of Geophysical Research* **99**: 24421–47.

Nerem, R. S., Schrama, E. J., Koblinsky, C. J. and Beckley, B. D. (1994b) 'A preliminary evaluation of ocean topography from the TOPEX/POSEIDON mission', *Journal of Geophysical Research* **99**: 24565–83.

Parsons, C. L. (1979) 'Geos 3 wave height measurements: an assessment during high sea state conditions in the North Atlantic', *Journal of Geophysical Research* **84**: 4011–20.

Pierson, W. J. and Salfi, R. E. (1979) 'A brief summary of verification results for the spectral ocean wave model (SOWM) by means of wave height measurements obtained by Geos 3', *Journal of Geophysical Research* **84**: 4029–40.

Priester, R. W. and Miller, L. S. (1979) 'Estimation of significant wave height and wave height density function using satellite altimeter data', *Journal of Geophysical Research* **84**: 4021–26.

Rapp, R. H. (1993) 'Geoid undulation accuracy', *IEEE Transactions on Geoscience and Remote Sensing* **31**: 365–70.

Ray, R. D. (1993) 'Global ocean tide models on the eve of TOPEX/POSEIDON', *IEEE Transactions on Geoscience and Remote Sensing* **31**: 355–64.

Robinson, I. S. (1985) *Satellite oceanography*. Chichester: Ellis Herwood.

Robinson, I. S. and Guymer, T. (1996) 'Observing oceans from space', in C. P. Summerhayes and S. A. Thorpe (eds) *Oceanography: an illustrated guide*. London: Manson, pp. 69–88.

Sandwell, D. T. and Agreen, R. W. (1984) 'Seasonal variation in wind speed and sea state from global satellite measurements', *Journal of Geophysical Research* **89**: 2041–51.

Sarmiento, J. L., Slater, R. D., Fasham, M. J. R., Ducklow, H. W., Toggweiler, J. R. and Evans, G. T. (1993) 'A seasonal three-dimensional ecosystem model of nitrogen cycling in the North Atlantic euphotic zone', *Global Biogeochemical Cycles* **7**: 417–50.

Shum, C. K., Werner, R. A., Sandwell, D. T., Zhang, B. H., Nerem, R. S. and Tapley, B. D. (1990) 'Variations of global mesoscale eddy energy observed from Geosat', *Journal of Geophysical Research* **95**: 17865–76

Siegenthaler, U. and Sarmiento, J. L. (1993) 'Atmospheric carbon dioxide and the ocean', *Nature* **365**: 119–25.

Smith, W. H. F. and Sandwell, D. T. (1994) 'Bathymetric prediction from dense satellite altimetry and sparse shipboard bathymetry', *Journal of Geophysical Research* **99**: 21803–24.

Smythe-Wright, D., Gordon, A. L., Chapman, P. and Jones, M. S. (1996) 'CFC-113 shows Brazil Eddy crossing the South Atlantic to the Agulhas Retroflection region', *Journal of Geophysical Research* **101**: 885–95.

Stanley, H. R. (1979) 'The Geos-3 project', *Journal of Geophysical Research* **84**: 3779–83.

Stewart, R. H. (1985) *Methods of satellite oceanography*. Berkeley: University of California Press.

Walsh, E. J. (1979) 'Extraction of ocean wave height and dominant wavelength from Geos 3 altimeter', *Journal of Geophysical Research* **84**: 4003–10.

Warnecke, G., McMillin, L. M. and Allison, L. J. (1969) 'Ocean current and sea surface temperature observations from meteorological satellites', NASA Technical Note D–5142, Washington, DC.

Yoder, J. A., McClain, C. R. Feldman, G. C. and Esaias, W. E. (1993) 'Annual cycles of phytoplankton chlorophyll concentrations in the global ocean: a satellite view', *Global Biogeochemical Cycles* **7**: 181–93.

12 Transient tracers and tracer release experiments
New tools for the oceanographer

Andrew Watson

Introduction

Unlike some of the other subjects in this book, my subject is of recent origin, and I cannot enlist in its cause any of the scientists who took part in the original *Challenger* Expedition. That is, not if the word 'tracer' is defined in the sense in which it is usually used today. If I were to use the term in its wider and more literal definition, it would be appropriate to almost any measurement of a property of sea water, for most such properties have been used to trace (i.e. study the movement of) the oceans at some time in the past. Temperature, salinity, density, potential vorticity, nutrients, plankton abundance and speciation have all been used as tracers, and I could take virtually the entire field of marine science as my subject. However, today oceanographers use the word for a more restricted range of variables, the main defining features apparently being that the measurements should require esoteric chemical skills incomprehensible to all but a select few, and that it should take hours to produce a single measurement.

Nowadays in oceanography the word tracer is usually applied to *transient* tracers. These are chemicals or isotopes, the concentration of which is changing on a time-scale much faster than an ocean mixing-time in the order of 1000 years – they are far from the steady state. This contrasts with properties such as temperature and salinity, the distributions of which are close to steady state. Historically, however, even among the more esoteric measurements, steady-state compounds were more important until the 1960s, when global pollution from human activities really began in earnest and created many new transient signals.

Transient tracers are normally compounds that have been recently introduced into the oceans as a result of industrial or military activity; they are pollutants by any other name, and include some of the most potentially damaging substances on earth. In most cases, the original reason they were measured in the seas was to track the course of global pollution. It was soon realized, however, that the measurements also tell us a great deal about natural processes occurring in the oceans. Pollutants from our more recent activities may be spreading into the ocean, but we can at least take advantage of these 'unplanned experiments' to learn about ocean circulation.

Steady-state tracers: ^{14}C in the abyss

Though I could choose many examples of the use of steady-state tracers, carbon-14 is perhaps the most familiar; everyone is aware of the use made of carbon-dating in archaeology. The process makes use of 'natural' ^{14}C, generated continually in the upper atmosphere by the action of cosmic rays. All carbon derived from atmospheric carbon dioxide therefore has a ^{14}C content, but once isolated from the atmosphere, this decays away with a half-life of 5700 years. Carbon in the deep sea has a residence time there in the order of 1000 years, and can be carbon-dated in the same way.

Today the concentration of ^{14}C in the atmosphere and oceans is no longer in steady state; it has been hugely disturbed, first by the burning of fossil fuels (which contain little ^{14}C and therefore dilute the ^{14}C in the atmosphere), and second (and to a much larger extent) by pollution from nuclear testing. Thus ^{14}C is now a transient tracer in 'young' water which has had recent contact with the surface. In old water, however, its concentration is still set by the steady-state considerations that applied until the industrial revolution.

In 1966, Walter Munk published his famous 'abyssal recipes' paper, in which he used the distribution of temperature, salinity and ^{14}C to deduce the balance between advection and diffusion that maintains the properties of the deep Pacific (Munk 1966). He used a one-dimensional representation of a Stommel–Arons type circulation in which cold dense water sinks to the bottom in the southern Pacific, and upwells uniformly everywhere else. The 1–D steady-state advection–diffusion equation for a conserved tracer such as salinity or temperature is homogeneous, because there are neither sources nor sinks. If the vertical advection rate w and diffusion rate k are assumed constant, it is readily solved analytically for the ratio w/k. The same equation applied to Carbon-14 is not homogeneous, because it also contains the decay term in ^{14}C; this equation can also be readily solved, but this time for the ratio w^2/k; simultaneous solution of both equations therefore enables both the advection and diffusion rates to be calculated. Munk derived these values. The vertical mixing rate he found was about 10^{-4} m^2 s^{-1} – too large to be attributable to molecular processes, and he suggested that it was caused either by shear-induced mixing by internal waves, or by interactions with the margins of the oceans. Thirty years later, his conclusions about internal mixing still encapsulate well what we know for certain about the abyssal mixing of the ocean.

Transient tracers

Pollution of the sea is as old as civilization. Perhaps the oldest 'transient tracer' to be used as such was not a chemical substance at all, but the derelict hulks that littered the oceans towards the end of the nineteenth century. Richardson (1985) made a fascinating study of these. During that period most ships were still made of wood, and they were numerous compared to the shrunken merchant fleets of today. Each year hundreds of ships were abandoned at sea, but the wooden hulks often did not sink for months or even years. As dramatized in Conrad's story *Lord Jim*, these hulks were a major hazard to ships' masters in the pre-radar era. Collision with a derelict could mean abandoning ship, thus creating another derelict. Because of this, for a period between 1883 and 1902, the US navy charted sitings of all known derelicts in the North Atlantic. Richardson plotted the drifts of those hulks, which very elegantly show up the surface circulation of the North Atlantic. A better description of the surface circulation of the North Atlantic has only recently been possible with the advent of satellite trackable buoys and other high-tech devices. Some might feel it is stretching the definition to say that these hulks were a 'transient tracer', but I think that they fit the description rather well, being (like the ones studied today) pollutants, and certainly transient.

Radioactivity, CFCs and carbon dioxide

Radioactivity, CFCs and carbon dioxide are being used today as transient tracers, and all are the subject of intense environmental concern. In the case of excess carbon dioxide, the impetus to measure its penetration into the oceans has come from that environmental concern and the desire to understand how CO_2 gets into the sea. By contrast, for CFCs the first measurements were done simply because it was possible to do them – the analytical technique came first and, in a sense, gave rise to the environmental concern. Radioactivity measurements were also made possible by exquisitely sensitive analytical techniques, but in this case spurred on by environmental fears. In all cases these substances enter the oceans from the atmosphere into the surface, and serve therefore to mark recently ventilated (i.e. recently at the surface) water masses.

Chronologically the first to be measured was the fall-out generated by the atmospheric hydrogen bomb tests in the 1950s and 1960s. The tests injected into the atmosphere unprecedented levels of tritium, carbon-14, strontium-90, caesium-137, iodine-131 and dozens of other radioactive isotopes. After the first H-bomb test in 1957, the sizes and numbers of warheads being tested increased very rapidly up until the test-ban treaty of 1963, when the input stopped abruptly. (Even in that time when environmental awareness was in its infancy and the Cold War at its height, the amounts of radioactivity being released were alarming enough to both sides for sense at last to prevail.) The result, for tritium and many other isotopes with relatively short lifetimes in the atmosphere, was a rather well defined transient input (Fig. 12.1) to the oceans, concentrated in the northern hemisphere mid-latitudes.

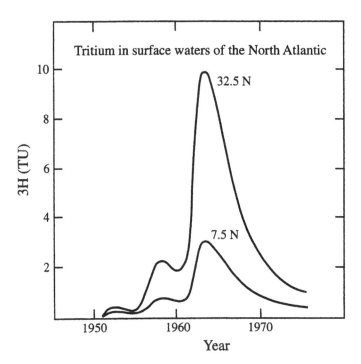

Figure 12.1 The input of tritium to the surface of the North Atlantic, at two latitudes, as a function of time, showing the peak in the early 1960s. (Data redrawn from Wunsch 1988.)

More recently, a second substantial source of radioactivity has been added to the North Atlantic. This is a contribution to oceanography that is uniquely and especially British, being the outfall from the reprocessing site at Windscale/Sellafield. Production of significant quantities of caesium-137 and strontium-90 began in the late 1960s, reached a peak in the early 1970s and then declined to lower levels. Smaller amounts of radioactivity have come from the French facility at Cap de la Hague.

Some kinds of volatile halocarbons are relatively stable products of the chemical industry in the developed countries, and are eventually released to the atmosphere where they are building up in concentration. The chlorofluorocarbons or 'freons' are the best known, because of their harmful effect on stratospheric ozone, but there are more than a dozen such compounds now routinely measurable in the atmosphere. The history of the build-up has been measured or reconstructed from the amounts released for many of these compounds, the stimulus to do so being not their utility to the marine sciences, but their effects on the ozone layer; Figure 12.2 shows examples. Because the various compounds have different rates of rise, their ratios at the surface have changed with time, so measurement of these ratios can serve to pinpoint more accurately the history of the water masses involved.

Lovelock was the first to measure these compounds, in both the air and the ocean. During a voyage of the RRS *Shackleton* from London to Montevideo, he found that the compounds were globally distributed, and observed how the concentration dropped off with depth in the ocean. In a subsequent paper (Lovelock et al. 1973), he remarked almost in passing that they might be useful as tracers of ocean mixing. Within a year of its publication, these measurements were coupled by Rowland and Molina (1975) with some hitherto esoteric calculations of the effect of chlorine on stratospheric ozone. Thus began the debate on CFCs and ozone that dominated the environmental sciences in the 1970s. It was not until the passion of that debate had begun to cool in the early 1980s that oceanographers picked up on the consequences of Lovelock's early work, and began to make systematic and accurate measurements of the CFCs in the oceans.

Since that time, the CFCs have become the most extensively measured of the transient tracers. Their popularity is due to the practicality of the measurement, which can be made with relatively cheap apparatus and in nearly real time on board a ship. This contrasts with tritium or ^{14}C, where samples must be returned to a specialist shore-based facility. With the CFCs, a density of sampling similar to that of other common chemical measurements (e.g. nutrients) can be obtained. Furthermore, several compounds determined in the same measurement can in principle have some utility as transient tracers; several groups now routinely determine and report CFC-113 and carbon tetrachloride in addition to the more standard CFC-11 and CFC-12 (Wallace, Beining and Putzka 1994, Haine, Watson and Liddicoat 1995). The inclusion of these tracers substantially extends the range of time-scales that can be addressed by the technique. Carbon tetrachloride is thought to have been first released into the atmosphere early in this century, while CFC-113 was not manufactured in quantity until the 1970s, so both extremely rapid ventilation and longer-term processes can be studied.

But it is often the case that, to obtain convenience in measurement, some other property must be sacrificed. The CFCs are relatively easy to measure, but they are probably not ideal tracers because they may not be fully conservative. It has recently become clear that carbon tetrachloride breaks down, probably by a biological process, in low oxygen regimes (Wallace et al. 1994). It is probable that all of these compounds are liable to similar fates, but with the more highly fluorinated species being the more stable (Krysell, Fogelqvist and Tanhua 1994). Their quantitative use as tracers is obviously heavily dependent on their stability (or, at least,

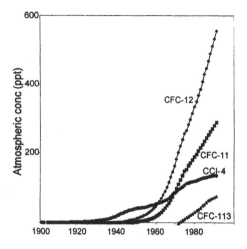

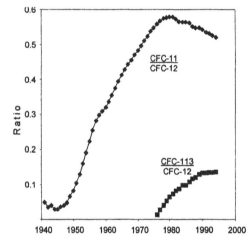

Figure 12.2 Upper: The rise in concentrations of four anthropogenically released chlorocarbons (measured in parts per trillion) in the atmosphere: CFC-11 ($CFCl_3$), CFC-12 (CF_2Cl_2), CFC-113 (CF_2ClCCl_2F) and carbon tetrachloride (CCl_4). Lower: The ratios of CFC-11 : CFC-12 concentrations rise monotonically until the early 1970s, and can be used to 'age' water up to that time. Beyond that time, this ratio declines, but the measurement of CFC-113 : CFC-12 can be used to age more recent waters.

on the accuracy with which any loss can be taken into account) so there is a serious need for information on what controls their breakdown.

Transient tracers are almost all input to the surface ocean from the atmosphere. Because they have been around for only a few decades, they can be used to study only rather recently ventilated water. However, ventilation – the process by which surface waters are subducted to depth – is a process of particular interest, first because the properties of water masses are most usually formed at the surface, and second because the interaction of the oceans with the atmosphere takes place through the process of ventilation. Of special interest in recent years has been the process by which the ocean takes up excess atmospheric carbon dioxide, and there are measurements (Chen 1982) in which the signal of increasing carbon dioxide penetrating the ocean can be seen directly. To see the effect requires some skill in interpretation, because, unlike the other tracers, carbon dioxide is already present in the sea in large quantities, and has a huge biological flux in addition to that carried passively by water movements. Because of these complications, most of our estimates of the flux of CO_2 from air to sea have been made using models, which in turn are calibrated, or at least checked for consistency, with more easily measured transient tracers. Most useful among these have been tritium and bomb-generated carbon-14 (Broecker, Peng, Ostlund and Stuiver 1985; Siegenthaler and Joos 1992). Surprisingly, as a transient tracer, ^{14}C behaves very differently from ^{12}C, as a result of its very different distribution between the air and the sea, and different time history of input.

What do transient tracers really tell us?

The first and still perhaps the most dramatic example of the utility of transient tracers was the measurements made during GEOSECS (Geochemical Ocean Sections) during 1972–4. This was the first and only time that a cruise programme of global extent was devoted mainly to the making of geochemical measurements. By luck or good judgement it came at a good time, ten years after the bomb transient had reached its peak being ideal to look at the fast-spreading signal.

Figure 12.3 shows the tritium distribution in the deep western boundary current as it flows around the continental slope of eastern North America. Note that this was only ten to fifteen years after the tritium transient had begun; the deep tongue of the core of the current is clearly evident, as is the pattern of deeper mixing in the surface waters as one goes north. It is a dramatic 'picture' of a process previously assumed to be happening on both theoretical grounds and from the interpretation of hydrographic data, but which was much more difficult to quantify using those methods. You can actually see the water penetrating the deep Atlantic, just as clearly as if you were watching a demonstration in a laboratory tank.

How much information does the picture actually contain, and how much has been learned by the effort expended in obtaining transient tracer data in general? This subject has been debated with some heat by, on the one side, the chemical oceanographers who make the measurements and are enthusiasts for transient tracers, and on the other hand, some rather sceptical theoreticians.

Theoreticians such as Wunsch (1988) have made attempts to use tracer data in a diagnostic way – that is, to deduce the velocity field from tracer data alone or in combination with other variables. Attempts to derive flows in any real detail always seem to fail. A prime example is the analysis made by Wunsch of tritium and helium-3 data collected by Jenkins (1987) in a study of the ventilation of the North Atlantic thermocline. The data represent one of the most densely sampled sets of transient tracer available, so one might think that this was a fair test. However, dense though the spatial sampling was, it still represents only a single section, at a single instant of time as the bomb-induced tritium–helium signal passed through.

Wunsch investigated models of the distributions of temperature, salinity, oxygen, tritium and helium-3 along this section, and attempted to deduce the flow based on conservation of these tracers and simple quasi-geostrophic dynamics. He found that even a simple 3-D model was compatible with an infinity of solutions. If the purpose of taking oceanographic measurements was solely to provide a data set that would yield a unique solution for the

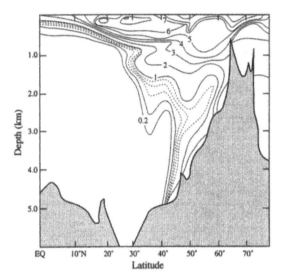

Figure 12.3 The penetration of bomb-derived tritium into the water of the North Atlantic, measured during the GEOSECS expedition in 1973. (Data from Ostlund and Fine, 1979.)

circulation of water in a given region, it would be difficult to escape the conclusion from Wunsch's analysis that, densely sampled as this region was, the data were seriously inadequate. From this, the physical oceanographer would infer that transient tracer data are a waste of time, being expensive to collect and too sparse ever to be adequate. The chemical oceanographer will reach the opposite conclusion: that there are not nearly enough transient tracer data. (These two positions are beautifully illustrated by the exchange between Wunsch and Jenkins following Wunsch's paper in the *Philosophical Transactions of the Royal Society*.)

Several related problems contribute to Wunsch's rather disappointing result. First, there is a fundamental difficulty in solving an *inverse* problem involving the advection–diffusion equation; the effect of mixing is to smear out any distribution as time progresses, so that information about the distribution in former times tends to be lost. In the inverse problem we are given the distribution at some late stage, when mixing has attenuated the information of former times. We are then required to solve the equation in an unstable direction, upstream or backwards in time, which has the effect of amplifying any errors present in the initial data. Second, there is the boundary condition problem: for a section in the middle of an ocean, the concentrations over most of the boundaries are unknown and must be added to the internal unknowns, increasing the under-determined nature of the problem.

The secret of avoiding such frustrations lies, I think, in the correct choice of problem. Because the tracer data will be always be sparse in space, and in the case of transient data in time as well, the only problems that can be solved with such data alone will be of very coarse resolution. The situation can be greatly helped by posing the problem across entire ocean basins, because then zero-flux boundary conditions can be specified, and in this sense the problem examined by Wunsch, far from being the fairest possible test, is in fact particularly ill suited to the nature of the tracer data. However, if we wish to derive circulations *a priori*, from conservations of tracers alone, we will have to be content with models not much more sophisticated than the geochemists' 'box' model. Something like this has been done by Bolin,

Bjorkstrom and Holmen (1983) who used twelve boxes to simulate the entire global ocean.

And yet, it seems self-evident that something important is revealed by the picture of tritium penetrating the ocean given by Figure 12.3; something is learned about the speed and shape of this penetration, which requires no model at all to understand, and which must be of value when detailed models of the North Atlantic circulation are evaluated. If, having constructed a model of the circulation of the North Atlantic, I use it to simulate the penetration of the tracer and find the results do not match, then I know that the model is wrong. Furthermore, the transient tracer distributions speak to the ventilation processes, which are important to understand but in general are not well handled by circulation models, since they involve much slower and more subtle water movements than do, for instance, the major surface currents. Thus the tracer data cannot by themselves be used to deduce the circulation in detail, but they can provide an important and difficult test that circulation models must pass.

Deliberately released tracers

Transient tracers have fundamental limitations. As mentioned above, one is the uncertainty over the boundary conditions. Even if we take as completely accurate the source function of a given tracer (invariably in the atmosphere), we may have difficulty specifying accurately the boundary conditions at the surface of the ocean for all times. A second limitation is that, because they almost always relate to surface waters, they cannot be used to study processes deep in the interior. Finally, the invasion of transient tracers into the sea is a sort of experiment, but it is unplanned; the scientist cannot make the measurements optimal because he or she is simply taking opportunistic advantage of events. A source that is widely distributed over the surface, for instance, is unlikely to be the best to study any one phenomenon.

There are a few circumstances in which it is possible to get around these drawbacks by releasing a tracer expressly for scientific study. This procedure also has its difficulties, notably that care must be taken not to cause needless pollution, and for this reason – and because scientists are impatient and do not like to wait decades for the results of a given experiment – large-scale experiments using this method will never be very common. Nevertheless, much can be learned from small and medium-sized releases. The technology has been developed to allow such sensitive detection that useful results can be obtained with a few grams of common inert gases such as sulphur hexafluoride (normally used as an inert gaseous insulator for the electricity supply industry). Several such release experiments have now been done, from which have emerged some solid advances in oceanography.

The origins of the technique go back to the late 1970s when John Shepherd and Wallace Broecker discussed the idea of a 'deep ocean tracer release experiment' to measure directly the rate of vertical or diapycnal (across density surfaces) transport at thermocline depths in the ocean, this being recognized as the main barrier limiting the penetration of carbon dioxide into the sea. The original concept was for a release large enough to measure vertical transport over an entire ocean basin. Perhaps fortunately, there was no consensus at the time about how such an experiment could be carried out. It took fifteen years of discussion, research and development before an experiment that was the true descendant of Shepherd and Broecker's musings was actually carried out; this was the North Atlantic Tracer Release Experiment (NATRE) (Ledwell, Watson and Law 1993). In the intervening period, ideas on ocean mixing moved on considerably and the final experiment was designed very differently from the original conception. It was not, for example, a huge release designed to cover an entire ocean basin; such a release is technically possible but expensive, unduly polluting, and of debatable value. Instead, it was made in a small and well characterized region (the eastern subtropical gyre of the North Atlantic, in the same 'subduction zone' as the measurements of tritium analysed by Wunsch and discussed above). This type of experiment can be done with amounts of tracer that are insignificant compared to the amounts being released by industry every day, so they do not materially contribute to the build-up of potentially harmful substances in the environment.

Released on a single, well-defined isopycnal at a known time, the subsequent spread of the tracer in the vertical plane gave us, within a few months, an unequivocal measurement of the rate of diapycnal mixing in the interior of the ocean at this location. This is not the same thing as Munk's analysis gives us (even were that analysis possible in the thermocline region, which it is not), for the steady-state tracer integrates over long times so that the mixing can occur elsewhere (e.g at the ocean boundary) and the resulting distribution be transported to the interior by isopycnal processes. Neither is it derivable from transient tracer distributions, for a transient tracer such as tritium or CFCs, being introduced over the surface of the ocean, is simultaneously introduced onto a wide range of density surfaces. Subsequently, we cannot deduce how much of it reaches a given location by rapid isopycnal transport along density surfaces, and how much by diapycnal mixing across them. By contrast, the deliberate release technique gets around the problem of specifying the boundary conditions, these being precisely known, provided only that the release is accurately targeted. (That is no trivial task, however; much of the technical expertise developed over the fifteen years between inception and execution of this experiment was devoted to ways of ensuring this.)

NATRE was designed primarily to measure diapycnal mixing rates, but some of the most interesting results have concerned the dynamics of horizontal mixing. The tracer patch has so far been mapped four times (at two weeks, six months, one year and two years after release). We monitored its spread from a series of fine streaks contained within an area of a few tens of kilometres on a side, to a series of ribbons a few kilometres wide but hundreds long, to a continuous plate of tracer-marked water, a thousand kilometres on a side. The process of the spreading into the ocean has been likened to the slow stirring of cream into a cup of coffee.

Using tracer release techniques to make oceanography a more experimental science

The technique of deliberate release was originally developed as a tool to study water-mixing and movement. However, it seems likely that an even more important development will be the use of the method to enable detailed studies on particular 'patches' of water. It is possible to release a small amount of sulphur hexafluoride into surface water, and so create a marked patch of water a few kilometres in size on the surface of the ocean. Though the tracer cannot be seen, apparatus can be set up on a ship that will detect whenever the vessel is in this water. By this means the ship can follow the marked water for periods of a week or two. It therefore becomes possible to perform experiments in which the detailed physics, chemistry and biology of the same water are followed over a period of time.

This may sound simple, but it has never before been possible, and it is an important step forwards. For example, if a terrestrial agronomist or ecologist wishes to study the effect of this or that nutrient, this is easily done experimentally on a controlled patch of ground. But only with the help of an added tracer is it possible to do this in the ocean. The first two such experiments, which have proved that iron is an important limiting nutrient in the equatorial Pacific, have recently been completed (Martin et al. 1994, Watson et al. 1994). The 'iron hypothesis' was proved in the most direct way possible, by simply adding inorganic iron to the surface ocean and recording the results. It could only be done, however, because it was possible to keep track of the water to which the iron had been added.

These experiments share a similar philosophy with the physical oceanographic tracer release experiments such as NATRE. By this I mean that both are designed as *experiments*, in which the natural system is perturbed by addition of a tracer and perhaps other substances also, and the consequences of this perturbation are followed. This contrasts with the more traditional approach in which the oceanographer simply observes what is going on in the unperturbed system. The 'passive observation' approach has the disadvantage that it is rarely possible to prove unequivocally a causal connection between variables observed in this way. Thus, for example, variable A may correlate with variable B, but this may be because A affects B, or B affects A, or they are both affected by some third variable C, or any combination of these relations.

The great advantage of the experimental approach is that mechanisms of cause and effect can be studied more precisely. Thus during the NATRE release, we could distinguish between locally occurring mixing and that which had occurred at a distance because the release was carefully controlled and targeted, whereas the iron hypothesis experiments explicitly tested the importance of iron, and iron alone, in affecting productivity in the open ocean. Setting up such experiments has proved to be arduous, team-effort science, but the rewards in terms of substantive advances in knowledge have been great and I believe we have only just begun to realize the potential of tracers used in this way.

The future

The use of specialist chemical measurements as tracers in oceanography is a young and still rapidly developing topic. It has become especially active as a result of the rise in emissions of various substances by human military or industrial activity, which have provided important 'tracers of opportunity' for studying the ventilation of the ocean. Most recently, it has become possible to use specific tracers expressly to perform experiments in the ocean.

Looking to the future, it is inevitable that many more potential transient tracers will be generated by human activity. These will probably follow a similar pattern to that of the bomb radioisotopes or the CFCs, where a potentially harmful substance is released with exponentially increasing amounts for a time, until enough environmental concern is voiced to bring about a slowing and then cessation of the release. Presumably, the amount of science to be extracted from measurements of the penetration of these substances in the ocean will progressively decline as they will in many respects be duplicating what has already been measured.

The use of purposefully released tracers addresses a qualitatively different area. Because the method gives for the first time a means of marking a particular volume of water and following it for a period of time, it may act to enable a range of experimental studies not previously possible. Only a handful of workers are currently employing this technique, but its use is likely to grow substantially over the coming decade or so if the technology can be made simpler to use. This method could then become almost a standard tool, like the use of a CTD or moorings, for marine scientists by the early years of this century.

References

Bolin, B., Bjorkstrom, A. and Holmen, K. (1983) 'The simultaneous use of tracers for ocean circulation studies', *Tellus* **B35**: 206–36.

Broecker, W. S., Peng, T.-H., Ostlund, G. and Stuiver, M. (1985) 'The distribution of bomb radiocarbon in the ocean', *Journal of Geophysical Research* **90**: 6953–70.

Chen, C. T. A. (1982) 'On the distribution of anthropogenic CO_2 in the Atlantic and Southern Oceans', *Deep-Sea Research* **29A**: 563–80.

Haine, T. W. N., Watson, A. J. and Liddicoat, M. I. (1995) 'Chlorofluorocarbon-113 in the Northeast Atlantic', *Journal of Geophysical Research* **100**: 10745–53.

Jenkins, W. J. (1987) 'H-3 and He-3 in the beta-triangle – observations of gyre ventilation and oxygen utilization rates', *Journal of Physical Oceanography* **17**: 763–83

Krysell, M., Fogelqvist, E. and Tanhua, T. (1994) 'Apparent removal of the transient tracer carbon-tetrachloride from anoxic seawater', *Geophysical Research Letters* **21**: 2511–14.

Ledwell, J. R., Watson, A. J. and Law, C. S. (1993) 'Evidence of slow mixing across the pycnocline from an open ocean tracer release experiment', *Nature* **364**: 701–3.

Lovelock, J. E., Maggs, R. J. and Wade, R. J. (1973) 'Halogenated hydrocarbons in and over the Atlantic', *Nature* **241**: 194–6.

Martin, J. H. and 43 others (1994) 'Testing the iron hypothesis in ecosystems of the Equatorial Pacific Ocean', *Nature* **371**: 123–9.

Munk, W. H. (1966) 'Abyssal recipes', *Deep-Sea Research* **13**:707–30.

Ostlund, H. G. and Fine, R. A. (1979) 'Oceanic distribution and transport of tritium', International Atomic Energy Publication IAEA-SM-232/67.

Richardson, P. L. (1985) 'Drifting derelicts in the North Atlantic 1883–1902', *Progress in Oceanography* **14**: 1–4, 463–83.

Rowland, F. S. and Molina, M. J. (1975) 'Chlorofluoromethanes in the environment', *Reviews of Geophysics and Space Physics* **13**: 1–35.

Siegenthaler, U. and Joos, F. (1992) 'Use of a simple model for studying oceanic tracer distributions and the global carbon cycle', *Tellus* **44B**: 186–207.

Wallace, D. W. R., Beining, P. and Putzka, A. (1994) 'Carbon-tetrachloride and chlorofluorocarbons in the South Atlantic Ocean, 19 degrees S', *Journal of Geophysical Research* **99**: 7803–19.

Watson, A. J., Law, C. S., Van Scoy, K., Millero, F. J., Yao, W., Friederich, G. E., Liddicoat, M. I., Wanninkhof, R. H., Barber, R. T. and Coale, K. H. (1994) 'Minimal effect of iron fertilization on sea-surface carbon dioxide concentrations', *Nature* **371**: 143–5.

Wunsch, C. (1988) 'Eclectic modelling of the North-Atlantic, II. Transient tracers and the ventilation of the Eastern Basin thermocline', in 'Tracers in the Ocean', ed. H. Charnock, J. E. Lovelock, P. S. Liss and M. Whitfield, *Philosophical Transactions of the Royal Society* **A325**: 201–36.

Part 4

The ocean ecosystem

Introduction

Tony Rice

The *Challenger* Expedition is generally, and justifiably, accorded enormous respect if not reverence by historians of oceanography, as this book testifies. But it has also had its share of critics, mainly concentrating on what has been interpreted as a technological conservatism (see Chapter 1) and a distressing failure to make the most of the opportunities the voyage provided for the study of oceanic physics. In contrast, the expedition's chemistry, and particularly biology, the subject of the chapters in this section, usually receive unbounded praise.

This is not surprising. After all, as Roger Wilson points out in Chapter 14, it was William Dittmar's careful analyses of the water samples collected from all over the world oceans during the expedition by John Buchanan that demonstrated the remarkable, and extremely important, constancy of the relationship between the dissolved constituents making up the familiar saltness of the seas. Similarly, Paul Tyler and his co-authors emphasize the significance of the *Challenger* biological collections and the resulting vast amount of taxonomic research that was based on them. They go on to demonstrate how many of Wyville Thomson's general conclusions about animal life in the deep sea, based on the *Challenger* results, have turned out to be correct despite the technological limitations under which the expedition's scientists worked and even, on occasion, when they were based on inappropriate data.

But important as these basic observations were in providing both a foundation for, and a stimulus to, the later developments in oceanography, they were, understandably, extremely limited in their scope. Essentially, the *Challenger* observations were static rather than dynamic. They provided a sort of coarse 'snapshot' of the chemistry, geology and biology of the oceans which, although totally novel and therefore of enormous significance, did almost nothing to explain the processes controlling and modifying them – in other words how the oceans actually worked. The secrets of these processes were to be uncovered slowly and painstakingly by subsequent generations of scientists, many of them much less well known than the *Challenger* scientists despite the importance of their work. Some of these later developments are dealt with in this section.

Roger Wilson, for example, describes the fascinating story of the study of the seemingly most mundane of the ocean's characteristics, its saltness. How do you measure it, how do you define it and, above all, how do you explain it? Suddenly, this possibly yawn-provoking topic for many readers becomes of crucial importance in explaining why life was able to arise and evolve on our planet, and also what might happen if we treat our environment with insufficient respect.

Similarly, Eric Mills traces the development of chemical methods for the analysis of inorganic nutrients in sea water to unravel the complexities of the annual cycle of phytoplankton growth in the oceans. Generally these methods involved the application, often with modifications, of techniques developed primarily for other purposes, or at least for other, often less challenging, environments. But the stimulus for these advances cannot be traced back to the *Challenger* Expedition other than via the most tenuous of links. The importance of phytoplankton as the main source of food for the animal populations in the whole of the deep ocean was simply not appreciated in the 1870s. Indeed, the study of the phytoplankton was a very minor player in the *Challenger* objectives and resulted in only one small section of the official reports, that on the diatoms by Conte Castracane (1886). Although, in an editorial note accompanying this report, John Murray writes that it had been hoped that accounts of other planktonic algal groups would have been forthcoming, 'The material brought home was . . . not extensive enough, nor in a sufficiently good state of preservation, to admit of the preparation of a satisfactory report'.

Instead, the serious study of the phytoplankton, and subsequently the chemical techniques described by Eric Mills, had to await the pioneering quantitative plankton research instigated by Victor Hensen at Kiel in the 1880s, taken up in Plymouth in the 1920s and subsequently in the USA from the 1940s. This fascinating story has been told superbly by Eric Mills (1989) in his book *Biological oceanography*.

Although phytoplankton ecology, which was not covered at all adequately by the *Challenger* scientists, made enormous strides over the next seventy or eighty years, deep-sea zoology, which was a primary objective of the expedition, made almost no progress over the same period. Consequently, the commonly held view of animal life in the deep ocean in the middle years of the last century, and expressed by N. B. Marshall in his book on deep-sea biology published in 1954, was almost indistinguishable from that held by Wyville Thomson and John Murray following the *Challenger* Expedition (see Chapter 15). In summary, this view was that a constant and monotonous deep sea, effectively isolated from the surface layers and, for that matter, from the rest of the planet, supported a rather small number of individual animals representing a similarly small number of species. Within twenty years after the mid-1950s these ideas were turned on their heads. We now know, or think we know, that the deep ocean is by no means as stable and monotonous as it was confidently regarded for almost a century after the *Challenger*. Instead, it exhibits a geological, physical, chemical and biological patchiness at spatial scales ranging from centimetres to hundreds of kilometres, and is subject to disturbance on the same scales at intervals ranging from hours to millennia. Perhaps in response to this environmental diversity, the species richness of the deep-sea floor may, in the opinion of many deep-sea biologists, rival that of tropical rain forests. These startling new discoveries arose from a number of origins, but mainly the availability of new technologies, some, such as satellites and manned submersibles, relatively complex and expensive, others rather simple and cheap. But whatever the source, they should warn us that our faith in our apparent knowledge of the deep ocean should be tempered with caution. Transported to a modern research vessel or oceanographic laboratory, *Challenger* Expedition scientists would no doubt be greatly impressed by late twentieth-century technology, but even more so by the new discoveries. In his thought-provoking personal view of the requirements of oceanography in the next twenty years, John Woods (Chapter 16) suggests that we need a quite different approach involving a quantum leap in our data gathering and analysis capabilities. If he is right, then by the year 2020 present-day oceanographers are likely to feel more estranged technologically than they currently do from the days of the *Challenger*. More importantly, will our present ideas about the largest environment on earth seem as curious and out-of-date as some of those of our *Challenger* forebears?

References

Castracane, F. (1886) 'Diatomaceæ', *Report on the scientific results of the voyage of HMS* Challenger. *Botany* **2**. London: HMSO.

Marshall, N. B. (1954) *Aspects of deep-sea biology*. London: Hutchinson.

Mills, E. L. (1989), *Biological oceanography: an early history*. Ithaca: Cornell University Press.

13 'Problem children of analytical chemistry'

Elucidating the seasonal cycle
of marine plankton
production through nutrient
analysis

Eric L. Mills

Introduction

The complex causation of the yearly plankton production cycle in the sea could only be determined once reliable chemical analyses of the inorganic nutrients available in sea water became available beginning in 1902. The first reliable analyses were made in Kiel by Emil Raben. By the mid-1930s, W. R. G. Atkins and H. W. Harvey at Plymouth applied colorimetric nutrient analyses whose origins were in the synthetic dye industry and in public health studies of drinking water. None of the techniques used in studying the plankton cycle (which Raben characterized as 'problem children of analytical chemistry') was devised for that purpose; each came, often after long delay, from pure or applied chemistry. Every major advance on which production studies depended was linked with the arrival of a chemist to work on biological problems. Through adaptation rather than revolution, chemists, working in new settings with marine biologists, brought modifications of established techniques to a new problem – the understanding of how marine production is controlled in nature.

Nutrients in the sea

Late in 1889 the German steamer *National* set out from Kiel on a voyage devoted to studying open-ocean plankton. The organizer of the Plankton Expedition, Victor Hensen, had promoted it as a means of obtaining new information about organisms he referred to as 'blood of the sea' – the phytoplankton cells that formed the food source of commercially important fish species (Mills 1989: ch. 1). But despite the expectations of Hensen and his colleagues, that the open ocean in the tropics would show luxuriant plant growth like the tropical forests, the tropical oceans proved to be plankton-poor, whereas the high latitude North Atlantic seas traversed by *National* had a large stock of plankton. Explaining these paradoxical observations became the dominant theme of research at Kiel for thirty years in the hands of Hensen's younger colleague Karl Brandt (1854–1931) (Mills: 1989, ch. 2, Mills 1990).

Brandt, appointed Professor of Zoology at Kiel in 1887, had been trained as a classical zoologist. His previous work provided no basis for work on the abundance of marine planktonic organisms. But the richness of the land had been studied by agriculturalists, agricultural chemists and bacteriologists for decades. Kiel was a university fostering interest in agriculture as well as the sea; Brandt turned for advice and ideas to his colleagues in agriculture and to the burgeoning literature on agricultural chemistry to begin his explorations of marine production.

In a dramatic extrapolation from the conditions governing agricultural production, Brandt speculated that, just as on land, the nitrogen cycle in the sea must control the abundance of green plants. But although the nitrogen cycle on land was becoming well known, nothing was known about it in the sea. Using the seas near Kiel as a microcosm of the open ocean, and seasonal changes of plankton abundance as phenomena similar to the geographical variations observed during the Plankton Expedition, Brandt single-mindedly set out to elucidate how plankton abundance was controlled by dissolved nutrients, especially those of the nitrogen cycle.

When Germany began to contribute survey data and research to the newly founded International Council for the Exploration of the Sea (ICES, established in 1902), Brandt became responsible for marine biological research carried out in a new Internationale Meereslaboratorium set up in Kiel. New research personnel enabled him to expand his work on marine plankton and its control by nutrient cycles. Among the first to be appointed was a chemist, Emil Raben (1866–1919), whose work was crucial to knowledge of dissolved nutrients in the ocean.

Raben, who was directed in 1902 to develop reliable methods for nutrients in sea water, said of one method, after fifteen years of work: 'we must consider that the quantitative determination of phosphoric acid in earlier times, as Ostwald said in a lecture on Schmidt of Dorpat, was a problem child of analytical chemistry' (Raben 1916: 7). The quantitative determination of all the major nutrients such as ammonia, nitrate, nitrite, silicate and phosphate was difficult for Raben throughout his career, and it remained difficult for a succession of later investigators in Europe, Britain and North America, who were anxious to put the study of marine production on as fully scientific a basis as chemical agriculture. But the problem of determining microgram levels of dissolved substances in a salt solution, sea water, was a formidable one.

When Raben began his work on nutrients in the sea in 1902, chemical analysis had been applied to agriculture for more than a century. Justus von Liebig's reductionist approach to agricultural chemistry, dating from his book *Organic chemistry in its application to agriculture and physiology* (published in London and Germany in 1840) provided the leitmotiv for Brandt's approach to marine production, which was based on the analysis of agricultural production in terms of weight and chemical composition (Brandt 1898: 46ff.; 1901: 494ff.).

Brandt and Raben drew on nearly a century of debate about the nutrients most important in plant growth, especially the often bitter controversy over whether ammonia, free nitrogen, or nitrate was the main source of nitrogen for plants (Aulie 1970, Rossiter 1975). By 1902 the main elements of the nitrogen cycle on land were clear, including the processes of nitrification, denitrification, and nitrogen fixation, which had been firmly established as bacterial processes during the last three decades of the nineteenth century.[1] It is understandable that Brandt was preoccupied with nitrogen in the sea and its transformations. A century of agricultural chemistry testified to the crucial role of nitrogen in plant production, and Brandt's colleagues and students had begun to show that the elements of the nitrogen cycle were similar to those on land (Mills 1989: ch. 2). Thus Raben concentrated on the sources of nitrogen in solution, giving much less attention to silicate, which diatoms required for growth, and even less attention to phosphorus, which appeared to be a ubiquitous and relatively unvarying element in the sea.

When Raben first began to develop or adapt techniques to determine quantitatively what

Brandt called 'trace plant nutrients' (Brandt 1927: 207), the analysis of sea water, usually in qualitative terms, went back more than seventy-five years in a scattered literature from Europe and Great Britain (see references in Raben 1905a, 1905b, 1910, 1916; on 'mineral waters' in general, see Hamlin 1990). In addition to the analyses done by his predecessors, he had a series of textbooks on analytical chemistry, new techniques from the explosively growing German chemical industries (Beer 1959, Travis 1993), the advice of colleagues from university agricultural institutes, and superb sources of information on current and past analytical chemistry, among them Liebig and Wöhler's *Annalen der Chemie und Pharmacie* (since 1832), Fresenius's *Zeitschrift für analytische Chemie* (since 1862) and the thoroughly cross-indexed *Berichte der deutsche chemische Gesellschaft* (since 1868). What he lacked was much assurance that the techniques used in agriculture worked in the sea, where the nutrients were greatly diluted compared to land, and where the presence of salt complicated analyses. Only the pioneering work of Konrad Natterer, chemist on the Austrian ship *Pola*, which worked in the Mediterranean Sea, Sea of Marmara and Red Sea, from 1890 to 1896 (Natterer 1892, 1893, 1894, 1895, 1898), suggested that conventional chemical techniques could be applied profitably to sea water.

How did nutrient analysis develop in early biological oceanography, and where did the methods come from? What roles did industrial chemistry, the literature of pure chemistry, or discovery and innovation arising from oceanography's own needs, play in the development of reliable techniques for determining the amounts of plant nutrients in sea water? Each major nutrient regarded as important in early biological oceanography provides a useful case-history of scientific change.

Ammonia in sea water

By the time Brandt had developed his denitrification hypothesis, many decades of work had demonstrated that the major substances involved in the nitrogen cycle on land were ammonia (either as gas or ammonium ion), nitrate and nitrite. Air or water-borne ammonia, in particular, had been rejected as a major source of nitrogen on land (see Aulie 1970), but its role in the sea was uncertain; J. J. T. Schloesing, for example, had suggested that, whereas the land was dominated by nitrate, the sea produced ammonia from the decomposition of organic matter and, because of sea winds, ammonia was transferred from the sea to land (Schloesing 1875a, 1875b, 1875c). Only Natterer, on the Austrian *Pola*, provided a little information on the abundance of ammonia in the sea, although it had been known since the 1850s that ammonia was virtually undetectable in rain falling over land (Aulie 1970). Analyses were badly needed to determine the abundance of ammonia in the sea and its relation with plants and the other nitrogen compounds.

The very early qualitative analyses of ammonia in sea water, dating back at least to Alexander Marcet in 1822, involved either evaporation or distillation of sea water to concentrate ammonia, which then could be sublimed off or titrated in the distillate. In general, the distillation techniques involved boiling sea water with magnesium oxide before the ammonia was determined by one technique or another (Raben 1910: 305ff.; also Johnstone 1908: 213–18 on Raben's techniques).

In 1868 Julius Nessler (1827–1905) refined a method dating from 1852 in which ammonia yielded a yellow-brown precipitate (Nessler 1868). His test was adopted quickly as a test for sewage pollution in drinking water. John Murray of the *Challenger* and Robert Irvine appear to have been the first to apply the reaction to ammonia in sea water, when they used the Nessler reaction in their studies of coral reef formation (Murray and Irvine 1889).

Thereafter, the Nessler reaction was used by nearly everyone until the 1930s or later as a quantitative colorimetric test for ammonia.[2] Raben himself used a distillation technique, then the Nessler reagent, as part of his determination of ammonia, nitrate and nitrite, as I discuss later. Raben's values for ammonia in the Baltic and North Seas provided the first significant body of data on ammonia in the oceans.

The great disadvantage of the Nessler reaction was that the ions in sea water interfered with the formation of the precipitate. Rolf Witting (1879–1944), director of the Finnish ICES investigations in Helsingfors (now Helsinki) from 1911, developed the first direct method of determining ammonia, eliminating the need for distillation, by precipitating the major ions (Witting 1914). After this innovation in 1914, the source of which is unknown, Witting's colleague Kurt Buch (born 1881, death date not known), who became Dozent in chemistry at Helsingfors in 1927 (later professor of chemistry at Åbo), continued to use a direct method of ammonia determination with the Nessler reagent, using and developing another of Witting's innovations, a spectrophotometer, to measure the intensity of the colour produced (see discussion later, p. 239) (Buch 1923, 1929).

Another direct method, source unknown, which eliminated the tedium, duration and susceptibility to contamination of the distillation technique was introduced by Herman Wattenberg (1901–44) of Kiel in 1929. Rochelle salts complexed the interfering ions, keeping them in solution while the Nessler reaction proceeded.

In 1929 Buch recognized that there were serious problems with distillation methods, for example contamination from ammonia in corks, seals and even laboratory air. Eight years later, Wattenberg emphasized that the existence of three methods (distillation and two direct techniques) indicated that there was no 'universally satisfactory method' of quantitatively measuring ammonia in sea water (Wattenberg 1937: 21). And in 1945, H. W. Harvey of Plymouth, in his book *Recent advances in the chemistry and biology of sea water*, pointed out that the direct methods were insensitive, and that the older distillation methods always gave values that were too high (Harvey 1945: 82). The difficulties of measuring ammonia reliably had not been solved after more than forty years of modified chemical techniques from the classical analytical literature of chemistry.

The determination of ammonia after 1945 is also a story of mixed success. Two direct methods appeared in the early 1950s, which, modified, remained useful into the 1980s. Kruse and Mellon, industrial chemists, publishing in the journal *Sewage and Industrial Waste*, showed in 1952 that water treated with a pyrazolone reagent and extracted into carbon tetrachloride gave a yellow solution that could be measured photometrically. Modified, their method was sensitive, but it included dissolved amino acids as well as ammonia (Strickland and Parsons 1960: 75–80). Nearly simultaneously, J. P. Riley, a marine chemist, devised a method of reacting ammonia in sea water with sodium hypochlorite and sodium phenate to produce a blue dye that was easy to measure photometrically (Riley 1953, Strickland and Parsons 1972: 87–9). Finally, taking their method from the chemical literature in the 1960s and an unpublished method used in the Scripps Institution of Oceanography, J. D. H. Strickland (1920–70) and T. R. Parsons (b. 1932) proposed a sensitive method in which ammonia was oxidized to nitrite and then coupled into an azo dye, giving a red colour for spectrophotometry (the latter is the Griess–Ilosvay reaction, which plays an important role in developments described in the next section).

My tale of ammonia determination in sea water is a complex one. Raben refined previously existing distillation techniques and used the nearly forty-year-old Nessler technique (with a colorimeter) to measure ammonia. His method, and his results, dominated the literature until direct methods were used in the 1920s and 1930s – their basis being conventional chemical techniques. Only the introduction of dyes in the 1950s was truly innovative (as far as ammonia measurement is concerned), although dye chemistry dates from the period *c.* 1865–90 and dyes had been used to indicate other nutrients far earlier, as the following sections show. In the early development of techniques, pure chemistry and the chemical tradition from agriculture played varying roles; later, only the introduction of the Kruse and Mellon technique can be construed as a transfer from industrial technology into science.

Measuring nitrite and nitrate in sea water

When Raben began his analytical work on sea water in 1902, both nitrite and nitrate were implicated in the nitrogen cycle on land. Solid evidence existed that nitrate was the ion providing a large part of plant nitrogen and that, under some conditions, for example in anaerobic soils, it could be reduced to nitrite; conversely, nitrite was a step in the production of nitrate during the oxidation (mineralization) of decomposing organic matter. It seemed very likely that nitrate would have similar roles in the sea.

The detection of nitrate and nitrite is a lengthy and complicated story during the early history of biological oceanography, largely because the major ion, nitrate, was difficult to detect without measuring the less abundant ion, nitrite, simultaneously. All the early methods and most modern ones were sensitive to both ions; thus the estimation of nitrate has always been too high, or has needed correction, because of interference by the nitrite ion. Ironically, a sensitive test for nitrite alone appeared very early and has been in use, slightly modified in industry and analytical chemistry, to the present day.

Peter Griess (1829–88), the eminent German dye chemist who spent the last quarter century of his life working in the Allsop & Sons brewery at Burton-on-Trent in the English Midlands, developed a test for nitrite in 1879. His method, modified by Lajos Ilosvay (1851–1936) in 1889, was to react sulphanilic acid, naphthylamine and nitrite to produce a rose-coloured azo dye (see Robinson and Thompson 1948a). The coloured product was stable, sensitive and specific for nitrite. It could be used as a quantitative assay by colorimetry (at first) or spectrophotometry (from the 1940s onwards).

Although the reliable Griess–Ilosvay reaction was well known as a test for the nitrite ion in food, water and sewage, it entered oceanography slowly. H. C. Geelmuyden, working in Christiania (now Oslo) in 1902–3, first applied the technique to sea water (Geelmuyden 1902, 1903). Raben, whose work was beginning then, followed a technique from water chemistry involving distillation and the addition of metaphenylenediamine, which gave a coloured product. Only in 1911–12 did he replace his early cumbersome technique with the quick, elegant Griess–Ilosvay reaction (Brandt 1927: 210; Raben 1914). With the work of Kurt Buch at Helsingfors in the 1920s, the method became well known (Buch 1923, 1929; see also Orr 1926) and remained a standard one in the literature of chemical oceanography.[3] Ironically, the Griess–Ilosvay sensitive test was available for the form of nitrogen least abundant in the sea. The results of Natterer (1890–6) in the Mediterranean and adjacent seas showed Raben and Brandt that nitrite was virtually absent, a result they confirmed in the Baltic and the North Sea after 1902. They and their successors were under no pressure to apply any test for nitrite, since it had proved to be a variable, scarce and unimportant component of sea water according to the early results. This accounts for the demise of the metaphenylenediamine test and Raben's adoption of a technique that measured the sum of nitrate and nitrite. The Griess–Ilosvay reaction, which was widely used to test for the contamination of food and water, began to play a major role in oceanography only in the 1950s when it was applied in a new test, not for nitrite but for nitrate, after a lengthy history of searches for a sensitive, rapid and reliable test for that ion.

In 1902, a variety of tests existed for nitrate (or 'nitric acid' as it was frequently called) in water, soil and chemicals. J. B. Boussingault (1802–87), who analysed water samples from the Dead Sea in 1856, used indigo with sulphuric acid as a test for nitrate, a method he attributed to Liebig (Boussingault 1856: 1235–7). Both he and J. B. Lawes, J. H. Gilbert and E. Pugh (1863), who refuted Liebig's belief that atmospheric ammonia was the major source of nitrogen for plants, used the indigo sulphate method and a complementary one, the 'protosulphate of iron test' ($FeSO_4$), which gave a coloured precipitate with nitrate.

Diphenylamine was synthesized about 1872 by Merz and Weith; within a year it was used by Kopp in industrial chemistry to detect the contaminants nitrous and nitric acid in commercial sulphuric acid (Meister 1872, Letts and Rea 1914). Natterer applied the method in marine chemistry for the first time to samples from the Mediterranean, which proved to be nearly free of nitrate, and Geelmuyden also used diphenylamine as a direct test for nitrate in Dutch coastal waters. Raben, who was aware by searching the chemical literature that several substances such as diphenylamine gave a coloured product with nitrate and nitrite, distrusted their results because he found that sea water interfered with the reaction (Raben 1905a: 91). Thus he refused to consider that diphenylamine might provide a simple, reliable assay and he turned to a more complex distillation–reduction technique.

Raben's technique of nitrogen determination, based on the best methods available (as he believed) from the analytical literature, was a synthesis of older techniques. Working first in the laboratory, then directly on board the German research vessel *Poseidon*, he used a distillation–catalysis process (Fig. 13.1a) to produce ammonia, which could be estimated using the Nessler reaction. The final amount of ammonia represented the total nitrogen in nitrites and nitrates. His method, which was subject to all the criticisms of distillation techniques and the insensitivity of the Nessler reaction, reflected Raben's belief that nitrite could be ignored safely in analyses of sea water, although he was aware that Natterer had tested for nitrite, and that a sensitive test for nitrite was available in a current text of water chemistry, Tiemann and Gärtner's *Untersuchungen und Beurteilung der Wasser* (Walther and Gärtner 1895). Natterer's technique was unduly sensitive to light and air, and thus unreliable. The other reaction allowed Raben to estimate nitrite, but he seldom used it, preferring to consider nitrite together with the more abundant nitrate (Raben 1905a: 89ff.).

Only Raben and Ringer and Klingen,[4] working in Holland, used the distillation technique extensively. Meanwhile, interest in the diphenylamine method, with its virtue of simplicity, continued at a low level. It was used to detect nitrites and nitrates in drinking water and milk, and improved (doubled sensitivity) by replacing diphenylamine with diphenylbenzidine (Tillmans 1911, Tillmans and Sutthoff 1911, Letts and Rea 1914, Smith 1917). Despite this, the technique languished in biological oceanography for nearly twenty-five years, until N. Tschigirine and P. Daniltchenko (1930) used the original method and W. R. G. Atkins (1884–1959) at Plymouth introduced an improved method using diphenylbenzidine to marine chemists and oceanographers in 1932. Atkins, in fact, proposed the diphenylbenzidine method to replace another, developed by H. W. Harvey (1887–1970) of Plymouth, which already dominated nitrate analysis in sea water and was well on its way to becoming a textbook technique. Despite Atkins's advocacy, the diphenylamine–diphenylbenzidine technique was a casualty of the forceful way Harvey's new technique, the colour reaction of reduced strychnine, was introduced into marine science.[5]

Harvey had been searching for a sensitive method that would allow nitrate to be estimated when it was least abundant, after the spring bloom in the English Channel. Curiously

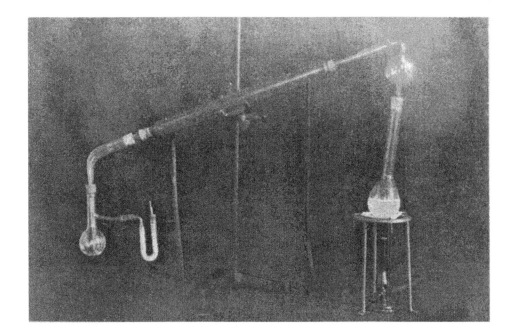

Figure 13.1 Apparatus used to determine the amount of ammonia, nitrite and nitrate in sea water. This equipment, based on Raben's at Kiel, was used in W. E. Ringer and F. J. M. P. Klingen's laboratory at Den Helder, The Netherlands, early in the twentieth century. (From Ringer and Klingen 1907.) (a) The distillation apparatus by which ammonia, then nitrite plus nitrate was distilled off before being determined using Nessler's reagent. (b) Chemical laboratory in the Dutch Institute for Sea Research. On the right hand corner of the table is a colorimeter with two Hehner cylinders set in place to be read. Beside it is a pan balance.

enough, he does not seem to have been aware at first of the use of diphenylamine-diphenylbenzidine reaction being examined by his older colleague Atkins, or he chose to ignore it (Harvey 1928: 42).[6] In 1924 Harvey found a reference in *Chemical Abstracts* to a sensitive colorimetric test for nitrate using an acidified solution of strychnine. This was based upon Tafel and Naumann's (1901) discovery that strychnidine (a reduced product of strychnine) gave a coloured product with nitrous acid, or with nitrate if acidified with sulphuric acid. Georges Denigès (1911), Professor of Biological Chemistry in the Faculty of Medicine and Pharmacy at Bordeaux, used this knowledge to devise a qualitative test for nitrate using reduced strychnine and sulphuric acid. In Harvey's hands this became a sensitive colorimetric quantitative test for nitrate in the sea, one that was the premier method in his papers and books until the 1950s. Like the other methods, it was sensitive to nitrite, that ion, according to Harvey (1928: 44), 'not being present in sufficient quantity materially to affect the result in the case of unpolluted waters from the open sea'. Harvey admitted that the reagent was difficult to prepare, requiring strychnine sulphate and acid completely free of nitrate and iron. Even then, as he said, 'the yield and time taken is somewhat capricious' (Harvey 1926: 72). In Harvey's hands the method worked very well, although others, such as T. G. Thompson and Martin W. Johnson at the University of Washington, found it so unreliable that they abandoned it (Cooper 1932: 161). L. H. N. Cooper (1905–85) at Plymouth defended the method by showing how to remove the greatest source of difficulty, nitric acid in the sulphuric acid. Even he struggled with the method, for, as he described later:

> Chance played its part in his [Harvey's] success because he had had little difficulty in preparing satisfactory reduced strychnine reagent. I did, and found that my failures coincided with a northeasterly or easterly air-stream over Plymouth bringing to the laboratory contaminated air from industrial England or the local power station. When the preparation was undertaken with an air stream from the southwest there was no difficulty. Subsequently successful preparations were made at Plymouth for a colleague bedevilled by impurities in the industrial air of Hamburg.
>
> (Cooper 1972: 336–7)

Undoubtedly it was Harvey's advocacy of the reduced strychnine method, first in his journal articles, then in his widely read and very influential books *Biological chemistry and physics of sea water* (1928) and *Recent advances in the chemistry and biology of sea water* (1945), that assured the survival and use of this difficult method for nearly thirty years. But it was never regarded as satisfactory (except perhaps by Harvey), for reasons explicitly listed in 1955 by Mullin and Riley (1955b). Nonetheless, they acknowledged that it had been used, as they said, '*faute de mieux*'.

Something better appeared in 1951 when Ernst Føyn (1951: 3–7) of the University of Oslo suggested that nitrate could be measured using the old but reliable and sensitive Griess–Ilosvay reaction if nitrate were first reduced to nitrite. Four years later, J. B. Mullin and J. P. Riley, marine chemists at the University of Liverpool, also proposed that reduction followed by the Griess–Ilosvay reaction would produce an easy, sensitive test for nitrate (which could be corrected for the nitrite content of sea water by using the Griess–Ilosvay reaction alone). Departing from Føyn's technique of reduction, which was slow and liable to reduce nitrate to ammonia, they experimented with reducing agents. Later (1963), Morris and Riley reduced the time of reaction from twenty-four hours to only a few minutes. Similar techniques became standard (Strickland and Parsons 1972: 71–6). Føyn's and the later reduction techniques allowed nitrate to be measured using the old but reliable and sensitive Griess–Ilosvay reaction, the only consistently used analytical reaction for nitrogen in the sea that has a history of use greater than twenty or thirty years.

This account indicates that the development of techniques to measure nitrite and nitrate in sea water was governed by varying independent factors. Qualitative and quantitative techniques for nitrate and nitrite were developed for use in industrial, water and food chemistry, beginning in the late nineteenth century. Natterer, who applied some of these methods to Mediterranean Sea water, found little or no nitrite and little indication of nitrate. Basing their ideas on his results, Raben and Brandt virtually ignored nitrite, to concentrate their efforts on the compounds of nitrogen most important in the terrestrial nitrogen cycle, ammonia and nitrate. Although they occasionally used the Griess–Ilosvay reaction, it was irrelevant because nitrite was never important in their conceptions of the marine nitrogen cycle. Nitrate (which included nitrite) became the concern of most analysts, beginning with Raben, but a simple colorimetric test did not become established early, partly because the standard test had a reputation for unreliability.[7] Despite Atkins's espousal of this method, Harvey's reduced strychnine test, with all its idiosyncrasies, became the standard method for twenty-five years, largely because of Harvey's promotion of it in two important books on production in the sea. L. H. N. Cooper, who compared the two methods, believes that Harvey accepted his conclusion that diphenylamine was less reliable than strychnine (Cooper, in a letter). Finally, when it became possible easily to reduce nitrate to nitrite, rather than completely to ammonia (a development that apparently occurred without direct influences from outside oceanography), a standard analytical tool from general chemistry, the old Griess–Ilosvay diazotization by nitrite and coupling to produce a strongly coloured red dye, could be used to determine both nitrate and nitrite. Throughout its history, the determination of nitrate in the sea presented both conceptual and technical difficulties that were overcome very slowly by applying standard techniques from analytical chemistry, most of them developed years or even several decades earlier for different purposes.

The analysis of silicate in sea water

Brandt was preoccupied with the role of nitrogen in the sea, but early in his explorations of marine production he recognized the possibility that other substances might limit plant growth. He said in 1902 that the 'question is [yet] to be established whether or not there are essential plant nutrients other than nitrogen compounds, for example phosphoric acid, carbonic acid or silicic acid, which will determine the maximum of production' (Brandt 1902: 53). Because it was the main constituent of diatom tests, silicate, in particular, required special attention. Murray and Irvine (1891) had suggested that silicate was low in the open sea; they proposed that phytoplankton and radiolarians liberated silicate from suspended clay particles. Brandt's early calculations convinced him that this was not correct, for there was plenty of silicate in solution, on average, for diatom growth. Nonetheless, he suspected that silicate could limit phytoplankton growth during the spring bloom (Brandt 1902: 69–72; 1905: 10). Reliable measurements were vitally needed but lacking in 1902.

Raben (1905a: 99–101; see also Johnstone 1908: 217–19) rapidly developed a gravimetric technique to determine silicates based on their insolubility in acid. After a course of drastic treatments with acid and hot water, the weight of the dried residue was the silicate content of the sample (a weight that frequently had to be corrected for the presence of other refractory substances).

Raben's analysis of silicate in sea water lay well within the tradition of German analytical chemistry, which concentrated on gravimetric methods throughout the nineteenth century (Ihde 1964: 281–2). Gravimetric determination of silicate, after removal of the soluble salts, went back to at least 1848, including the classic analyses of Georg Forchhammer (1794–1865) and William Dittmar (1833–92), who analysed the sea water samples collected by the *Challenger*.[8] Raben found it satisfactory at least until 1914, after which he published no information on the analysis of silicate. A colorimetric test for silicate existed before Raben's work began, for in 1898 Jolles and Neurath showed that molybdic acid gave a yellow product with silicate.[9] Apparently, Raben was not aware of this work; the use of molybdate compounds was restricted to pure chemists until F. Diénert and F. Wandenbulcke in 1923, as a result of Diénert's work on the circulation of subterranean waters in soils, showed that silicate reacted with ammonium molybdate and sulphuric acid to produce a yellow compound. W. R. G. Atkins (1923b) quickly adapted this method for sea water at the beginning of his studies of plankton production in the English Channel. It became a standard technique, rapidly replacing the gravimetric method, and was adapted for new colorimeters and spectrophotometers during the 1930s (Robinson and Spoor 1936, Wattenberg 1937: 15–18). Like other colorimetric methods for nutrients, the molybdenum yellow test had its flaws, for the colour was unstable, could be affected by the salts in sea water, and was not fully proportional to the silicate present at high concentrations (Mullin and Riley 1955a).

The reduction of molybdates to a blue compound was noted in 1924 by M. L. Isaacs, examining the silicon content of tissues. J. D. H. Strickland, working on his Ph.D. thesis at the University of London, explored the use of the molybdenum blue test to detect silicon in industrial materials. His work, published in 1952, was taken up by F. A. J. Armstrong at Plymouth, who established a new, more stable, and more sensitive test for silicate in sea water (see Armstrong 1951). Later, the stability of the blue compound was increased by using reductants giving a longer-lasting blue product (Mullin and Riley 1955a, Strickland and Parsons 1960, 1972).

Raben derived his method by refining techniques well known to chemists for many years. After his time the only major change in detecting silicate occurred when a sensitive colorimetric technique, also adapted from other branches of analytical chemistry going back several decades, was applied to the analyses of sea water by Atkins. This method was refined nearly thirty years later by applying chemical research on silicon in industrial materials to the reaction used with sea water. In short, all techniques for the measurement of silicate in sea water had their ancestry in more general branches of analytical chemistry. Innovation was by adaptation, not invention.

Phosphate

Phosphate (nowadays characterized as reactive phosphorus) ranked last among the trace plant nutrients that Brandt and Raben set out to study during the first decade of the twentieth century. In Raben's words (1916: 3), 'phosphoric acid, is a never-absent constituent of all natural waters and as a result cannot be absent in the sea'. He began his analyses in 1904, but did not publish his results until 1916, partly because, as Brandt (1902: 65–8) noted, combined nitrogen was less abundant in sea water than phosphate and thus was more likely to be limiting. After their initial uncertainty about its role, once Raben's analyses had accumulated, it became evident that although nitrate could virtually disappear from the surface waters of the Baltic, phosphate never disappeared, although it decreased after the spring bloom. This result, which was an artifact of Raben's method of analysis, persuaded them that phosphate was the least significant of the major nutrient ions. This is the main factor responsible for the scanty attention they paid to an important dissolved nutrient.

Standard analyses for phosphate existed long before Raben's time because of agricultural chemists' intense interest, beginning in the 1840s, in phosphate fertilizers (Ihde 1964: 449). Even sea water had been analysed for phosphate by the 1840s, although the early analyses were mainly qualitative, not quantitative (Raben 1916 gives a lengthy list of references). Precipitation was used very early, for example by Forchhammer (who published years of analyses in 1859) and C. Schmidt of Dorpat (1871, 1875, 1878), the intention of a variety of methods being to precipitate insoluble calcium or ferric phosphates that could be weighed (Matthews 1916: 123).

Raben used a complex and lengthy procedure involving the reaction of phosphate with ferric chloride and ammonia, its separation from silicate, and repeated precipitation from solution. Once adequately purified and dried, the precipitate was weighed. His method combined elements of qualitative analysis and quantitative techniques well known in texts of analytical chemistry, such as the ammonium molybdate precipitation method given by Treadwell (1905).

A new method of measuring phosphate was proposed by I. Pouget and D. Chouchak, agricultural chemists at the École supérieur des sciences in Algiers, in 1909 (Pouget and Chouchak 1909, 1911). They noted that phosphomolybdic acid forms insoluble compounds with alkaloids like strychnine; the resulting strychnine phosphomolybdate is opalescent and may be measured colorimetrically. D. J. Matthews (1873–1956), hydrographer at the Plymouth Laboratory, after using a method similar to Raben's, learned of the Pouget and Chouchak technique and began using it in 1915.[10] His results were the first to be published giving values of phosphate in English Channel water.

Matthews's results were considerably lower than Raben's from the Baltic and North Seas, as Atkins (1923a: 21) noted when he began his work on marine production in 1922. Atkins approached the problem of phosphate analysis by applying a method developed by Denigès for the analysis of phosphate in fresh water (a test widely used in the early twentieth century to indicate sewage pollution). Denigès (1920, 1921) had modified the well known reduction of a molybdate complex by stannous chloride to give a blue product (as in tests for silicate current in the 1920s) as a test for phosphate in fresh water. As Atkins showed, it proved a sensitive test in sea water, provided arsenate did not interfere in the reaction, a factor that he believed accounted for Raben's high phosphate values (Atkins and Wilson 1926, 1927; see Robinson and Thompson 1948b for reaction).

Atkins's application of the Denigès method was used virtually without modification for thirty-five years, although with corrections for an error caused by the salt in sea water, recognized in 1934 (see review by Cooper 1938). Later, in the late 1950s and after, modifications using well known chemical techniques made the reaction more reliable, stabilized the coloured product (which had tended to fade), avoided the problem of salt error, and increased its sensitivity by a factor of three. The result, in the words of Strickland and Parsons, 'probably represents the ultimate in sea-going techniques' (Murphy and Riley 1958, 1962, Strickland and Parsons 1972: 49–52).

Phosphate analysis, which was hindered by the scant attention paid to it by Brandt and Raben, proceeded nearly as fast (or as slowly) as the analysis of silicates, largely because the analytical chemistry of phosphate in agriculture and pollution detection was well developed early in the twentieth century. Colorimetric techniques were available in the 1920s to replace the outmoded and insensitive gravimetric techniques; they were rediscovered, used and improved once it became evident that phosphorus was important in marine production and that the early measurements by Raben and Matthews were anomalously inconsistent. Newly modified sensitive techniques revealed the problems of the old analyses and led to their solution.

Apparatus

The chemical analytic methods used by Raben, Matthews, Atkins, Harvey and many later analysts cannot be divorced from the instruments they required, which were, almost without exception, the standard ones used in pure chemistry, slightly modified for the special problems faced by oceanographers.

Raben's analyses required only a good balance (for the gravimetric techniques) and a colorimeter of standard design (Figs 13.1b, 13.2a). His measurements followed the tried and true techniques in a standard text dating from 1891 (Krüss and Krüss 1891). This drew on a long tradition of colorimetric comparisons in analytical chemistry going back to the 1830s (Ihde 1964: 565–6). Raben and all his successors were concerned that their analyses satisfy criteria later summarized by Wattenberg (1937: 6): (1) that the reaction must be sensitive to the lowest occurring concentration of the nutrient, (2) that the variations of concentration in nature must be greater than the error of the method, (3) that the colour must vary directly with the concentration of the substance to be measured, (4) that the colour developed must be independent of the concentration of the reagents, and (5) that the reaction must be specific to the nutrient being measured. To a greater or lesser extent, the methods used in the early twentieth century satisfied these criteria.

However, as Wattenberg suggested, the measurement of nutrients in sea water, sometimes under less than ideal conditions on a ship, brought special problems. The low concentration of nutrient salts in the sea was attacked by using new long colorimeter tubes; for example, the colorimeter tubes used by Harvey and co-workers at Plymouth were up to 60cm long, whereas the Hehner cylinders used by Raben and some later analysts were usually no longer than 20–30cm. In addition, Raben's technique had not taken into account the colour or turbidity of the sea water, which could be considerable. This problem was solved by subtracting a blank value for sea water or by other special techniques (Fig. 13.2b) (Harvey 1929: 69–70, Wattenberg 1937: 78). Finally, the colorimeters were made simpler and easier to handle so they could be used reliably at sea.

The heyday of simple colorimetric nutrient analyses by marine chemists undoubtedly lay between 1916 and 1937. Matthews, building on Raben's work, eliminated gravimetric analysis when he applied a colorimetric technique to the measurement of phosphate. Seven years later, Atkins did the same with the analysis of silicate. By 1937, when Wattenberg reviewed the variety of methods available for nutrient analysis of sea water, at least half a dozen types of colorimeters were in use in marine laboratories, their use dictated by custom, preference and cost (Wattenberg 1937: 6–11). Gravimetric analysis had entirely disappeared.

Although colorimeters had been greatly refined by 1937, they still required the use of the trained human eye. When Wattenberg's review appeared, the replacement of the human eye was in progress. The precursors of this protracted revolution appeared late in the nineteenth century, when the first spectrophotometers were built and sold, but Rolf Witting at Helsingfors was the first to use a spectrophotometer in oceanography, in 1914. His work was carried on by Kurt Buch, who developed the details of a spectrophotometric technique using the Nessler reaction for ammonia (Buch 1915, 1920, 1923, 1929). It was the complexity of the Nessler reaction and his difficulty in making a standard colour comparison that led Buch to use the spectrophotometer, which he called 'the only theoretically perfect measuring

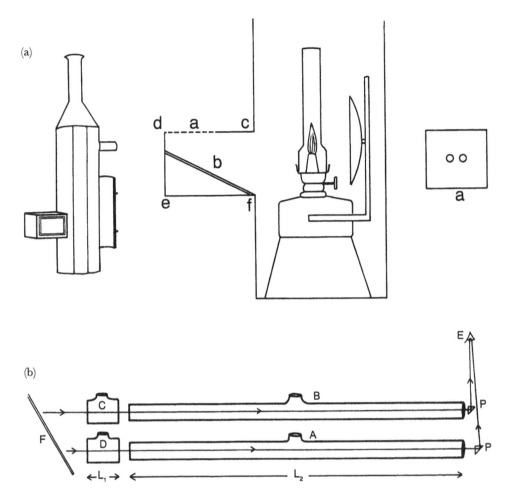

Figure 13.2 (a) Raben's colorimeter. The light source was an oil lamp. Its light was reflected by a mirror
b through a milk glass plate *a* set in a metal box. The glass plate, covered with black paper,
had two holes into which Hehner cylinders containing the sample and a colour standard
were set. (From Raben 1905a: 85.) (b) H. W. Harvey's colorimeter, used for determining
phosphate in discoloured water. The glass tubes A and B were 25–60cm long, 3cm diameter, closed at the left end and with lenses at the right. Tube A was filled with the sea water
sample and molybdate reagent with reducing agent, tube B contained only sea water and
molybdate and thus had only natural colour. Bottle C represents a series of colour standards (each representing known amounts of phosphate) which were put in place until the
light passing through A and B was equal. (From Harvey 1929: 70.)

instrument'. He used a commercial instrument dating to the nineteenth century that provided a continuous spectrum by means of prisms (Buch 1929: 40; Martens and Grünbaum
1903), but, reviewing techniques for nitrogen compounds and phosphate in sea water in
1927, he continued to mention standard colorimetric techniques because they were simple
and easy to use at sea.

Using a spectrophotometer, Buch described the extinction curves of light in sea water of
different salinity and phosphate contents, and showed that different waters, the age of the

samples, the preparation of the reagents and the time elapsed after the reagents were added, affected the results. All these complexities were hidden or difficult to explore using conventional colorimetric techniques (Buch 1929: 38–40).

Following Buch, Kurt Kalle (1931–5) did extensive work with a photometer in the 1930s, as did Wirth and Robinson (1933), using the Zeiss–Pulfrich photometer, a simpler instrument than Buch's, which used coloured filters to produce 40μm segments of the visible spectrum (Fig. 13.3). And at Lowestoft, a group under Michael Graham built a new photometer, specially designed for use at sea, using a photocell and galvanometer, to achieve 'an impersonal method of determining phosphate, that can be used at sea under ordinary conditions' because 'after one year's work bringing the samples ashore, we were not satisfied as to keeping, nor as to our colour judgement . . . after a sea trip' (Graham 1936: 205).

Graham's photometer, which was designed with the advice of Atkins, Cooper, Harvey and Gilson at Plymouth, should have been an oceanographic precursor of the switch from simple colorimetry to electrical spectrophotometry that occurred in chemistry during the 1940s (Ihde 1964: 565–6). However, even with a home-grown photocell spectrophotometer to their credit, the Plymouth group and many other marine chemists continued to use conventional colorimetric techniques for at least another decade. H. W. Harvey admitted grudgingly that 'the Pulfrich photometer is coming into increasing use, but it does not appear to give more exact results than visual comparison when made under the best conditions' (Harvey 1945: 80). As Wattenberg (1937: 8) pointed out, the human eye is very

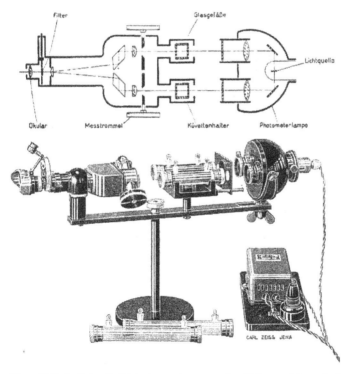

Figure 13.3 A Zeiss–Pulfrich 'step-photometer' widely used in colorimetric nutrient analyses during the 1930s. The filters gave a series of 40μm segments throughout the visible spectrum. 'Okular' = ocular; 'Messtrommel' = graduated scale; 'Glassgefasse' = glass vessel; 'Kuvettenhalter' = cuvette holder; 'Photometerlampe' = photometer lamp; 'Lichtquelle' = light source. (From Wattenberg 1937: 9.)

sensitive to small colour differences at low concentrations of nutrients; but when the intensity of colour is greater, errors increase, so the spectrophotometer may be more useful under those special conditions.

It is not difficult to account for the slow, unspectacular entry of photocell photometers into oceanography, for, as Atkins found in his pioneering work on light in the sea, which began in 1922, the early photocells were difficult to use, unpredictable and short-lived (Poole 1960: 7–8). But photometers of any kind, available since the 1890s, were not regularly used for nutrient analysis by the Plymouth group, which dominated biological oceanography from the 1920s through the 1950s, until they used one of their own design in the late 1940s (Harvey 1948: 337–40). What caused the delay? Part of the answer is certainly the success of standard practice. Colorimetry had become a refined art at Plymouth and in other marine chemical laboratories. And Wattenberg, who carefully described the available photometers in 1937, cautioned his readers that as the complexity of instruments increased so did the possibility of error. Although the eye can compensate for air bubbles, shading, slight variations, in sample tubes and the like, 'the photocell indiscriminately registers these disturbances' (Wattenberg 1937: 11).

Actually, for a time Harvey was alone in promoting the use of photocells in photometers for nutrient analysis. As his younger colleague L. H. N. Cooper (1905–85) told me in a letter in 1983:

> Everyone was attracted to the possibilities of photoelectric cells, everyone that is except Atkins. With these he had as much experience as anyone in the world except his associate Horace Poole. Poole strongly supported Atkins' opinion that they were much too unreliable to be used for chemical analysis. I knew of the experimental evidence for their opinions and agreed with them. If any development at Plymouth deserved to be called an innovation it was Harvey's acceptance of the selenium photoelectric cell. He went against the advice of a close colleague. Harvey was of course right. Atkins and I persisted in holding on to a well founded opinion when new technology had made it invalid. Our resistance to change was much like that of geologists who have rejected the evidence based on new-fangled geophysical technologies.

Even when the hegemony of electrical photometers was established in the 1950s, Harold Barnes (1955: 580), reviewing the analysis of sea water, recommended keeping a set of Hehner cylinders available for use 'in an emergency'. This kind of scientific conservatism was reinforced, especially during the 1930s and the war years, by the high cost of buying new analytical instruments such as photometers, a factor that was especially important in the financially starved Plymouth Laboratory throughout its most productive period. The unreliability of early photocells also undoubtedly delayed the Plymouth group's transition to photometers. Atkins's lengthy experience with photocells slowed the development of the Plymouth group's first absorption photometer, rather than speeding its use, until relatively reliable selenium photocells could be used in a balanced circuit, which allowed easy correction of changes in the properties of the cells and of fluctuations in the light source. The most significant question is why the Plymouth group did not use optical absorption photometers earlier, not why their switch to electric ones was slow. E. I. Butler of the Plymouth Laboratory told me that the simplicity, reliability and repairability at sea of Harvey's colorimeter, and its easy use to determine nitrate, phosphate and silicate, prolonged its use. When the Murphy and Riley method of phosphate determination came into use after 1962 (requiring a spectrophotometer), Harvey's method was abandoned (Butler, in a letter).

The impact of chemists on oceanography

The most evident conclusion I can draw from the preceding examples is that new chemical technology entered biological oceanography very slowly, and from several sources. Knowledge of the production cycle based on nutrient analyses followed the development of relevant techniques only after long delays, which had a variety of causes.

When nutrient analyses were being developed after 1902, the German dye industry had been in existence for more than thirty years (Beer 1959, Travis 1993). A host of synthetic dyes and a vast knowledge of their chemical properties existed, yet synthetic dyes were very little used in nutrient analysis until the 1950s, when they were introduced to detect ammonia and nitrate. The Griess–Ilosvay reaction, dating from the late nineteenth century, was available throughout, but it detected only the seemingly least important form of oxidized nitrogen, nitrite. Because nitrite was considered unimportant by Raben, Harvey and others, the Griess–Ilosvay reaction, certainly the most reliable one available for any nutrient throughout the early twentieth century, had a minor position until Føyn, apparently independently, realized in 1950 that it could be used to measure the major ion, nitrate.

It is hard to know why organic dyes were not used earlier to measure the always problematic ammonia in sea water. The dye techniques that were applied in the 1950s followed nearly a century of analyses of varying complexity and reliability using the inorganic Nessler's reagent. No one was ever very satisfied with the early results, whether they were obtained using distillation techniques or direct methods after precipitating the interfering salts. The existence of a well accepted test, the Nessler reaction, seems to have inhibited any search for a new test until the 1950s.

The slow development of a simple colorimetric test for nitrate can be attributed to Raben's mistake in believing that diphenylamine was unreliable. This judgement committed him to a decade of nitrate analysis using the tedious and inaccurate distillation technique, despite the fact that a far simpler technique was available and being used in pure and applied chemistry.

Finally, I have shown that the use of spectrophotometers in biological oceanography occurred very late. Only Kurt Buch, beginning in 1915, appears to have used optical spectrophotometry until it spread in the late 1940s, even though reliable instruments were available throughout the century. Instead, the colorimeter ruled, mainly because it was well known and understood by the early analysts. Both optical and photocell spectrophotometers, which were expensive and impersonal, had no evident advantages to many chemists, who had been trained to make fine distinctions between coloured solutions by eye.

In general terms the story I have outlined supports the now current belief that science and technology frequently develop independently but may interact through transfers from technology to science or science to technology (B. Barnes 1982; see also Barnes and Edge 1982: 147–54; for complexities see Bijker, Hughes and Pinch 1987 and Laudan 1984). It is quite clear in the case of biological oceanography that rapid transfers from technology (notably from industrial or public-health chemistry) did not occur; instead there were long delays between the introduction of a technique or an instrument and its use in oceanography. Many examples in the preceding pages leave no doubt of this. What the general model misses, correct though it may be in its own context, is the role of people as transmitters of scientific and technical information. As Derek Price perceptively noted, 'the research front is basically contained in people' (Price 1969: 99; see also Barnes and Edge 1982: 164–76).

Every major advance in the analytical chemistry of nutrients in the sea, on which production studies depended, was linked with the arrival of a chemist to work on biological problems. The first of these was Emil Raben, Brandt's appointee in 1901, who adapted classical techniques for use with sea water. Raben's training in classical chemistry and his connections with the agricultural chemists in the Kiel Institute of Agriculture were the basis of his success in the new science. At Plymouth, W. R. G. Atkins arrived in 1921 after a career as a chemist at Trinity College, Dublin, the Woolwich Arsenal, and the Imperial Department of Agriculture (where he was indigo research botanist in India) (Poole 1960: 1–5). He brought a distinguished record in chemistry and, no doubt, a thorough knowledge of dye chemistry. Atkins's chemical knowledge complemented the work of the Director, E. J. Allen, on the nutrition of phytoplankton and led directly to new tests for phosphate and silicate. H. W. Harvey, who joined Atkins in Plymouth in 1921, was also a chemist, trained at Cambridge in the School of Agriculture (Cooper 1972: 332–7). Although he was originally appointed as the hydrologist of the Plymouth Laboratory, he worked in the same flexible way as everyone in that small research group, taking on Atkins's early concern with nutrient analyses and nutrient cycling very soon after his arrival.

Several other examples of the innovating influence of chemists could be developed – Kurt Buch at Helsingfors, L. H. N. Cooper at Plymouth, R. J. Robinson at the University of Washington, J. P. Riley at Liverpool, and others. More recently, J. D. H. Strickland, whose Ph.D. research on compounds of silicon led to improvements in the estimation of silicate, worked as a chemist at the Woolwich Arsenal, then with the United Kingdom Atomic Energy Authority, before moving to Canada, where, at the Canadian government's Pacific Biological Station in Nanaimo, he and T. R. Parsons (trained as an agricultural chemist and biochemist) refined recent nutrient analyses and promulgated them in two influential handbooks.

The chemists brought their 'new' knowledge to oceanography from the chemical litera-ture, from texts and by reading abstracts (for example *Chemical Abstracts* and the summaries of foreign papers in the *Journal of the Chemical Society of London*), as well as in their training. Personal communication seems to have played a small role in the introduction of new techniques. Raben, for example, probably never left Germany, and Harvey travelled little, only once working with an ICES group on problems of nutrient analysis in Oslo in 1928 (Fig. 13.4). Harvey, in particular, seldom cited the German literature, although he clearly read the chemical journals that contained abstracts of foreign work in English. Price's (1965: 562–5) suggestion that technology and science affect each other after a generation's lag, representing the time it takes for new knowledge to be incorporated into textbooks and taught, applies very well to the transfer of knowledge and techniques from applied chemistry to oceanography, personified in the chemists who influenced biological oceanography during the early twentieth century.

It is possible to answer my original questions about the origin and development of nutri-ent analyses in biological oceanography in an abstract way, by developing a model in which technological innovations from industrial chemistry, optics and electronics entered ocean-ography after delays caused by developments in oceanography itself and by the time taken for chemists, carrying new but not frontier knowledge, to be recruited into the field. This would be correct, but for my purposes, too bloodless a way of describing the process. The most satisfying way of ending this survey of the history of nutrient analyses is to acknow-ledge that it was the rate at which chemists entered and formed new research groups devoted to the study of the ocean, not the abstraction 'new technical knowledge', that governed the halting changes in the study of plankton nutrients between 1901 and the 1950s.

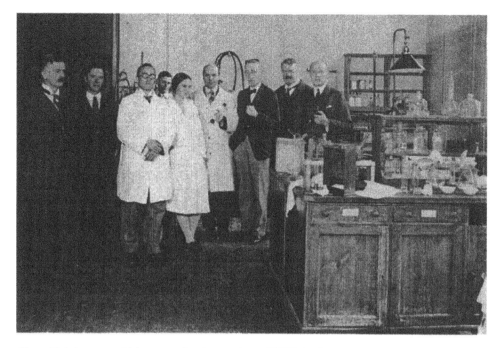

Figure 13.4 A group which met under the auspices of ICES in October 1928 in Oslo to standardize methods of nutrient analysis. From left to right: H. H. Gran, H. W. Harvey, Ernst Schreiber, Alf Klem, Birgitte Ruud Føyn, Kurt Buch, Hermann Wattenberg, Johan Hjort and Torbjørn Gaarder. (Photograph courtesy the late Professor Trygve Braarud; published with permission of the Department of Biology, Section of Marine Botany, University of Oslo.)

Acknowledgements

I am especially grateful to the late Leslie Cooper and to E. I. Butler, both of the Plymouth Laboratory, for the information they gave me in interviews and for their comments on an early version of this paper. The research and writing were supported by grants from the Social Sciences and Humanities Research Council of Canada.

Notes

1 The history of these developments has not been documented completely. An outline can be compiled from the accounts by Aulie (1970), Rossiter (1975) and Waksman (1952).
2 Føyn (1951) was apparently the last to recommend use of the Nessler reagent, although after precipitation not distillation.
3 See, for example, Strickland and Parsons (1972: 77–80). The modern technique uses a modification introduced by Shinn (1941) which improves the speed, stability and clarity of the test.
4 Ringer and Klingen (1909) concluded that Raben's distillation method for NO_2 and NO_3 did not reduce the ions fully to ammonia and that it gave only approximate results.
5 It was mentioned in standard works later – e.g. by Wattenberg (1937) and by Harvey himself (1945) – but seldom used.
6 L. H. N. Cooper, in an interview in 1983, told me that either would have been characteristic of the relation between Atkins and Harvey.
7 Based largely on the fact that diphenylamine was converted to diphenylbenzidine in the course of

the reaction, according to Atkins (1932: 170. This was the basis of the Letts and Rea suggestion that the test would be more reliable using diphenylbenzidine.

8　Raben (1905a: 101; 1905b; and 1910: 317–18) gives a lengthy list of early analysts and their results.

9　Atkins (1923b: 157–8; 1926); Atkins and Wilson (1926). Robinson and Thompson (1948c) describe the reaction.

10　Matthews (1916) had heard of Raben's work but had not seen it in print – in fact their papers were published nearly simultaneously.

References

Armstrong, F. A. J. (1951) 'The determination of silicate in sea water', *Journal of the Marine Biological Association of the United Kingdom* **30**: 149–60.

Atkins, W. R. G. (1923a) 'The phosphate content of fresh and salt waters in its relationship to the growth of the algal plankton', *Journal of the Marine Biological Association of the United Kingdom* **13**: 119–50.

Atkins, W. R. G. (1923b) 'The silica content of some natural waters and of culture media ', *Journal of the Marine Biological Association of the United Kingdom* **13**: 151–9.

Atkins, W. R. G. (1926) 'Seasonal changes in the silica content of natural waters in relation to the phytoplankton', *Journal of the Marine Biological Association of the United Kingdom* **14**: 89–99.

Atkins, W. R. G. (1932) 'Nitrate in sea water and its estimation by means of diphenylbenzidine', *Journal of the Marine Biological Association of the United Kingdom* **18**: 167–92.

Atkins, W. R. G. and Wilson, E. G. (1926) 'The colorimetric estimation of minute amounts of compounds of silicon, of phosphorus, and of arsenic', *Biochemical Journal* **20**: 1223–8.

Atkins, W. R. G. and Wilson, E. G. (1927) 'The phosphorus and arsenic compounds of sea water', *Journal of the Marine Biological Association of the United Kingdom* **14**: 609–14.

Aulie, R. P. (1970) 'Boussingault and the nitrogen cycle', *Proceedings of the American Philosophical Society* **114**: 435–79.

Barnes, B. (1982) 'The science–technology relationship: a model and a query', *Social Studies of Science* **12**: 166–72.

Barnes, B. and Edge, D. (eds) (1982) *Science in context. Readings in the sociology of science*. Cambridge, Mass.: MIT Press.

Barnes, H. (1955) 'The analysis of sea water. A review', *The Analyst* **80**: 573–92.

Beer, J. J. (1959) *The emergence of the German dye industry*. Illinois Studies in the Social Sciences **44**. Urbana, Ill.: University of Illinois Press.

Bijker, W. E., Hughes, T. P. and Pinch, T. J. (eds). (1987) *The social construction of technological systems. New directions in the sociology and history of technology*. Cambridge, Mass.: MIT Press.

Boussingault, J. B. (1856) 'Recherches sur les variations que l'eau de la Mer Morte semble subir dans sa composition', *Académie des Sciences de Paris, Comptes Rendues* **42**: 1230–8.

Brandt, K. (1898) 'Beiträge zur Kenntnis der chemischen Zusammensetzung des Planktons', *Wissenschaftliche Meeresuntersuchungen, Abteilung Kiel*, N.S. **3**: 43–90.

Brandt, K. (1901) 'Life in the ocean', in *Smithsonian Institution, Annual Report for 1900*. Washington, D.C.: Smithsonian Institution.

Brandt, K. (1902) 'Ueber den Stoffwechsel im Meere. 2 Abhandlung', *Wissenschaftliche Meeresuntersuchungen, Abteilung Kiel*, N.S. **6**: 23–79.

Brandt, K. (1905) 'On the production and the conditions of production in the sea', Appendix D, *Conseil international permanent pour l'Exploration de la Mer, Rapports et Procès-Verbaux* **3**: 1–12.

Brandt, K. (1927) 'Stickstoffverbindungen im Meere I', *Wissenschaftliche Meeresuntersuchungen, Abteilung Kiel*, N.S. **20**: 201–92.

Buch, K. (1915) 'Bestimmungen des Ammoniakgehaltes im Wasser der Finland umgebenden Meere', Helsingfors, *Öfversigt af Finska Vetenskaps-Societetens Förhandlingar* **57**, Afdeling A., No. 21.

Buch, K. (1920) 'Ammoniakstudien an Meer- und Hafenwasserproben', Helsingfors, *Merentutkimuslaitoksen Julkaisu, Havforskningsinstitutets Skrift* No. 2.

Buch, K. (1923) 'Methodisches über die Bestimmung von Stickstoffverbindungen im Wasser', Helsingfors, *Merentutkimuslaitoksen Julkaisu, Havsforskningsinstitutets Skrift* No. 18.

Buch, K. (1929) 'Über die Bestimmungen von Stickstoffverbindungen und Phosphaten im Meerwasser', *Conseil permanent international pour l'Exploration de la Mer, Rapports et Procès-Verbaux* **53**: 36–52.

Cooper, L. H. N. (1932) 'The determination of nitrate in the sea by means of reduced strychnine', *Journal of the Marine Biological Association of the United Kingdom* **18**: 161–6.

Cooper, L. H. N. (1938) 'Salt error in determinations of phosphate in sea water', *Journal of the Marine Biological Association of the United Kingdom* **23**: 171–8.

Cooper, L. H. N. (1972) 'Hildebrand Wolfe Harvey, 1887–1970', *Biographical Memoirs of Fellows of the Royal Society* **18**: 331–47.

Denigès, G. (1911) 'Recherche rapide des nitrates et des nitrites dans les eaux a l'aide d'un nouveau réactif hydro-strychnique', *Bulletin de la Société Chimique de France*, 4 Ser., **9**: 544–6.

Denigès, G. (1920) 'Réaction de coloration extrèmement sensible des phosphates et des arséniates', *Académie des Sciences de Paris, Comptes Rendues* **171**: 802–4.

Denigès, G. (1921) 'Détermination quantitative des plus faibles quantités de phosphates dans les produits biologiques par la méthode céruléomolybdique', *Société de Biologie de Paris, Comptes Rendues* **84**(17): 875–7.

Diénert, F. and Wandenbulcke, F. (1923) 'Sur le dosage de la silice dans les eaux', *Académie des Sciences de Paris, Comptes Rendues* **176**: 1478–80.

Forchhammer, J. G. (1859) *Om Sövandets bestandele og deres fordeling i havet.* Copenhagen: J. H. Schultz.

Føyn, E. (1951) 'Nitrogen determinations in sea water', Bergen, *Fiskeridirektoratets Skrifter. Serie Havundersokelser* **9**(14).

Geelmuyden, H. C. (1902) 'Om kvantitativ bestemmelse af sövandets kvaelstoffholdige bestanddele', Christiania, *Videnskabs-selskabet Skrifter, Mathematisk-naturvidenskabelig Klasse* **1**(6).

Geelmuyden, H. C. (1903), 'Über die quantitative Bestimmung der stickstoffhaltigen Bestandteile des Meereswassers nebst Bemerkungen colorimetrische Methoden', *Zeitschrift für analytischen Chemie* **42**: 276–92.

Graham, M. (1936) 'The "Lowestoft photometer" ', *Conseil permanent international pour l'Exploration de la Mer, Journal du Conseil* **11**: 205–10.

Griess, J. P. (1879), 'Bemerkungen zu den Abhandlung der H. H. Weselsky und Benedikt: "Ueber einige Azoverbindungen" ', *Berichte der deutschen chemischen Gesellschaft* **12**: 426–8.

Hamlin, C. (1990) *A science of impurity: water analysis in nineteenth century Britain.* Bristol: Adam Hilger.

Harvey, H. W. (1926) 'Nitrate in the sea', *Journal of the Marine Biological Association of the United Kingdom* **14**: 71–88.

Harvey, H. W. (1928) *Biological chemistry and physics of sea water.* Cambridge: Cambridge University Press.

Harvey, H. W. (1929) 'Methods of estimating phosphates and nitrates in seawater', *Conseil permanent international pour l'Exploration de la Mer, Rapports et Procès-Verbaux* **53**: 68–74.

Harvey, H. W. (1945) *Recent advances in the chemistry and biology of sea water.* Cambridge: Cambridge University Press.

Harvey, H. W. (1948) 'The estimation of phosphate and of total phosphorus in sea waters', *Journal of the Marine Biological Association of the United Kingdom* **27**: 337–59.

Harvey, H. W. (1955) *The chemistry and fertility of sea waters.* Cambridge: Cambridge University Press.

Ihde, A. J. (1964) *The development of modern chemistry.* New York: Harper & Row.

Ilosvay, L. (1889) 'Sur les réactions des acides azoteux et azotique', *Bulletin de la Société Chimique de France*, 3 Ser., **2**: 347–51.

Isaacs, M. L. (1924) 'Presence of silicon in tissues. A micro-method for the determination of silicon', *Bulletin de la Société de Chimie Biologique* **6**: 157–68.

Johnstone, J. (1908) *Conditions of life in the sea. A short account of quantitative marine biological research.* Cambridge: Cambridge University Press.

Jolles, A. F. and Neurath, F. (1898) 'Eine colorimetrische Methode zur Bestimmung der Kieselsäure im Wasser', *Zeitschrift für angewandte Chemie* **11**: 315–16.

Kalle, K. (1931–5) 'Meereskundliche chemische Untersuchungen mit Hilfe des Zeiss'schen

Pulfrich-Photometers', *Annalen der Hydrographie* 1931: 313–17; 1933: 124–8; 1934: 65–74; 1935: 58–65, 195–204.

Kruse, J. and Mellon, M.G. (1952) 'Colorimetric determination of free ammonia with a pyridine-pyrazolone reagent', *Sewage and Industrial Waste* **24**: 1098–1100.

Krüss, G. and Krüss, H. (1891) *Kolorimetrie und quantitative Spektralanalyse in ihrer Anwendung in der Chemie.* Hamburg and Leipzig.

Laudan, R. (ed.) (1984) *The nature of technological knowledge: are models of scientific change relevant?* Dordrecht and Boston: D. Reidel.

Lawes, J. B., Gilbert, J. H. and Pugh, E. (1863). 'On the sources of the nitrogen in vegetation', *Journal of the Chemical Society* **216**: 100–86.

Letts, E. A. and Rea, F. W. (1914) 'An extremely delicate colorimetric method for detecting and estimating nitrates and nitrites', *Journal of the Chemical Society* **105**: 1157–61.

Liebig, J. von (1840) *Organic chemistry in its applications to agriculture and physiology.* London: Taylor and Walton.

Martens, F. F. and Grünbaum, F. (1903) 'Über eine neue Konstruktion des Königischen Spektralphotometers', *Annalen der Physik*, Ser. 4, **12**: 984–1003.

Matthews, D. J. (1916) 'On the amount of phosphoric acid in the sea water off Plymouth Sound', *Journal of the Marine Biological Association of the United Kingdom* **11**: 121–30, 251–7.

Meister, O. (1872) 'Correspondenzen 78. O. Meister, aus Zürich, am 23 März', *Berichte der deutschen chemischen Gesellschaft* **5**: 283–7.

Mills, E. L. (1989) *Biological oceanography: an early history, 1870–1960.* Ithaca, N.Y.: Cornell University Press.

Mills, E. L. (1990) 'The ocean regarded as a pasture: Kiel, Plymouth and the explanation of the marine plankton cycle, 1887 to 1935', in W. Lenz and M. Deacon (eds) *Ocean sciences: their history and relation to man. Deutsche Hydrographische Zeitschrift, Ergänzungsheft* **B22**: 20–9.

Morris, A. W. and Riley, J. P. (1963) 'The determination of nitrate in seawater', *Analytica Chimica Acta* **29**: 272–9.

Mullin, J. B. and Riley, J. P. (1955a) 'The colorimetric determination of silicate, with special reference to sea and natural waters', *Analytica Chimica Acta* **12**: 162–76.

Mullin, J. B. and Riley, J. P. (1955b) 'The spectrophotometric determination of nitrate in natural waters, with particular reference to sea-water', *Analytica Chimica Acta* **12**: 464–80.

Murphy, J. and Riley, J. P. (1958) 'A single-solution method for the determination of soluble phosphate in sea water', *Journal of the Marine Biological Association of the United Kingdom* **37**: 9–14.

Murphy, J. and Riley, J. P. (1962) 'A modified single solution method for the determination of phosphate in natural waters', *Analytica Chimica Acta* **27**: 31–6.

Murray, J. and Irvine, R. (1889) 'On coral reefs and other carbonate of lime formations in modern seas', *Proceedings of the Royal Society of Edinburgh* **17**: 79–109.

Murray, J. and Irvine, R. (1891) 'On silica and the siliceous remains of organisms in modern seas', *Proceedings of the Royal Society of Edinburgh* **18**: 229–50.

Natterer, K. (1892) 'Chemische Untersuchungen im östlichen Mittelmeer. Berichte der Commission für Erforschung der östlichen Mittelmeeres. I. Reise S. M. Schiffes *Pola* im Jahre 1890', Vienna, *Denkschriften der Österreichischen Akademie der Wissenschaften* **59**: 83–104; 'II. Reise S. M. Schiffes *Pola* im Jahre 1891', Vienna, *Denkschriften der Österreichischen Akademie der Wissenschaften* **59**: 105–118.

Natterer, K. (1893) 'Chemische Untersuchungen im östlichen Mittelmeer. III. Reise S. M. Schiffes *Pola* im Jahre 1892', Vienna, *Denkschriften der Österreichischen Akademie der Wissenschaften* **60**: 49–82.

Natterer, K. (1894) 'Chemische Untersuchungen im östlichen Mittelmeer. IV. Reise S. M. Schiffes *Pola* im Jahre 1893', Vienna, *Denkschriften der Österreichischen Akademie der Wissenschaften* **61**: 23–63.

Natterer, K. (1895) 'Tiefsee-Forschungen im Marmara-Meer auf S. M. Schiff *Taurus* im Mai 1894', Vienna, *Denkschriften der Österreichischen Akademie der Wissenschaften* **62**: 19–117.

Natterer, K. (1898) 'Berichte der Commission für oceanographische Forschungen. Expedition S. M. Schiff *Pola* in das Rothe Meer (nördliche Hälfte) 1895–96. IX Chemische Untersuchungen', Vienna, *Denkschriften der Österreichischen Akademie der Wissenschaften* **65**: 445–572.

Nessler, J. (1868) 'Ueber Bestimmung des Ammoniaks und der Salpetersäure in sehr verdünnten Lösungen', *Zeitschift für analytischen Chemie* **7**: 415–16.

Orr, A. P. (1926) 'The nitrite content of sea water', *Journal of the Marine Biological Association of the United Kingdom* **14**: 55–61.

Poole, H. H. (1960) 'William Ringrose Gelston Atkins, 1884–1959', *Biographical Memoirs of Fellows of the Royal Society* **5**: 1–22.

Pouget, I. and Chouchak, D. (1909, 1911) 'Dosage colorimétrique de l'acide phosphorique', *Bulletin de la Société Chimique de France*, 4 Ser., **5** (1909): 104–9; **9** (1911): 649–50.

Price, D. J. de S. (1965) 'Is technology historically independent of science? A study in statistical historiography', *Technology and Culture* **6**: 553–68.

Price, D. J. de S. (1969) 'The parallel structures of science and technology', in W. H. Gruber and D. G. Marquis (eds) *Factors in the transfer of technology*. Cambridge, Mass.: MIT Press, pp. 91–104.

Raben, E. (1905a) 'Über quantitative Bestimmung von Stickstoffverbindungen im Meerwasser, nebst einem Anhang über die quantitative Bestimmung der im Meerwasser gelösten Kieselsäure', *Wissenschaftliche Meeresuntersuchungen, Abteilung Kiel*, N.S. **8**: 81–101.

Raben, E. (1905b) 'Weitere Mitteilungen über quantitative Bestimmungen von Stickstoffverbindungen und von gelöster Kieselsäure im Meerwasser', *Wissenschaftliche Meeresuntersuchungen, Abteilung Kiel*, N.S. **8**: 277–87.

Raben, E. (1910) 'Dritte Mitteilung über quantitative Bestimmungen von Stickstoffverbindungen und von gelöster Kieselsäure im Meerwasser', *Wissenschaftliche Meeresuntersuchungen, Abteilung Kiel*, N.S. **11**: 303–319.

Raben, E. (1914) 'Vierte Mitteilung über quantitative Bestimmungen von Stickstoffverbindungen im Meerwasser und Boden, sowie von gelöster Kieselsäure im Meerwasser', *Wissenschaftliche Meeresuntersuchungen, Abteilung Kiel*, N.S. **16**: 207–29.

Raben, E. (1916) 'Quantitative Bestimmung der im Meerwasser gelösten Phosphorsäure', *Wissenschaftliche Meeresuntersunchungen, Abteilung Kiel*, N.S. **18**: 1–24.

Riley, J. P. (1953) 'The spectrophotometric determination of ammonia in natural waters with particular reference to sea-water', *Analytica Chimica Acta* **9**: 575–89.

Ringer, W. E. and Klingen, F. J. M. P. (1907) 'Ueber die Bestimmung von Stickstoffverbindungen im Meereswasser', *Verhandelingen uit het Rijksinstuut voor het Onderzoek der Zee* **2**: 1–27.

Robinson, R. J. and Spoor, H. J. (1936) 'Photometric determination of silicate in sea water', *Industrial and Engineering Chemistry, Analytical Edition* **8**: 455–7.

Robinson, R. J. and Thompson, T. G. (1948a) 'The determination of nitrites in sea water', *Journal of Marine Research* **7**: 42–8.

Robinson, R. J. and Thompson, T. G. (1948b) 'The determination of phosphates in sea water', *Journal of Marine Research* **7**: 33–41.

Robinson, R. J. and Thompson, T. G. (1948c) 'The determination of silicate in sea water', *Journal of Marine Research* **7**: 49–55.

Rossiter, M. W. (1975) *The emergence of agricultural science. Justus Liebig and the Americans, 1840–1870*. New Haven: Yale University Press.

Schloesing, J. J. T. (1875a) 'Sur l'ammoniaque de l'atmosphère', *Académie des Sciences de Paris, Comptes Rendues* **80**: 175–8.

Schloesing, J. J. T. (1875b) 'Sur les échanges d'ammoniaque entre les eaux naturelles et l'atmosphère', *Académie des Sciences de Paris, Comptes Rendues* **81**: 1252–4.

Schloesing, J. J. T. (1875c) 'Sur les lois des échanges d'ammoniaque entre les mers, l'atmosphère et les continents', *Académie des Sciences de Paris, Comptes Rendues* **81**: 81–4.

Schmidt, C. (1871, 1875, 1878) 'Hydrologische Untersuchungen von Prof. Dr. Carl Schmidt, Dorpat', *Bulletin de l'Académie impériale des sciences de St. Petersbourg* **16**: 117–203; **20**: 130–69; **24**: 177–258, 419–36.

Shinn, M. B. (1941) 'Colorimetric method for determination of nitrite', *Industrial and Engineering Chemistry Analytical Edition* **13**: 33–5.

Smith, L. (1917) 'Use of diphenylamine and diphenylbenzidine for colorimetric estimations', *Journal of the Chemical Society* **112**(2): 217.

Strickland, J. D. H. (1952) 'The preparation and properties of silicomolybdic acid. I-III', *Journal of the American Chemical Society* **74**: 862–7, 868–71, 872–6.

Strickland, J. D. H. and Parsons, T. R. (1960) *A manual of sea water analysis*. Ottawa: Fisheries Research Board of Canada, Bulletin Number 125.

Strickland, J. D. H. and Parsons, T. R. (1972) *A practical handbook of seawater analysis*, 2nd edn. Ottawa: Fisheries Research Board of Canada, Bulletin Number 167.

Tafel, J. and Naumann, K. (1901) 'Die elektrolytische Reduction des Strychnins und Brucins', *Berichte der deutschen chemischen Gesellschaft* **34**: 3291–9.

Tillmans, J. (1911) 'Detection and estimation of nitric acid in milk by means of the diphenylamine-sulphuric acid test', *Journal of the Society of Chemical Industry* **30**: 44.

Tillmans, J. and Sutthoff, W. (1911) 'Ein einfaches Verfahren zum Nachweis und zur Bestimmung von Salpetersäure und salpetriger Säure im Wasser', *Zeitschrift für analytische Chemie* **50**: 473–95.

Travis, A. S. (1993) *The rainbow makers. The origins of the synthetic dyestuffs industry in Western Europe*. Bethlehem, Pennsylvania: Lehigh University Press.

Treadwell, F. P. (1905) *Kurzes Lehrbuch der analytischen Chemie*. Leipzig and Vienna: F. Deuticke.

Tschigirine, N. and Daniltchenko, P. (1930) 'De l'azote et ses composés dans la Mer Noire', *Travaux de la Station Biologique de Sebastopol* **2**: 1–16.

Waksman, S. A. (1952) *Soil microbiology*. New York: John Wiley.

Walther, G. and Gärtner, A. (1895) *Tiemann–Gärtner's Handbuch der Untersuchung und Beurtheilung der Wässer*, 4th edn. Braunschweig: Friedrich Wieweg.

Wattenberg, H. (1937) 'Critical review of the methods used for determining nutrient salts and related constituents of sea water. 1. Methoden zur Bestimmung von Phosphat, Silikat, Nitrit, Nitrat und Ammoniak im Seewasser', *Conseil permanent international pour Exploration de la Mer, Rapports et Procès-Verbaux* **103**: 1–26.

Witting, R. (1914) 'Zur Methodik der Bestimmung von geringen Ammoniakmengen mit besonderer Berücksichtigung der Meerwasseranalysen', Helsingfors, *Öfversigt af Finska Vetenskaps-Societetens Förhandlingar* **56**: Afdeling A, No. 15.

Wirth, H. E. and Robinson, R. J. (1933) 'Photometric investigation of Nessler Reaction and Witting method for determination of ammonia in sea water', *Industrial and Engineering Chemistry Analytical Edition* **5**: 293–6.

14 Why is the sea salty?
What controls the composition of ocean water?

T. Roger S. Wilson and J. Dennis Burton

Introduction

The most obvious and far-reaching question of marine chemistry is the one that many children have posed to their puzzled elders 'Why is the sea salty?' Once the reply advanced from 'Because it is . . .' to 'Because water evaporates and falls again on the land as rain, while the salt stays behind in the sea' (the first point is to be found in the book of Ecclesiastes), progress in understanding came to a halt. Even after experimental oceanography became established as a science late in the nineteenth century, it took about one hundred years of research – during which time the question had been refined to 'Why is the composition of sea water as it is?' – to answer it in any detail.

Some aspects of the question can be expressed in ways that make clear its central relevance for many currently important environmental issues: 'What is the impact of human activities on marine ecosystems?'; 'How are the ocean and atmosphere linked in terms of composition, and what are the global climatic implications of their interaction?' In this chapter some of the main thrusts in the development of the study of the chemistry of the ocean are outlined.

Early priorities: the concept of salinity

The voyage of HMS *Challenger* (1872–6) represented an important stage in the development of oceanography as an organized science. Not least among the impressive achievements of this expedition was its provision of the experimental evidence that underpins the concept of salinity. At first sight, this seems rather an odd concept; everyone knows that the sea is salty, but most people also know that this 'salt' is a complicated mixture of dissolved constituents. How can it be that this complexity can usefully be rendered down into a single number, the 'salinity'?

Systematic collection of water samples by John Young Buchanan, the chemist on HMS *Challenger,* and the subsequent quantitative analysis of the six most abundant salt constituents in seventy-seven of these samples by William Dittmar at Glasgow University, showed an amazing consistency of composition, confirming the earlier findings of Marcet and of Forchhammer (Deacon 1997). Although the total amount of dissolved salt varied from place to place, Dittmar's more accurate analyses revealed hardly any systematic variability in the relative proportions of the dissolved elements. Though the total salt content can vary from place to place, within certain limits the major elements in sea water are always present in the same relative proportions.

Geological evidence (from evaporite deposits and other sediments formed from sea water constituents) indicates that ocean salt composition has not varied hugely in the last 10^8 years. This salt composition is very different from the typical composition of river water, and so there must be control processes that alter the composition of river water into that of sea water. However, the immediate and most useful consequence of Dittmar's work was the insight that salt could be regarded as a kind of index or tracer of mixing within the ocean interior, and the variation in the total quantity of salt could be used to study both mixing and transport processes. Because of the importance of understanding these basic physical processes, all major efforts were bent in this direction. The processes that caused the composition of sea water to remain so fascinatingly constant were largely left aside by oceanographers, while the problem of finding exactly how much salt was in a given sample of sea water became paramount. Thus, for two generations, marine chemistry became a hand-maiden to physics, and, as knowledge was acquired about the chemical factors involved in the productivity of the ocean, to marine biology. (See Chapter 13.)

Oceanographic expeditions are expensive, and a particular part of the ocean may be visited very infrequently. It is vitally important that the salinity measurements made at widely different times from different vessels belonging to different nations should be consistent. This produced an urgent need for a method of international intercalibration that would provide the whole oceanographic community with an agreed definition of salinity and a precise standardized methodology for measurement.

Measurement of the absolute concentration of the dissolved material in a sample of sea water poses intractable technical problems. In practice, what is needed is a readily measurable quantity, closely related to the total salt content, which can serve as a tracer in the way described above, and can be used to compute density values for physical oceanographic calculation. For what can now be seen as historical reasons, the quantities used have been converted to salinity values, which approximate to the total salt concentrations.

During the exploratory phase of the International Council for the Exploration of the Sea in 1899–1901, the Danish scientist Martin Knudsen worked under the auspices of an international committee to establish appropriate methods and definitions in this area. Salinity was defined in terms of an elaborate gravimetric method of measurement and related to a much more readily measurable property, the concentration of halide (in parts per thousand). Because of time constraints the committee used only nine samples, and because the Baltic was over-represented in that sample set they obtained a relationship between salinity, S and chlorinity, Cl:

$$S = 1.805 \, Cl + 0.03$$

differing slightly from that derived by Dittmar. Nevertheless, this way of arriving at salinity values was in use for more than fifty years, a fact which gives the method a considerable historical significance in the development of ocean studies.

As often happens, the stimulus for eventual change was provided by advancing technology. In the 1950s, a number of salinometers based on the developing technique of conductivity measurement were devised. The early instruments needed large thermostat baths but portable inductive salinometers were soon developed. By steady working, any reasonably careful person could measure salinity to within 0.003 parts per thousand (ppt or ‰) on about 150 samples per day. This was a considerable advance on the fifty samples a day that a skilled analyst (if not seasick) could measure to 0.01 ppt with the old chlorinity titration method.

The quantity and quality of the data that flowed from these salinometers, subsequently to be enormously increased as the conductivity technology was adapted to permit *in situ* measurement, raised the necessity to review the work of Dittmar and to revisit the questions addressed by Knudsen and his co-workers. In particular, the concept of constancy of composition needed to be evaluated at higher precision levels. This work was initiated at the National Institute of Oceanography in the UK by Roland Cox, using samples collected all over the world under the auspices of a UNESCO-funded International Joint Panel. John Riley of the University of Liverpool and several of his students were also involved. In a series of papers published in the mid-1960s, the major constituent composition of sea water was more closely defined.

Following on from that work, Cox and his co-workers determined the conductivity to chlorinity relationship for these samples, and also examined their density and refractive index properties. This classic work, which defined the relationship between salinity and a conductivity ratio, referred to IAPSO Standard Sea Water, and established reliable means to relate these measured quantities to the basic parameters required in the study of water movement, has underpinned and validated all oceanographic measurements made in the latter decades of the twentieth century.

The use of a standard sea water, defined in terms of its source and preparation but not formally defined in terms of its composition, is very convenient in practical terms (Culkin and Smed 1979). However, it suffered from the conceptual weakness that it was not possible to relate the measurements made using this procedure to a compositionally defined standard. In the late 1970s, an international group working under the auspices of UNESCO developed the Practical Salinity Scale (PSS78, currently in use), which provided a practical definition of salinity in terms of electrical conductivity relative to the conductivity of a closely specified and reproducible potassium chloride solution. To retain continuity with the older standard, and to take continued advantage of its practical suitability, the conductivity ratio (K15) of each batch of Standard Sea Water is compared with that of the defined potassium chloride solution at 15°C, to give a certified value of K15. The fact that the quantity 'salinity' is now formally referred to a conductivity ratio rather than to the dissolved solid content of the sample is somewhat obscured by the great care that has been taken to retain numerical comparability with historical data as the underlying approach has altered over time. The only difference of which many workers are actually aware is that salinity values based on the 1978 Practical Salinity Scale are dimensionless ratios, whereas those based on the older scales were formally dimensioned in 'parts per thousand' in deference to the historical link to a measure of total dissolved solids, a link that was always more conceptual than practical.

For calibration of laboratory salinometers, it is only necessary to fill the cell with Standard Sea Water and either adjust the instrument to read the certified value of K15 or note the correction to be applied to subsequent readings. Tables were originally used to determine salinity from the measured conductivity ratio over a wide range of salinities, temperatures and pressures. More recently, as the algorithms relating the measured quantities to those required have been closely defined, the system was adapted to cope with the vast increase in data quantity that followed from the development of reliable *in situ* conductivity measuring instruments.

Early priorities: micronutrient elements

As with the major composition, some basic aspects of micronutrient cycles and their relation to marine primary productivity began to be appreciated in the nineteenth century, but the major developments began early in the twentieth century and were greatly stimulated by the introduction of rapid colorimetric methods for the determination of phosphate, nitrate and dissolved silicon (see Chapter 13). A considerable body of information was collected using these methods. The behaviour of the important marine micronutrients, nitrate, phosphate and silicate, was investigated and the understanding of their links to primary productivity was refined. In contrast to the major constituents, the micronutrients in sea water show large variations in space and time. The range is typically from less than 1 to more than 600 micrograms per litre for nitrate, from less than 0.1 to over 100 micrograms per litre for phosphate, and from less than 20 to about 5000 micrograms per litre for dissolved silicon. Plant growth depletes the nutrient concentration in the upper (euphotic) water layer. Where the thermocline persists throughout the year, as in much of the open ocean, plant biomass production is limited by the quantity of nutrients supplied to this layer, primarily by diffusion from below, the deeper waters being enriched by the regeneration of the nutrients into dissolved inorganic forms by the microbial breakdown of sinking organic material. Where seasonal mixing takes place, as on the continental shelf in temperate latitudes, the surface layer is replenished by this winter overturn and a seasonal bloom appears with the increasing light levels of spring.

The proportions to carbon in which the micronutrients nitrate and phosphate are utilized (the Redfield ratios) vary generally over quite a restricted molar ratio range close to 106C : 16N : P. The process of carbon fixation to form organic material may therefore be represented chemically by the relation (Redfield, Ketchum and Richards 1963):

$$106CO_2 + 122H_2O + 16HNO_3 + H_3PO_4 = (CH_2O)_{106}(NH_3)_{16}H_3PO_4 + 138O_2$$

Thus, the progress of any particular bloom may be affected by the exhaustion of one particular micronutrient while the others are still present. A succession of various phytoplankton species may occur, since the requirements of different species can differ in other ways. For instance, small-celled organisms are favoured at low nutrient concentrations, and the presence of dissolved silicon in relative abundance favours diatom production. As microbiological breakdown occurs subsequently the nutrients are released back into the water column, but if the cells have in the meantime sunk out of the euphotic zone it may be a considerable period before the micronutrients are again utilized. The system is at steady state: if the nutrient supply to depleted waters were to increase then global production would rise, causing increased transfer of organic material to the deep ocean and its sediments. Thus, the system would balance itself at a new higher level of production.

Subsequent research has led to an appreciation of many more complex factors in the relationship between micronutrient concentrations and primary productivity. The marine nitrogen cycle in particular involves many chemical forms including ammonium, urea, dinitrogen (N_2) and organic nitrogen compounds. Phytoplankton species vary considerably in their ability to utilize these different forms of nitrogen. Developments in the study of trace metals in sea water have led to an appreciation that certain low-concentration elements can also play a significant role by limiting the utilization of primary micronutrients. The organic chemistry of sea water and the interactions between organic and inorganic constituents are other important research areas closely linked to the biota.

Marine geochemistry

The early role of chemistry in the marine sciences was to a large extent determined by major questions arising in physical and biological oceanography, mainly questions concerning density and fertility. Chemists were much involved, but fundamental curiosity about the chemistry of the ocean did not generally dominate as the impetus for their investigations. A further reason for this lies in the relationship between marine chemistry and the primary discipline. It is difficult to point to any truly important impact of marine chemistry on pure chemistry, in contrast with the immense influences of marine researches on the disciplines of biology or geology. Marine chemistry depends critically upon the application of the concepts of fundamental chemistry, but there has been little incentive for the pure chemist to work on the inherently messy material that sea water represents, with its high electrolyte concentrations and its multiphase compositional complexity.

The ideas and approaches of geochemistry, dealing as it does with 'the distribution and migration of the chemical elements within the earth in space and in time' (Mason 1952: 1), provide a coherent, unifying framework for the chemical study of the ocean. From this standpoint the ocean is important as a major reservoir in the water cycle that drives the weathering of the continents and the transport of weathered material, and also as a holding reservoir for the products of weathering. In essence, the ocean acts as a gigantic reactor processing material entering from diverse sources and dispatching constituents at different rates according to their properties; the main immediate destination is the seabed with volatiles passing to the atmosphere. (See Chapter 6.)

This way of looking at the ocean was explicit in the work of some early scientists, notably Forchhammer. In the twentieth century it developed distinctively with the work of V. M. Goldschmidt and others. Mass balance calculations showed that the major anion-forming elements in the ocean, such as chlorine and sulphur, were present in amounts greatly exceeding those compatible with derivation from continental weathering; it was recognized that they derived mainly from magmatic volatiles. It was found also that the majority of elements, for which continental weathering is the major source, had been supplied in amounts far in excess of those remaining in the ocean, and that the removal processes operated more efficiently for some elements than for others; these differences were broadly explicable in terms of the electronegativity of the elements. A widely used index of the differences in geochemical reactivity among the elements is the mean oceanic residence time, the average time elapsing between the entry of an atom of the element and its removal from the ocean. This is calculated by dividing the total ocean content of that element by its flux into (or out of) the ocean.

The introduction of the piston corer, extensively deployed by the Swedish Deep-Sea Expedition, 1947–8, led to a much greater knowledge of the composition of marine sediments. Understanding of the sedimentary record was subsequently advanced by the development of radiochemical dating methods using carbon-14 and nuclides in the uranium, thorium and actino-uranium series. As oceanography expanded in the 1950s and 1960s, particularly in North America, major advances ensued in understanding the ocean as a geochemical system, through the work of scientists including Edward Goldberg, Karl Turekian and Wallace Broecker. An example of the developing perspectives is afforded by the work undertaken around this time on ferromanganese concretions (manganese nodules) from the sea floor. These deposits had been discovered during the *Challenger* Expedition but, following early work on their composition, they had received little further attention. It was now realized that they contained manganese and iron minerals, which accumulate very slowly as a result of *in situ* reactions, and could provide valuable information on the scavenging of trace metals from the ocean (Goldberg 1954). Additional stimulus to investigations of the concretions was provided by their potential as an economic resource.

What was now emerging was an appreciation of the role played in the control of oceanic composition by what Turekian (1977) was to describe as the 'great particle conspiracy' by which particles derived from the continents and transported by rivers and winds, and those produced *in situ* by organisms and by inorganic processes, remove constituents from solution and transfer material to sediments. Many aspects of the composition of sea water can be understood in terms of a dynamic balance between the rate of input and the rate of removal by particles (Broecker 1971, 1974).

The oceanic particle conspiracy operating in the water column cannot, however, explain adequately the removal of all of the more concentrated major constituents. The distinguished inorganic and physical chemist, Lars Gunner Sillén, who was invited to address the First International Oceanographic Congress in New York in 1959, drew attention to the similarity in the ratios of some of the major cations in sea water to those calculated for model equilibrium systems containing primary igneous rock minerals, such as feldspars, and the clay minerals, which are products of their weathering and which are abundant in marine sediments (Sillén 1961).

Independent geochemical arguments were advanced supporting the idea that the aluminosilicates in the ocean were implicated in geochemical reactivity (Garrels 1965, Mackenzie and Garrels 1966). Bicarbonate ions derived from weathering on the continents account for the major part of the ocean's alkalinity. To maintain the oceanic alkalinity against the continuous addition of bicarbonate by rivers, the inputs must be balanced by removal. One readily observable process is the use by some surface-living organisms of calcium and bicarbonate ions to form skeletal calcium carbonate, some of which is eventually deposited and buried in the ocean sediments. This reaction essentially reverses the continental weathering of limestone, removing alkalinity from the ocean and restoring to the atmosphere carbon dioxide used in weathering. But some of the bicarbonate in the ocean is present as a result of weathering of igneous aluminosilicates and this suggested that there must be compensating 'reverse weathering' reactions involving the accompanying cations and clay minerals.

Evidence for a major role for these particular reactions has proved elusive. Marine sediments provide potential sites for the postulated reactions. In the 1960s and 1970s an increased appreciation developed of the significance of post-depositional changes in the chemistry of sediments and the potential for them to influence the overlying water column through diffusion along concentration gradients at the sediment–water interface set up by differences in composition between the sediment pore waters and the adjacent ocean bottom waters. Once the difficult technical problems of obtaining accurate information on the concentrations of dissolved constituents in the sediment pore waters, under the *in situ* conditions of temperature and pressure, were overcome, evidence was obtained that these solutions were depleted in potassium, consistent with the reverse weathering hypothesis. Overall, however, the question of oceanic mass balance was rendered more complicated, since for some elements such as silicon post-depositional reactions gave rise to a global increase in the supply of the dissolved element to ocean waters.

By the late 1970s, the occurrence of additional source–sink reactions at the seabed, involving interaction of sea water with the oceanic crust, had been discovered. These reactions take place under a range of conditions of temperature and proportions of basalt and sea water (Thompson 1983), but the most dramatic compositional changes arise in sea-floor areas with a high geothermal heat flux, mainly at the mid-ocean ridge axes, where sea water can convect through some kilometres' depth in the crust, react with basalt at temperatures up to in excess of 350°C, and return to the ocean at hydrothermal vents. Among the spectacular changes in the composition of the sea water are the essentially total removal of magnesium and sulphate, and increases in the concentrations of dissolved silicon, calcium and potassium; the hydrothermal solutions are acidic and reducing and spectacularly enriched in dissolved manganese and iron. On mixing with cold oxygenated sea water, whether within the basalt or in the hydrothermal plumes resulting from fluid discharge directly into ocean-bottom waters, metalliferous precipitates are produced. Paradoxically, hydrothermal plumes can act as net sinks for some trace metals through the scavenging action of the particles on deep-ocean water entrained into the plume. Quantifying the rates of input and removal resulting from the hydrothermal reactions is difficult because of uncertainty about the volume of sea water undergoing high-temperature reactions in a given time. This can be arrived at only by indirect calculations using various approaches based either on heat fluxes or on geochemical constraints (Elderfield and Schultz 1996). Early calculations were made on the basis that the whole of the convective heat transport in newly formed oceanic crust was due to the circulation of water undergoing high-temperature interaction, that is, water heated to about 350°C. This assumption leads to estimates of chemical fluxes that are maximal; it neglects the role of diffuse flow at much lower temperatures and this may be the dominant component in the hydrothermal circulation.

The picture that has emerged is of a variety of mechanisms that contribute to the maintenance of the composition of sea water between limits that have remained remarkably constant. Dynamic factors, including the rate of formation of new oceanic crust and the flux of biogenous material to the seabed, are crucial, but Sillén's observations still serve to provoke ideas concerning equilibrium factors as long-term determinants of some basic characteristics of sea-water composition.

Trace metals

In the 1970s a coherent picture of the distribution of trace metals in the ocean started to emerge. This was made possible by the development of sensitive techniques, particularly atomic absorption spectrophotometry, and in part by rigorous attention to the prevention of contamination during sampling and analysis. In some cases the application of these ultra-clean techniques showed that previous measurements had been erroneously high by at least an order of magnitude. With these developments it gradually became clear that there are three principal categories of trace metal behaviour (Whitfield and Turner 1987). The trace alkali metals lithium, rubidium and caesium mirror the behaviour of their more abundant counterparts. Such elements have long residence times and a low geochemical reactivity, and they vary in concentration essentially in proportion to salinity. Other elements that behave in this way include thallium, molybdenum and uranium. A second group are to varying degrees removed from surface water and transported to deep water by sinking particles. They thus mirror the behaviour of the micronutrients. They may be biologically utilized because of the metabolic demands of organisms or strongly bound to ligands characteristic of sinking organic material. This group includes cadmium, zinc, inorganic germanium, nickel, inorganic selenium and barium. In general they have intermediate residence times in the ocean. The third group are strongly scavenged from sea water and show wide horizontal and vertical variation. Examples are aluminium, gallium, manganese, lead and perhaps cobalt. These elements have a very low residence time, perhaps less than 1000 years.

The measurement of the low deep-water concentrations characteristic of elements that are rapidly scavenged from sea water can be particularly difficult. The work of Schaule and Patterson (1981) on lead is a classic of its kind, overcoming considerable contamination problems to determine that this element varies in concentration from 5pmol/kg in the deep

Pacific to as much as 75pmol/kg in surface waters subject to atmospheric input of this element with a major anthropogenic component. Higher values (20 to 130pmol/kg) have been reported for the Atlantic.

Gases

The dissolved gas content of ocean water is fundamentally related to atmospheric composition, and thus to climate. Although, as indicated above, free oxygen is consumed in the regeneration of organic material and also by the input of reduced material in hydrothermal plumes, most of the ocean is oxygenated. This was not so in the remote past, and the presence of oxygen is a consequence of photochemical reactions including the photosynthetic activity at the surface of our planet. In addition to the photosynthesis of land plants, oxygen is produced in the surface ocean. Oxygen and other gases undergo exchange across the ocean surface and are carried to abyssal depths by the circulation of ocean waters. Surface waters acquire concentrations of oxygen, nitrogen and argon (exemplifying the inert gases) that are close to the saturation levels for the particular conditions of temperature and salinity. In subsurface mixing processes without *in situ* reaction the gas content is conserved, but the microbiological breakdown of organic material *in situ* consumes oxygen, and to a much smaller extent denitrification adds nitrogen in certain localized areas and in the sediments. There is recent evidence that in the deep sediments nitrogen is produced by the microbial oxidation of ammonia and diffuses into bottom waters, but again this is a quantitatively less important process.

Concentrations of oxygen in the euphotic zone may considerably exceed saturation levels as a result of photosynthetic activity. Below the euphotic zone, however, the effects of respiratory processes become apparent. They lead to the formation of a characteristic oxygen minimum layer in the open ocean in tropical and temperate latitudes where there is a permanent density stratification (the permanent pycnocline) within the top 1km of the water column. Permanent total depletion of dissolved oxygen occurs in bottom waters isolated from exchange with surface waters over long periods, as in the Black Sea. It may also arise even in relatively well-ventilated waters if the oxygen demand is abnormally high. In some coastal basins and fjords, seasonal changes in the water circulation may lead to alternation between oxic and anoxic regimes. Variations in the location and extent of anoxic water bodies in the Baltic Sea provide striking examples of the complex interaction between water circulation, nutrient cycles and organic decomposition processes, further complicated by waste discharges.

The ocean and atmosphere are linked in many ways, and our increasing need to understand the factors influencing climate has given these links a new significance. The importance of the atmosphere in weather prediction is easy to grasp, but the longer time-scale implicit in climate studies requires that the ocean–atmosphere system has to be considered as a unit. Just as predictions of global warming cannot ignore the thermal capacity of the sea, so considerations of the carbon dioxide greenhouse gas budget must include the ocean carbon system. Vast fluxes of carbon dioxide flow from the warm tropical ocean into the atmosphere, and these are balanced by equally large fluxes into the cold surface ocean in high latitudes. Some of this high latitude cold water sinks into the deep ocean, and there it is enriched further in carbon dioxide by the decay of falling organic particles already mentioned. Simultaneously, the solution of sinking particles containing carbonate minerals tends to oppose the pH drop resulting from this organic breakdown. Renewing itself on a time-scale of about 1000 years, the deep water advects upward in tropical regions, thinning the surface mixed layer and replacing the carbon dioxide that originally escaped from the atmosphere.

Enrichment of the atmosphere with fossil fuel carbon dioxide thus has the initial effect of reducing the flux of carbon dioxide out of the tropical ocean and increasing the flux into the high latitude surface waters. In both cases, the water becomes slightly more acid. As it sinks, the high latitude ocean water will gradually make the deep water more acid (i.e. it will lower the pH). This water will be more aggressive towards calcium carbonate minerals, so that more of the falling particles will dissolve and on a much longer time-scale some of the sediments at the seabed will dissolve. Thus, paradoxically, water that has been enriched in fossil fuel carbon will become further enriched in carbon from calcium carbonate minerals. When this water reaches the ocean surface again after many centuries, it will be significantly enriched in carbon dioxide, but interactions with carbonate will mean that the partial pressure of carbon dioxide is not increased proportionally. Thus, the added carbon dioxide will have been substantially removed from the atmosphere by the oceans and 'titrated' with carbonate. The ultimate fate of the fossil fuel carbon that we release into the atmosphere is to make a layer of ocean sediment that contains rather less calcium carbonate than would have been deposited otherwise.

Unfortunately, the rate at which removal from the atmosphere can occur is limited by the rate at which deep water formation can carry the carbon dioxide away from the surface mixed layer. This imposes a delay of several hundred years on the removal process, and ensures that the present rate of anthropogenic release, although small relative to the natural fluxes, is capable of driving the atmospheric concentration upward to levels that have not been equalled for at least 1.5×10^5 years. It is this increase that carries the possibility of significant climate alteration. Paradoxically, the warming of the surface ocean that this entails would act to reduce the density of surface ocean water, and thus to reduce the sinking flux that is necessary to transfer the additional carbon dioxide into the deep ocean.

The interaction of ocean and atmosphere may also be mediated by the release of relatively minor gases by the biota. The volatile compound dimethylsulphide is released to the atmosphere as a result of its production in surface waters by some kinds of phytoplankton; it is the most abundant of a number of sulphur-containing gases that are released from the surface ocean in this way. In the atmosphere, some of it is converted into sulphur dioxide, that forms sulphuric acid aerosol particles that can act as a nucleus for cloud condensation. Clouds modify the infra-red albedo of the earth, reflecting more energy back into space. It has been postulated (Charlson, Lovelock, Andreae and Warren 1987) that changes in the flux of dimethylsulphide as a result of changes in the temperature of the upper ocean could feed back to modify the temperature through a change in cloud cover.

Another hypothesis links inputs of iron by wind-blown mineral dust to the recycling of carbon dioxide. Some oceanic surface waters, unlike those of the stratified ocean, do not become strongly depleted in the principal micronutrients needed for phytoplankton growth. Experimental work has shown that the addition of dissolved iron to such waters promotes an increased utilization of these relatively abundant nutrients. The intensive and large-scale experiments at sea undertaken to examine such effects (Coale et al. 1996, Martin et al. 1994) have involved the use of seagoing resources in a way that contrasts tellingly with the *Challenger* Expedition. Perhaps the most important strand that can be picked out from the web of research is the understanding that life on earth is underpinned by a multiplicity of unseen, more or less subtle biogeochemical processes, each of which contributes to the stability of the whole. This insight was given almost poetic form by James Lovelock's 'Gaia Hypothesis', in which the earth's biosphere is seen as acting as if it were a huge organism maintaining stable conditions on the planet in order to ensure its own survival (Lovelock 1979, Lovelock and Margulis 1974). This has been misinterpreted as a comforting concept for the human race, a sort of eco-cradle that will protect us in spite of our own worst efforts to unbalance the system. Alas, the truth must be very different. Each of the processes that form the agents of Gaia can only operate within a limited range. If the system is overstressed, a rearrangement will occur and the system will come to a new balance. The indications are that this might be a very uncomfortable process, and not all species would survive. In other words, Gaia is tolerant only up to a point. After that, the homeostatic process might act to remove the most destabilizing factor, which in the present case is us. The lesson is that we do well to respect our planet and its life-support system. We cannot live without it, but it might manage to live quite well without us. Ultimately, we are constrained within the limits of our system, and must learn its ways with care in order to avoid fatal transgressions. Our present behaviour is analogous to tampering with the life-support system that we do not understand. What we know is only a fraction of what we need to know ultimately, in order to survive.

References

Broecker, W. S. (1971) 'A kinetic model for the chemical composition of sea water', *Quaternary Research* **1**: 188–207.

Broecker, W. S. (1974) *Chemical oceanography*. New York: Harcourt Brace Jovanovich.

Charlson, R. J., Lovelock, J. E., Andreae, M. O. and Warren, S. G. (1987) 'Oceanic phytoplankton, atmospheric sulphur, cloud albedo and climate', *Nature* **326**: 655–61.

Coale, K. H. et al. (1996) 'A massive phytoplankton bloom induced by an ecosystem-scale iron fertilization experiment in the equatorial Pacific Ocean', *Nature* **383**: 495–501.

Culkin, F. and Smed, J. (1979) 'The history of standard seawater', *Oceanologica Acta* **2**: 355–64.

Deacon, M. B. (1997) *Scientists and the sea, 1650–1900: a study of marine science*. Aldershot: Ashgate.

Elderfield, H. and Schultz, A. (1996) 'Mid-ocean ridge hydrothermal fluxes and the chemical composition of the ocean', *Annual Review of Earth and Planetary Science* **24**: 191–224.

Garrels, R. M. (1965) 'The role of silica in the buffering of natural waters', *Science* **148**: 69.

Goldberg, E. D. (1954) 'Marine geochemistry I. Chemical scavengers of the sea', *Journal of Geology* **62**: 249–65.

Lovelock, J. E. (1979) *Gaia: a new look at life on earth*. Oxford: Oxford University Press.

Lovelock, J. E. and Margulis, L. (1974) 'Atmospheric homeostasis by and for the biosphere: the Gaia hypothesis', *Tellus* **26**: 2–10.

Mackenzie, F. T. and Garrels, R. M. (1966) 'Chemical mass balance between rivers and oceans', *American Journal of Science* **264**: 507–25.

Martin, J. H. et al. (1994) 'Testing the iron hypothesis in ecosystems of the equatorial Pacific Ocean', *Nature* **371**: 123–9.

Mason, B. (1952) *Principles of geochemistry*. New York: Wiley; London: Chapman & Hall.

Redfield, A. C., Ketchum, B. H. and Richards, F. A. (1963) 'The influence of organisms on the composition of sea-water', in M. N. Hill (ed.) *The sea*. New York: Interscience, vol. 2, pp. 97–116.

Schaule, B. K. and Patterson, C. C. (1981) 'Lead concentrations in the northeast Pacific; evidence for global anthropogenic perturbations', *Earth and Planetary Science Letters* **54**: 97–116.

Sillén, L. G. (1961) 'The physical chemistry of sea water', in M. Sears (ed.) *Oceanography*. Washington, D.C.: American Association for the Advancement of Science, pp. 549–81.

Thompson, G. (1983) 'Hydrothermal fluxes in the ocean', in J. P. Riley and R. Chester (eds) *Chemical oceanography*, vol. 8. London: Academic Press, pp. 271–337.

Turekian, K. K. (1977) 'The fate of metals in the oceans', *Geochimica et Cosmochimica Acta* **41**: 1139–44.

Whitfield, M. and Turner, D. R. (1987) 'The role of particles in regulating the composition of sea-water', in W. Stumm (ed.) *Aquatic surface chemistry*. New York: John Wiley and Sons, pp. 457–93.

15 Deep-sea biology in the 1990s
A legacy of the *Challenger* Expedition

P. A. Tyler, A. L. Rice and C. M. Young

Introduction

The biological contribution of the *Challenger* Expedition is most tangibly commemorated by the vast collections of material brought home (see Chapter 1), now mainly in the Natural History Museum in London, and the fifty volumes of the official scientific reports. More than 18,000 of the 30,000 or so pages in these reports are devoted to the marine biological collections and contain an enormous amount of systematics, which have formed the basis of the subsequent taxonomic study of most deep-sea and many shallow-water marine groups.

Within the taxonomic treatments, most of which are of very high quality, there are many observations on the biology of the groups concerned, including the morphology of the reproductive organs and even the reproductive strategies the animals were thought to employ. These will be the main subject of this comparison of the views of the authors of the *Challenger* reports and those we hold now. But before dealing with this specific subject, it is worth comparing the views of the *Challenger* scientists on more general aspects of the benthic biology of the deep sea with current opinions (see, for instance, Gage and Tyler 1991) to see to what extent these have changed during the intervening century or so.

The biology of the deep-sea floor, then and now

The detailed study of the *Challenger* marine collections by the taxonomic specialists inevitably took many years. But even before the material had been dispatched to the specialists, some general conclusions about deep-sea biology had already been drawn by the expedition scientists on the basis of what they had seen during the voyage itself. Consequently, in the 'General introduction to the zoological reports', published in 1880, the expedition's leader Charles Wyville Thomson was able to make a number of general statements about the deep-sea fauna; some of these have stood the test of time, others have not.

First, Thomson pointed out that 'animals of all the marine invertebrate groups, and probably fishes also, exist over the whole of the floor of the ocean'. In other words, there is no depth limit to animal life in the seas. This was not really a *Challenger* discovery, for earlier voyages, and particularly those of HMS *Porcupine* (see Chapter 1), had demonstrated that animals were to be found in very deep water off the coasts of Europe. But the *Challenger* expedition had extended this observation to the world ocean, trawling up animals from all depths wherever they had sampled, including twenty-five hauls from deeper than 4500m, the deepest from 5700m. Thomson's reservation about fishes (and also the swimming decapod crustaceans) was because the *Challenger* scientists had no means of preventing their trawls and dredges from catching animals in mid-water on their way to and from the sea floor. Consequently, it was possible that at least some of these groups in the bottom samples had actually been collected in the water column at much shallower depths. We now know that Thomson's general conclusion was correct, for all the main animal groups, including fishes, are to be found on, or close to, the bottom even in the very deepest ocean trenches in excess of 11,000m depth.

But although there was a wide range of animals in the deep samples, Thomson noted that the number of individuals was usually very small. So the second general conclusion was that animal abundance decreases dramatically with increasing depth. Again, we now know that he was right, for abyssal benthic communities are in general a hundred to a thousand times less abundant per unit area than their continental shelf or upper continental slope counterparts. In fact, the *Challenger* sampling gears and fishing techniques were not reliable enough to demonstrate this relationship in any detail, and it was not until the development of sophisticated quantitative sampling gears and the use of deep-sea photography and submersibles in the 1960s that this became possible. Nevertheless, the general principle seemed clear to the *Challenger* scientists, and they were correct.

Third, Thomson believed that, although there were some notable exceptions, the sea-floor animals generally become smaller with increasing depth. This conclusion also seems to be supported by modern data, though it is difficult to reconcile it with the relatively simple sampling gears available on the *Challenger*. Many megafaunal taxa, particularly at the genus or family level, actually seem to be represented by larger species with increasing depth. The smaller–deeper relationship results rather from an increase in the relative importance of the smaller fractions of the benthic communities, and particularly the meiofauna – the very small crustaceans, polychaete and nematode worms, and foraminiferans – inhabiting the deep-sea sediments. But these small creatures were very poorly sampled from the *Challenger*. The relatively coarse-meshed nets used on the trawls and dredges would certainly not have retained them effectively, and most of the small animals collected were brought up in the 'hempen tangles', teased-out lengths of rope attached to the back of the dredge frames (see Chapter 1). In fact, it was again not until the development of gears, and particularly corers, in the 1960s, with the specific intention of sampling the small mud-dwelling animals, that this smaller–deeper relationship could be demonstrated convincingly. Once more, Thomson and his *Challenger* colleagues were right in their general conclusion, but probably for the wrong reasons.

The increasing interest in, and ability to sample, the very small infaunal animals on the deep-sea floor has also highlighted the one major error in the *Challenger* biological conclusions, but one that was perpetuated for more than eight decades after the expedition. For Thomson also believed that with increasing depth there was a decrease in the *variety* of animal life, what we would now call species richness or even biodiversity. The *Challenger* samples showed this clearly. For example, the fifty-seven seabed samples obtained from depths greater than 3660m contained, on average, about twenty-five specimens belonging to about ten different species. But these were mainly the big sediment-surface-dwelling animals such as the echinoderms (sea-cucumbers, sea urchins, seastars and brittle stars), molluscs and decapod crustaceans that the *Challenger*'s trawls and dredges were quite good at collecting. These same techniques, with rather minor modifications and improvements, were used by all subsequent deep-sea biological expeditions up to the 1950s and beyond (see Rice 1986). Not surprisingly, the scientists who worked on these samples came to much the same conclusion as their predecessors. Consequently, the main textbook of the middle years of this century (Marshall 1954), summarizing the current state of knowledge of deep-sea biology, stated that 'with increasing distance from land [and, by implication, with increasing depth] there is an increasing tendency for the deep-sea floor to be populated by fewer individuals belonging to fewer species', a view very similar to that expressed by Thomson seventy years earlier. Within twenty years the relatively simple developments in sea-floor sampling technology referred to above had provided data on the smaller animals that demonstrated a previously unsuspected species richness in the deep ocean (Hessler and Jumars 1974, Sanders 1968). These early results have been amply confirmed subsequently, so that many biologists now believe that the deep-sea floor may be one of the most species-rich environments on the planet (see Grassle and Maciolek 1992). Clearly, in this respect the *Challenger* scientists, and many of their successors, were quite wrong.

Deep-sea scientists from the *Challenger* era to the 1950s had no reason to question the concept of a species-poor deep-sea floor. It seemed to fit well with the equally widespread view of the deep ocean as a physically and chemically monotonous environment lacking the temporal and spatial variability seemingly typical of species-rich environments such as coral reefs or rain forests. Indeed, this view persisted well after the acquisition of the samples indicating a species-rich deep-sea floor, making its explanation difficult.

A temporally stable deep-sea floor also seemed incompatible with data indicating that at least some deep-sea animals reproduce seasonally like their shallow-water relatives. However, since the early 1980s large areas of the deep sea have been found to be subjected to a highly seasonal and spatially patchy supply of organic matter, in total contrast to the ideas held for the previous century (see Tyler 1988). But this situation had been more or less forecast by one of the *Challenger* naturalists, H. N. Moseley. In an article in *Nature* in 1880 Moseley noted that:

> Life must be very monotonous in the deep sea. There must be an entire absence of seasons, no day and night, no change of temperature. Possibly there is at some places a periodical variation in the supply of food falling from above which may give rise to a little annual excitement amongst the inhabitants.

This proved to be one of the most perceptive statements by a member of the *Challenger* Expedition, but remained untested for over a century until technological developments allowed us to determine the rate and magnitude of particle flux from the sea surface to deep waters.

There were also some very perceptive contemporary observations on the reproductive biology of the organisms collected by the *Challenger*. Best known among these are the detailed observations of Théel (1882) on gonad development in the holothurians. Over 100 years later Smiley (1988) presented a model of holothurian reproduction, the rebuttal of which could be found in the figures of Théel (Sewell, Tyler, Young and Conand, 1997). Young (1994) has produced an elegant review of the contribution to reproductive biology, either of gametogenesis or larval development, made by the authors of the *Challenger* volumes.

To understand reproductive processes in the deep sea properly, it is necessary first to identify the species, second to determine the nature of the reproductive patterns, and third to determine the dynamics of those patterns, especially the rates at which they occur and the energy which they use. The first two of these aspects were covered for some species in the *Challenger* reports (Young 1994), but the very nature of the expedition (single collections at spatially rather than temporally varying stations) did not allow for the determination of the dynamics of the species collected. Fifty years after the *Challenger* Expedition Orton (1920) suggested that the deep sea was likely to be typified by year-round reproduction, as in this environment the temperature showed very little change and this was perceived, at that time, to be the single significant controlling factor for reproductive periodicity.

The new age of deep-sea biology, ushered in by the Gay Head–Bermuda transect study (Sanders, Hessler and Hampson 1965), gave an additional impetus with the use of new types of sampling equipment. The use of the fine-meshed WHOI epibenthic sledge (ibid.) demonstrated the high species diversity within the deep sea and the development of the USNEL box corer (Hessler and Jumars 1974) allowed quantitative determination of the macrofaunal biomass at the deep-sea bed. By this time, also, the use of submersibles was allowing direct observation and experimentation at the deep-sea bed. These technical developments, *inter alia*, allowed the dynamics of the deep sea to be more comprehensively understood.

During this era, and up to the present day, much of the interest in deep-sea ecological dynamics has been at the community level. In parallel with this community approach there were a few autecological reproductive studies. Of initial, as well as controversial, interest was the apparent seasonal reproductive pattern observed in deep-sea Antarctic amphipods (George and Menzies 1967, 1968). These data were considered equivocal and Rokop (1974) established a sampling time-series designed to test the hypothesis of seasonal reproduction in the deep sea. Of thirteen invertebrate species sampled in a twelve-month, five-sample series from 1250m in the San Diego Trough, two, the brachiopod *Frieleia halli* and the scaphopod *Cadulus californicus*, had distinct seasonal reproductive cycles (Rokop 1977). Rokop's study had been based on observations of gonad development. An indirect method had been used already by Schoener (1968) in her examination of the ophiuroids from the Gay Head–Bermuda transect. Schoener analysed the population structure of adults and juveniles sampled at different times of the year. In the summer months there was a significant increase in the number of newly recruited juveniles compared to other times of the year, suggesting a seasonal reproductive pattern.

Population structure and gonadal development were used in the analysis of time-series samples taken between 1977 and 1985 from two stations at 2200m and 2900m respectively in the Rockall Trough, northeast Atlantic. The analysis of these samples has been complemented by samples from stations down to 4000m depth in the Porcupine Seabight and Abyssal Plain. Orton had forecast the most common form of reproductive pattern is the asynchronous production of gametes throughout the year. In species exhibiting this pattern the ovaries in all samples contain oocytes of all sizes from small primary oocytes to those of maximum development. The largest oocytes observed tend to be large (600μm to 4mm,) suggesting that post-fertilization development is by lecithotrophy. Although originally thought to limit distribution, lecithotrophy is now interpreted as a suitable mechanism for wide dispersal, owing to the low metabolic rate at deep-sea temperatures and the high food reserves stored in the egg (Shilling and Manahan, 1994). Examination of the population structure of these species suggests no apparent division into size-related cohorts, which is interpreted as indicating recruitment of juveniles to the adult population throughout the year.

More surprising than the apparent dominance of this continuous reproductive pattern, as Orton had predicted, was the seasonal pattern of oocyte development found in a few species in the northeast Atlantic. In the echinoderms *Ophiura ljungmani*, *Ophiocten gracilis*, *Plutonaster bifrons*, *Echinus affinis*, the anemone *Amphianthus inornata* and the protobranch bivalves *Ledella pustulosa* and *Yoldiella jeffreysi*, the development of oocytes is synchronous within individuals. Not only is there synchronous development of oocytes between individuals of a single species, but there is also synchronous development of oocytes between the species. Oogenic proliferation in the echinoderms starts in the early months of the year (or very late the previous year in *Echinus affinis*) and continues through to the summer months (Fig. 15.1). In

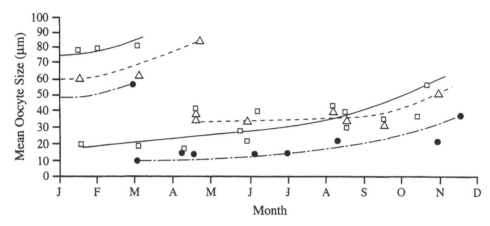

Figure 15.1 The pattern of seasonal reproduction, as determined by mean oocyte size, in *Echinus affinis* (open squares), *Plutonaster bifrons* (open triangles) from 2200m depth and *Ophiura ljungmani* (closed circles) from 2900m depth in the northeast Atlantic Ocean.

the late summer, autumn and winter the oocytes grow steadily to reach a maximum diameter of around 120μm in the late winter followed by a synchronous spawnout. Examination of the population structure of the adults reveals that it is possible to discern size-related cohorts that are indicative of periods of intense recruitment to the adult population.

The obvious question is 'What drives the seasonal reproduction and recruitment in the deep sea?' In the deep northeast Atlantic the temperature variation below 2000m is minimal and tidal currents are less than 10cm s^{-1}. There is evidence for a seasonal variation in the eddy kinetic energy (Dickson et al. 1986) although the ecological implications of this are difficult to define. Of more relevance is the input of organic material into the deep sea. The conceptual model of Rowe and Staresinic (1979) suggested that the main inputs are in the form of large food falls of either macrophyte or animal remains, material carried into the deep sea in turbidity currents or small organic particles sinking from surface primary production. The last form of input was for many years thought to consist of a slow rain of particles, but is now known to be intense, rapid and highly seasonal in certain parts of the world ocean (Billett, Lampitt, Rice and Mantoura 1983, Deuser 1986, Tyler 1988). This input may drive the seasonal reproductive processes in two ways: (a) the egg size of season-ally breeding deep-sea species is consistent with planktotrophic development and the sinking of particulate organics from the spring bloom may provide a food source for developing larvae; and (b) the organic flux arrives at the seabed coincident with the start of the laying down of yolk in the developing oocytes of the adults. If Moseley had had the equipment and time he might have been able to justify his concept of a 'little annual excitement'!

One aspect that the scientists on the *Challenger* were not able to achieve was the visualiza-tion of the sea floor at great depths. The advent of deep-sea cameras and submersibles has allowed us to explore the deep sea in this way and, as a result, we are able to conduct *in situ* experiments as well as take advantage of serendipitous observations. Two excellent examples of the latter in the context of reproduction were observed at bathyal depths off the Baha-mas. In May 1988 two of us (C. M. Young and P. A. Tyler) were conducting a series of experiments on reproductive activity at 500m depth. After one dive Craig Young was con-vinced he had seen a number of 'pairs' of the sea urchin *Stylocidaris lineata*. Analysis of the onboard video confirmed his observations. Aggregations of deep-sea megabenthic species had been observed before (Grassle et al. 1975) and these were thought to be feeding aggregations analogous to those of the great herds of bison that roamed the plains of North America. We were not convinced that pairing was a feeding aggregation, but thought it more likely to be a reproductive behaviour. Using the twelve-bucket carousel of the *Johnson Sealink* submersible, we were able to collect pairs of *Stylocidaris lineata* as well as those that occurred alone. Analysis of the pairs after the dive showed that they consisted of female/male, male/ male and female/female and all were ripe (Fig. 15.2). Statistical analysis demonstrated that the pairs we saw were a function of random pairing. As long as one individual of *Stylocidaris lineata* paired with another, it had a 50 per cent chance of pairing with the opposite sex. In the low population densities found in deep-sea megabenthos, a 1 in 2 chance of finding a mate is considerably better than finding no mate at all! May is the season of maximum reproductive development of *S. lineata*. When we repeated the experiment in February when the adults are not ripe, pairing was rarely observed (Young, Tyler, Cameron and Rumrill 1992).

Pairing, also as a reproductive behaviour but for different reasons, was observed on the same cruise in the bathyal holothurian *Paroriza pallens* (Fig. 15.3a). Pairing had also been observed in *P. pallens* in the northeast Atlantic (Fig. 15.3b) (Tyler, Young, Billett and Giles

(a)

(b)

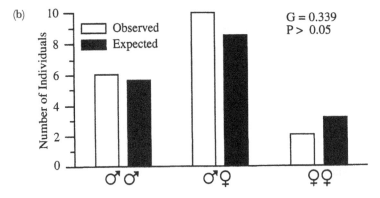

Figure 15.2 (a) *Stylocidaris lineata*; (b) analysis of collected pairs from 500m depth off the Bahamas in May when the adults are ripe. (For details see Young et al. 1992.)

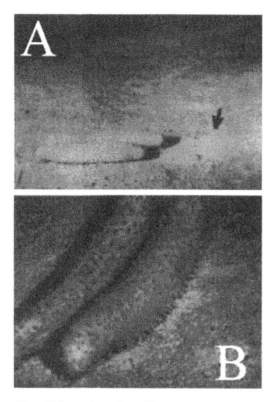

Figure 15.3 Paroriza pallens. Observations of pairing (a) from the *Johnson Sealink* submersible at 900 m in the Bahamas and (b) from the submersible *Cyana* at 800 m depth in the northeast Atlantic Ocean. The maximum length of this species is about 27 cm. (For details see Tyler et al. 1992.)

1992). *P. pallens* is a simultaneous hermaphrodite that appears to produce eggs all year round. Examination of the gonads of the pairs showed them to be well developed. Since self-fertilization is of limited benefit to a species, pairing in this case allows cross-fertilization and outbreeding. Théel (1882), in his *Challenger* report on the holothurians, commented on their large gonads and large oocyte size. This observation was reinforced by Hansen (1975), who showed that some of the eggs of abyssal holothurians can be up to 4.4mm diameter. The cosmopolitan occurrence of individual species and their large egg size hint at the spectacular dispersal capabilities attached to lecithotrophic development.

Another example of serendipity is the settlement of a larva of the deep-sea stalked barnacle *Poecilasma kaempferi* on the frame of a deep-sea camera system deployed at 1526m depth on the Goban Spur in the northeast Atlantic (Lampitt 1990). The camera was deployed for almost ten months and the larva settled on a vane in view of the camera early in this period. As the camera took a picture every four hours, it kept a record of the growth of *P. kaempferi*, which proved to be more rapid than expected for a deep-sea invertebrate. Analysis of the gametogenic biology of *P. kaempferi* (Green, Tyler, Angel and Gage 1994) showed that the gonad development was also much more rapid than would be expected at this temperature and depth in the deep sea. Initiation of gametogenesis occurs at day 70, maturity is reached by day 155 and the first egg lamellae are laid down by day 170. For *P. kaempferi* to complete the gametogenic cycle in 100 days was 250 days faster than previously recorded for a deep-sea species other than meiofauna and the aplacophoran mollusc *Prochaetoderma yongei* (Scheltema 1987). The reason for the rapid growth and development in *Poecilasma kaempferi* may be two-fold. First, there may be a phylogenetic constraint. All stalked barnacles grow and reproduce rapidly (known to Darwin 1851). Second, *P. kaempferi* lives, normally, as an epizooite on the carapace of deep-sea decapod Crustacea, especially the stone crab *Neolithodes grimaldi* (Williams and Moyse 1988). As with all Crustacea, *N. grimaldi* must moult in order to grow. When the exoskeleton is discarded, the barnacle is also discarded and thus loses its habitat. If *N. grimaldi* moults every two years (an estimate) the barnacle must produce enough larvae to ensure detection and colonization of a new host before the present host sheds its exoskeleton.

Rapid growth and reproduction is not confined to the barnacles in the deep sea. Any wood entering the deep sea is colonized and digested by wood-boring molluscs. Ballard's pictures of the *Titanic* show no wood and his text suggests that only the hardest teak had not been eaten away, though oak timbers from the SS *Central America*, sunk in 1857, still survive (Herdendorf, Thompson and Evans 1995). We (Young and Tyler) have deployed spruce and oak panels at three-monthly intervals at 500m depth off the Southwest Reef in the Bahamas. These have been retrieved after three- and six-month periods of immersion and the blocks X-rayed, the colonizing *Xylophaga* species measured and the specimens analysed for gonad development. Growth of *Xylophaga* can be up to 0.089mm d^{-1} and reproductive development starts at 1mm adult diameter. The reason for this rapid growth and reproduction is not dissimilar from that for *Poecilasma kaempferi*. As the name implies, *Xylophaga* has the ability to digest wood and as a result 'eats' its environment. A point is reached at which all the wood is digested and the population of that block dies. In the meantime the adults have had to produce sufficient larvae to locate and colonize other pieces of wood – a very unusual resource in the deep sea. The pattern we have observed implies a year-round process of gamete production. But Turner (1973) suggested that there may be a seasonal component to this production related to the spring thaw washing down trees into the ocean, which become waterlogged, sink and provide a seasonal substratum for *Xylophaga* species. But these wood-boring molluscs have been found in all regions of the world's oceans, including the abyssal plains and the mid-oceanic ridges, sites where one might not expect to find wood as a resource. This ubiquitous distribution may be a function of their prolific reproductive output.

An area that exercised the minds of the scientists on the *Challenger*, and which has been the subject of much debate since, is the 'zonation' of organisms in the deep sea. As one goes down the continental slope, species appear and disappear with the increase in depth. There has been much speculation as to the cause of this zonation, but very little experimental evidence. We have followed our interest in reproductive biology to examine the effect of pressure on early embryogenesis to determine if this is a factor in zonation. Our experimental design is to spawn ripe sea urchin adults and fertilize the eggs. The fertilized embryos are then subjected to different pressures from 1atm to 200atm at 50atm increments and at temperatures of 5° to 20°C at 5°C increments. Our data (see Young and Tyler 1993, Young, Tyler and Fenaux 1997, Tyler and Young 1998, Tyler, Young and Clarke 2000) show that embryos of shallow-water sea urchins will tolerate pressure down to the equivalent of 2000m depth. Conversely, the embryos of the deep bathyal echinoid *Echinus affinis* will tolerate pressures only greater than 1500m depth (Fig. 15.4), suggesting that they are barophilic. The wide range of the pressure tolerance of embryos of shallow-water species suggests that other factors control their zonation. By contrast, the lower pressure tolerance

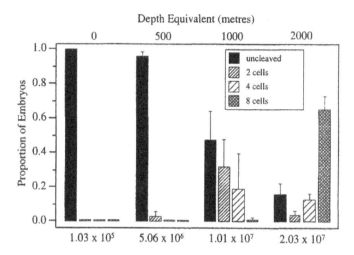

Figure 15.4 Echinus affinis. Pressure tolerances of early embryos at 1, 50, 100 and 200atm pressure, which demonstrate the barophilic requirement of embryogenesis in this species. (For details see Young and Tyler 1993.)

of the embryos of *Echinus affinis* is close to the upper limit of their zonation, which may be interpreted as causal or coincidental. Additional *in situ* and *in vitro* experimentation will answer such questions.

One of the main legacies that the *Challenger* cruise left us was the excitement of discovering the biological processes in the deep sea. Reading the volumes of the report one can imagine the excitement that occurred on the deck when previously unknown organisms were brought up in the deep-sea trawls. That legacy of excitement is still with us today. One has only to view the first videos of the discovery of hydrothermal vents and cold seeps to hear the eagerness in the scientists' voices. Samples collected by any method from the deep sea will yield their interest to the enthusiastic observer. It is that thrill of the sample arriving on deck and the information it may provide that is the real legacy of the *Challenger* expedition.

References

Billett, D. S. M., Lampitt, R. S., Rice, A. L. and Mantoura, R. F. C. (1983) 'Seasonal sedimentation of phytodetritus to the deep-sea benthos', *Nature* **302**: 520–2.

Darwin, C. (1851) *A monograph of the sub-class Cirripedia, with figures of all the species of the Lepadidae; or pedunculated cirripedes.* London: The Ray Society.

Deuser, W. G. (1986) 'Seasonal and inter annual variations in deep-water particle fluxes in the Sargasso Sea and implications for the transfer of matter to the deep oceans', *Deep-Sea Research* **33**: 225–46.

Dickson, R. R. , Gould, W. J., Griffiths, C., Medler, K. J. and Gmitrowicz, E. M. (1986) 'Seasonality in currents of the Rockall Channel', *Proceedings of the Royal Society of Edinburgh* **88**B: 103–25.

Eckelbarger, K. J. and Watling. L. (1995) 'The role of phylogenetic constraints in determining reproductive patterns in deep sea invertebrates', *Invertebrate Biology* **114**: 256–69.

Gage, J. D. and Tyler, P. A. (1991) *Deep sea biology: a natural history of organisms at the deep sea floor.* Cambridge: Cambridge University Press.

George, R. Y. and Menzies, R. J. (1967) 'Indication of cyclic reproductive activity in abyssal waters', *Nature* **215**: 878.

George, R. Y. and Menzies, R. J. (1968) 'Further evidence for seasonal breeding cycles in the deep sea', *Nature* **220**: 80–1.

Grassle, J. F. and Maciolek, N. J. (1992) 'Deep-sea species richness: regional and local diversity estimates from quantitative bottom samples', *The American Naturalist* **139**: 313–41.

Grassle, J. F., Sanders, H. L., Hessler, R. R., Rowe, G. T. and McLennan, T. (1975) 'Pattern and zonation: a study of the bathyal megafauna using the research submersible Alvin', *Deep-Sea Research* **22**: 643–59.

Green, A., Tyler, P. A., Angel, M. V. and Gage, J. D. (1994) 'Gametogenesis in deep- and surface-dwelling stalked barnacles in the NE Atlantic Ocean', *Journal of Experimental Marine Biology and Ecology* **184**: 143–58.

Hansen, B. (1975) 'Systematics and biology of deep sea holothurians', *Galathea Report* **13**: 1–262.

Herdendorf, C. E., Thompson, T. G. and Evans, R. D. (1995) 'Science on a deep-ocean shipwreck', *Ohio Journal of Science* **95**: 4–212.

Hessler, R. R. and Jumars, P. A. (1974) 'Abyssal community analysis from replicate box cores in the central North Pacific', *Deep-Sea Research* **21**: 185–209.

Hessler, R. R. and Sanders, H. L. (1967) 'Faunal diversity in the deep sea', *Deep-Sea Research* **14**: 65–78.

Lampitt, R. S. (1990) 'Directly measured rapid growth in a deep-sea barnacle', *Nature* **345**: 805–7.

Marshall, N. B. (1954) *Aspects of deep-sea biology*. London: Hutchinson.

Moseley, H. N. (1880) 'Deep-sea dredging and life in the deep sea', *Nature* **21**: 591–3.

Orton, J. H. (1920) 'Sea temperature, breeding and distribution in marine animals', *Journal of the Marine Biological Association of the United Kingdom* **12**: 339–66.

Rice, A. L. (1986) *British oceanographic vessels 1800–1950*. London: The Ray Society.

Rokop, F. J. (1974) 'Reproductive patterns in deep-sea benthos', *Science* **186**: 743–5.

Rokop, F. J. (1977) 'Seasonal reproduction in the brachiopod *Frieleia halli* and the scaphopod *Cadulus californicus* at bathyal depths in the deep sea', *Marine Biology* **43**: 237–46.

Rowe, G. T. and Staresinic, N. (1979) 'Sources of organic matter to the deep-sea benthos', *Ambio Special Report* **6**: 19–24.

Sanders, H. L. (1968) 'Marine benthic diversity: a comparative study', *The American Naturalist* **102**: 243–82.

Sanders, H. L., Hessler, R. R. and Hampson, G. R. (1965) 'An introduction to the study of the deep-sea benthic faunal assemblages along the Gay Head–Bermuda transect', *Deep-Sea Research* **12**: 845–67.

Scheltema, A. (1987) 'Reproduction and rapid growth in the deep-sea aplacophoran mollusc *Prochaetoderma yongei*', *Marine Ecology: Progress Series* **37**: 171–80.

Schoener, A. (1968) 'Evidence for reproductive periodicity in the deep sea', *Ecology* **49**: 81–7.

Sewell, M. A., Tyler, P. A., Young, C. M. and Conand, C. (1997) 'Ovarian development in the Class Holothurioidea: a reassessment of the "tubule recruitment model"', *Biological Bulletin* **192**: 17–26.

Shilling, F. M. and Manahan, D. T. (1994) 'Energy metabolism and amino acid transport during early development of Antarctic and temperate echinoderms', *Biological Bulletin* **187**: 398–407.

Smiley, S. (1988) 'The dynamics of oogenesis and the annual ovarian cycle of *Stichopus californicus* (Echinodermata: Holothurioidea)', *Biological Bulletin* **175**: 79–93.

Théel, H. (1882) 'Report on the Holothurioidea dredged by HMS *Challenger* during the years 1873–76. Part I', *Reports on the scientific results of the voyage of HMS Challenger, Zoology* **4**(13). London: HMSO.

Thomson, C. W. (1880) 'General introduction to the zoological series of reports', *Reports on the scientific results of the voyage of HMS Challenger, Zoology* **1**: 1–50. London: HMSO.

Turner, R. D. (1973) 'Wood-boring bivalves, opportunistic species in the deep sea', *Science* **180**: 1377–9.

Tyler, P. A. (1988) 'Seasonality in the deep sea', *Oceanography and Marine Biology: An Annual Review* **26**: 227–58.

Tyler, P. A. and Young, C. M. (1998) 'Temperature and pressure tolerances in dispersal stages of the genus *Echinus* (Echinodermata: Echinoidea): prerequisites for deep-sea invasion and speciation?', *Deep-Sea Research* **II–45**: 253–77.

Tyler, P. A., Young, C.M. and Clarke, A. (2000) 'Temperature and pressure tolerances of embryos of the Antarctic echinoid *Sterechinus neumayeri* (Echinodermata: Echinoidea): potential for deep-sea invasion from high latitudes', *Marine Ecology Progress Series* **192**: 173–80.

Tyler, P. A., Young, C. M., Billett, D. S. M. and Giles, L. A. (1992) 'Pairing behaviour, reproduction and diet in the deep-sea holothurian genus *Paroriza* (Holothurioidea: Synallactidae)', *Journal of the Marine Biological Association of the United Kingdom* **72**: 447–62.

Williams, R. and Moyse, J. (1988) 'Occurrence, distribution and orientation of *Poecilasma kaempferi* Darwin (Cirripeda: Pedunculata) epizoic on *Neolithodes grimaldi* Milne-Edwards and Bouvier (Decapoda: Anomura) in the northeast Atlantic', *Journal of Crustacean Biology* **8**: 177–86.

Young, C. M. (1994) 'The tale of two dogmas: early history of deep-sea reproductive biology', in C. M. Young and K. J. Eckelbarger (eds) *Reproduction, larval biology and recruitment of the deep sea benthos*. New York: Columbia University Press, pp. 1–25.

Young, C. M. and Tyler, P. A. (1993) 'Embryos of the deep-sea echinoid *Echinus affinis* require high pressure for development', *Limnology and Oceanography* **38**: 178–81.

Young, C. M., Tyler, P. A. and Fenaux, L. (1997) 'Potential for deep-sea invasion by Mediterranean shallow water echinoids: pressure and temperature as stage-specific dispersal barriers', *Marine Ecology Progress Series* **154**: 197–209.

Young, C. M., Tyler, P. A., Cameron, J. L. and Rumrill, S. G. (1992) 'Seasonal breeding aggregations in low density populations of the bathyal echinoid *Stylocidaris lineata*', *Marine Biology* **113**: 603–12.

16 The *Challenger* legacy
The next twenty years

John Woods

Introduction

I have been asked to end this book on the *Challenger* legacy with a look into the future. The planning horizon for oceanography is about twenty years, set by the lead time to develop and introduce new technology, and to fund great endeavours. I shall therefore limit myself to the next twenty years, and concentrate on major challenges that need to be addressed now, which will take twenty years or longer to resolve. And because *Challenger* was about the global perspective, I shall not address the future of coastal oceanography.

My essay reflects the achievements of the *Challenger* Expedition, starting with the basic exploration of the world ocean, then moving on to key scientific issues that remain unresolved at the end of the twentieth century and to the new tools that will help us address them in the next twenty years. After discussing the practical exploitation of the new under-standing and methods, I shall close by asking who will pick up the baton and lead our subject in the twenty-first century as Britain did in the nineteenth.

Basic exploration

Bathymetry

The most basic question one might ask about the ocean is how deep is it? Soundings were a routine task on the *Challenger*, and have been continued on research ships ever since. The soundings are contoured in the new CD-ROM (Jones, Tabor and Weatherall 1994) made for the General Bathymetric Chart of the Ocean (GEBCO) (see Chapter 5) by the British Oceanographic Data Centre (BODC). We know that the compilation is not complete; it lacks military classified data for some limited regions. Does the coverage meet our needs? We have long suspected that it does not, if only because previously unsounded sea-mounts are discovered from time to time, most spectacularly through the dimpling of the sea surface detected by Seasat's altimeter. Geologists need to know the shape of the sea floor, and can diagnose processes from it, as they have from GLORIA surveys (see Chapter 5). Recently it has become clear that an accurate map of the ocean floor is also important for simulating the ocean circulation. Standard GCMs based on the Modular Ocean Model (MOM) code from Princeton University's Geophysical Fluid Dynamics Laboratory did not do a very good job of describing the response of ocean currents to orography, but new codes such as that from the Miami laboratory with grid points on isopycnic rather than isobaric surfaces, do much better and have revealed the sensitivity of global heat flux to assumed bathymetry. It will soon be possible to specify precisely what needs to be known about the shape of the sea floor, and which areas are most important for each subject, whether it is geology, ocean circulation or military operations. The existing database of soundings, largely collected on an opportunity basis, will be found wanting. The Scientific Committee on Ocean Research (SCOR) has established a working group to foster international collaboration in systematic mapping of the ocean's bathymetry. Mature technology exists to do that globally within twenty years: fifty GLORIA-years could provide a coarse survey to be expanded in key areas by TOBI (Towed Ocean Bottom Instrument) and other high-resolution acoustic mappers (see Chapter 5). It will be costly but valuable, and it has the virtue of needing to be done only once.

Manned submersibles

The geologist, who will be one of the principal beneficiaries of a new global bathymetry, will continue to use manned submersibles to inspect promising sites. The discovery of hydrothermal vents and associated benthic fauna in recent years encourages us to believe that there is plenty more to be found by searching with the human eye. Discoveries will lead to expeditions, such as the RIDGE with its international associates (BRIDGE in Britain; FRIDGE in France, and so on).

Ocean drilling

The geologist will also want to continue drilling into the sediments and rocks below the deep ocean, building on the great success of the International Ocean Drilling Programme. It seems likely that there will be new drilling ships from Japan and possibly Europe to complement the US JOIDES ship *Resolution*. During the next twenty years the drilling programme is likely to diverge into two streams. The first will be concerned with making a few very deep holes through the crust to address problems arising in the theory of plate tectonics, and perhaps penetrating the Moho,[1] the original dream of the founding fathers, including Walter Munk. The second stream will be concerned with sampling the climatic record held in the sediments, a scientific goal that demands many holes that stop at the underlying bedrock, a procedure that has been called 'pogo stick' geology.

The climate of the ocean

The scientists on the *Challenger* were conscious of the need to understand how climate had changed in the past, a factor of central importance in Charles Darwin's theory of natural selection published in 1859. And they were no doubt aware of James Rennell's speculation that one of the factors in climate change was variation in the course of the Gulf Stream (Rennell 1832). But they were not then in a position to discover the processes responsible for such contemporary changes in the climate of the ocean. It was not until recently that the subject has been brought to life by two developments. First, the painstaking work of Syd Levitus and others who have accumulated and analysed oceanographic data held in many different archives to produce empirical climatologies for recent decades. And second, the first steps in mathematical simulation of ocean climate fluctuation over centuries by integrating climate models. Thanks to these pioneering efforts it is now clear that the global ocean circulation undergoes natural fluctuations over decades, which modulate the regional distribution of heat flux to the atmosphere with significant consequences for climate. These natural changes are so large that they have all but masked the effects of man-induced climate change during the twentieth century. The next twenty years will see a major effort to describe and understand those changes in ocean climate, motivated by a combination of scientific curiosity and the practical benefits of prediction if that proves possible.

The World Ocean Circulation Experiment (WOCE)

This task will demand continuing progress in the march towards realistic mathematical simulation of global ocean circulation. Much has been achieved since the mid-1980s, with the first steps towards resolving the eddies originally discovered by John Swallow (see Chapter 10). More attention is also being paid to processes in the seasonal boundary layer, the buffer zone between the fluxes to the atmosphere above and the permanent thermocline below. Within the next decade we shall have the first systematic survey of the ocean from the data collected during the World Ocean Circulation Experiment. The results of individual components of WOCE are already forcing us to change our ideas about ocean circulation, but the real benefits will come from the global synthesis of the whole data set, which is now becoming available. The WOCE data set already dwarfs all previous expedition data and will be the dominant source of information until the Global Ocean Observation System (GOOS) comes on stream, which is unlikely to be until well after the twenty years covered by this chapter. During the next decade there will be no single WOCE Office beavering away at the data from that great endeavour like that established in Edinburgh after the return of *Challenger*. The WOCE industry will be dispersed around the globe, a hundred Ph.D. students linked by the Internet and nourished by a few processing centres taking responsibility for data quality. The WOCE harvest will be collected during the coming twenty years. It will be brought home in models run on Teraflops computers (capable of 10^9 floating point operations per second), fulfilling our predictions underpinning the design of WOCE that such machines would be available to oceanographers by AD 2000.

Given the rethinking already being forced on us by individual components of WOCE and by eddy-resolving models, it is not too ambitious to predict that within twenty years we shall have a new paradigm for the global circulation of the ocean. That paradigm will have an impact on future research on climate and all aspects of oceanography. It will be as important as the revolution in geology wrought by the evidence for sea-floor spreading and the results from the Ocean Drilling Programme. Indeed, it will be more important because plate tectonics occurs in other planets but ocean circulation characterizes our own. The new paradigm will foster changed scientific priorities. How representative is the WOCE data given the changing climate of the ocean? What are the limits to predictability of ocean circulation? What is the relationship between the regional structure of the climates of the ocean and the atmosphere? These will lead to new experiments designed to clarify the processes responsible for the observed and simulated variability in the climate of the ocean.

The distribution of life in the sea

The *Challenger* expedition had one overriding ambition: to explore the distribution of life in the world ocean. Since then other great expeditions and marine laboratories and continuous plankton surveys have had the same goal, but all their collections put together fail to yield an unaliased, or fully documented, empirical description of the seasonally varying geographical distribution of even the most prolific species. The problem is exacerbated by sharp changes in space and time. Patchiness on scales of one to one hundred kilometres observed from space through gaps in the clouds is attributable to a combination of processes that are not yet well understood. The seasonal cycle involves rapid changes that vary in phase and amplitude from one year to the next, so that carefully planned expeditions can easily miss their target. In short, the biological oceanographer is confronted with technical problems that have so far defeated attempts to map the distribution of life in the sea reliably and without missing key aspects of the overall picture. Such under-sampling impedes not only the empirical approach but also verification of theories expressed through mathematical simulation.

How can we hope to understand the distribution of life in the sea in terms of the basic laws of physiology and behaviour? What is known about the population dynamics of plankton, fish, mammals? Some progress has been made in tracking seals and whales, but there are not enough samples to describe populations. The fisheries recruitment problem continues to be resolved by imaginative conjectures, rather than well tested scientific theories. Our maps of plankton tell us where they have been encountered, but nothing about the extent of breeding populations.

The *Challenger* Expedition revealed for the first time the rich biodiversity on the ocean floor. Some oceanographers claim that it rivals the biodiversity of tropical forests. The geographical distribution of life on the deep ocean floor is not yet documented, but it is believed to reflect the better understood distribution of life in the surface layer, the detritus from which provides its food supply. The prolific biodiversity poses problems for theorists seeking to explain the distribution of life in the sea by mathematical simulation; their models include equations not for individual species but for functional groups of species. There is no theory for the selection of these guilds, or for the consequence of suppressing biodiversity in models.

New approaches in marine biology

Enough about the problems of the present: what are the prospects of a breakthrough in the next twenty years? As usual the prospects for progress depend on introducing new technology. Four developments give rise to optimism: molecular biology, mathematical simulation, complexity theory and remote sensing.

After a slow start, biological oceanographers are beginning to apply the methods of molecular biology, which have revolutionized almost every other branch of biology and which are now doing the same for medicine. The encouraging first results suggest that within twenty years it will become possible to map the distribution of breeding populations and to explore how they adapt to changing environmental stress encountered during the seasonal cycle and as they drift through different climatic regimes.

Mathematical simulation, which is revolutionizing our understanding of the ocean's climate, promises to do the same for its ecosystem when computers with sufficient power become available. The WOCE landmark for simulating circulation was the Teraflops machine; the entry level for simulating the distribution of plankton is the Petaflops machine, capable of one billion computations per microsecond. That should arrive within twenty years. Meanwhile, progress is being made by adding biological code to low-resolution general circulation models, and by one-dimensional models. The Lagrangian ensemble method of integrating models of the upper ocean ecosystem has opened the way to explicit demography in plankton population dynamics. The ability to simulate the impact of mutation and natural selection provides a tool for analysing observed distributions derived from the new molecular biology markers.

These advances will lead to rapid advances in theoretical biological oceanography in the next twenty years. The challenge will be to relate the new population information to ocean climate. This will be helped by the third technical advance, this time in mathematics. The development of complexity theory provides a framework applicable equally to the physics of the ocean environment and the population dynamics of the plankton. For example, it opens the way to designing experiments to find out how much of spatial patchiness is attributable to mesoscale dynamics and how much to population dynamics, and how much of inter-annual variability is caused by the weather and how much to predator–prey interaction.

Those experiments will require much higher sampling rates than can be achieved by traditional netting or continuous plankton recording. The introduction of high-frequency multi-band acoustics and advanced optical sensing promises to establish in observations at sea a better match to the relative information density of each component of the ecosystem derived from mathematical simulation. These experimental data collected *in situ* can then be used most effectively by assimilation into prognostic models. In fact the design of the experiments will be based on observing system simulation experiments. The same technique will ensure that the best information is extracted from satellite ocean colour observations by the SeaWiFS and other missions scheduled by the Committee for Earth-Observing Satellites (CEOS) for the next twenty years (IOC 1995). (See also Chapter 11.)

Together, these technical advances, which are already with us or soon to be so, will provide the basis for rapid progress in biological oceanography during the next twenty years.

Technology

Now we ask more generally what new tools we might expect for oceanography in the next twenty years. The challenge is simply stated. It is greatly to increase the rate at which we can sample the ocean. The overwhelming problem is that data in our archives undersample the changing distributions in the ocean. Existing methods will not resolve that problem. To meet the requirements of practical problems such as climate prediction, we need to increase sampling density, not by a small percentage increase, or even by a modest factor, but by orders of magnitude. There is no way that existing methods can be scaled up to achieve that demand. It is unrealistic to plan such an increase in the number of research ships, let alone the number of scientists to work on them. The only way to scale up is by automation, by introducing robotic systems that do not need men on board. Rapid progress is being made in developing unmanned measuring systems and new means to deploy them. That will revolutionize the way we do oceanography in the next twenty years.

Artificial satellites provided the first step towards systematic global sampling of the ocean by instruments operating automatically on an unmanned vehicle. Visible and infra-red images gave a new insight into the complex patterns of slicks, temperature and coccoliths at the sea surface, but on average 80 per cent of the globe is covered by cloud, so working images are often composites from cloud-free pixels collected over a month or so. Temperature mapping by IR was greatly improved with the ATSR (Along-Track Scanning Radiometer) on ERS-1, from ±2K to ±0.2K. Visible measurements in several wavebands allowed the possibility to map the distribution of chlorophyll. All-weather monitoring came in 1978 with the radars on Seasat. Looking to the next twenty years, there is good and bad news. The good news is that space agencies plan a succession of missions that will, according to the CEOS portfolio, maintain coverage by at least one of each type of measuring system demonstrated in the past twenty years. The bad news is that there are no plans to introduce any novel instruments. We remain in a late 1970s time warp.

Remote sensing is useful inside the ocean, too. Photography from submersibles and bottom landers will continue to play a role, as will other optical devices, including fluorometers and counters for molecular probes. I have already referred to the role played by high-frequency acoustic sampling of plankton. At much longer ranges, acoustic tomography will reach maturity and find its niche for mapping transient mesoscale patterns, and acoustic thermometry promises to be one of the most useful techniques for monitoring global warming. Acoustic mapping of the sea floor, following the path blazed by GLORIA and TOBI, will continue to be a key tool for the geologist over the next twenty years. The need for greatly increased rates of data collection will be solved by introduction of optical fibres linking instruments to exchanges on mid-ocean islands such as Bermuda. New applications can also be found for old technology. The Japanese have pioneered the use of redundant copper-covered telecommunication cables to transmit data from deep-sea seismographs to provide early warning of tsunamis.

All of those tools used in remote sensing from space or inside the ocean are well established today. What will be quite new in the next twenty years will be the widespread use of unmanned autonomous submersibles to carry instruments to all locations in the world ocean. Prototype vehicles were tested in the sea during the 1990s. Many are relatively short-range vehicles used to deploy acoustic or optical sensors around a research vessel, without the need for a cumbersome and costly tow cable. Others, such as the Autosub being developed at the Southampton Oceanography Centre, are designed for long-range operation. The economic case for these vehicles is well known: they will be able to sample the ocean more cheaply than a research vessel; and the sampling rate can be increased by deploying more vehicles with very little increase in manpower. They will be able to operate under rough seas and roam into regions where it is difficult to navigate ships, for example under the ice. As this new technology becomes mature over the next twenty years, it will transform the way we perform experiments at sea. Together with the development of autonomous instruments to serve as the scientific payload, autonomous vehicles such as Autosubs will deliver the greatest revolution in oceanography since the *Challenger* was equipped with a steam winch.

Operational oceanography

Since its beginning in ancient times, oceanography has always been closely related to practical applications. Over the past twenty years these have begun to include what is now known as operational oceanography, which includes analysis of purpose-made observations and, where possible, prediction of future changes. Most of these operations have been concerned with relatively small regions such as the North Sea. That is the case for storm surge prediction based on models developed at the Proudman Oceanographic Laboratory on Merseyside and used routinely at the Meteorological Office; these forecasts provide vital information for the operation of the Thames Barrier, and can therefore be described as very cost-effective. Until recently the forecasting of sea state has also been a local operation, but research on wind waves over the past twenty years has led to a global forecasting system based at the European Centre for Medium Range Weather Forecasting, which provides much better predictions in every region of the world ocean, and is paving the way to improved computation of fluxes through the sea surface, for example carbon dioxide. A recent development has been the use of water-quality models to guide the design of projects for dealing with urban waste disposal at sea (for example the Massachusetts Bay project), or to deal with pollution emergencies (for example during the Gulf War and the *Braer* shipwreck). High-resolution models of the North Sea are being developed for future operational use in the formulation of policy for and management of water quality and eventually fisheries. They extend beyond the shelf seas down the continental slope into the deep ocean, and will provide information for offshore operators working in those difficult environments. These advances will open the way to environmental impact assessments for the deep ocean floor, a pre-requisite for licensing waste disposal and for achieving public support for such actions as the disposal of oil platforms in the deep ocean.

The portfolio of what can be described or forecast operationally will increase substantially during the next twenty years. One of the most interesting developments will be in the tropical Pacific, where the research results and observing know-how developed during the Tropical Ocean Global Atmosphere (TOGA) research project are leading to an operational forecasting system for the El Niño–Southern Oscillation. That should reach maturity during the next twenty years, with great benefits to countries around the Pacific Rim, and as far afield as southern Africa.

Most of the customers for such environmental information are likely to be concerned with limited geographical areas. As was found in the 1980s with wind wave prediction, it is not sufficient to rely on local models fed by local observations to meet the customers' needs. It is important to take account of fluxes through the open ocean boundaries of such local models. That requires support from a basin scale, or even a global model fed with global observations. If investment in a full global ocean observing system (GOOS) and associated modelling is shared between many local services, it may be as cost effective as weather forecasting. A substantial part of the system is already in place, namely the ocean-observing satellites, which together cost about $5 million per day. A complete GOOS might cost double that. There will be substantial progress towards that goal during the next twenty years.

Funding, collaboration and leadership

Oceanography, as Margaret Deacon (1997) reminds us, is an ancient science that has advanced in brief spurts from time to time over the centuries. It is of course all a matter of money. Cicero's dictum on war, '*Nervos belli, pecuniam infinitam*' (The sinews of war, unlimited money) applies equally to oceanography. In the nineteenth century Britain had enough money for the *Challenger* Expedition, and earlier for the voyage of HMS *Beagle*. Britain also had the necessary machinery for reaching imaginative decisions: the Navy and Royal Society were together strong enough to win over the Treasury.

In the twentieth century there was a spurt in the 1960s, which has lasted for a couple of decades. There are signs that it may be ending, at least in the country that launched the *Challenger* Expedition and in the most successful of its former colonies. In this century other nations have tried (Denmark, Germany, France, Canada), but only the USA has risen to the pre-eminence achieved by Britain with the *Challenger*, most recently through Seasat and Topex. Russia had the money and the commitment led by Gorshkov's navy, but chose the wrong target. Like socialist planning in other fields it went for volume rather than innovation or quality. Nevertheless, it has left a legacy of data that provide a valuable record of ocean variability through the decades of the Cold War. Now others are becoming rich and taking up the challenge. And increasingly, the grand challenges in oceanography (ODP, WOCE, JGOFS, RIDGE, GLOBEC, GOOS) are being promoted as international joint ventures. But some country has to take the lead in each case. What countries will be the leaders in the next twenty years?

America still has the money, but has lost its political commitment to science and cherishes aerospace more than the navy. Its plans for satellite oceanography have no innovative drive of the kind that gave us Seasat; the plan is for more of the same over the next twenty years, as it is for the other great American success, the Ocean Drilling Programme. America's immense strength in underwater acoustics is being frittered away as it was in Britain in the 1970s. Like GLORIA and TOBI, acoustic tomography and thermometry are not being exploited vigorously to their full potential. NOAA is promoting ENSO prediction like a drowning man clutching the final straw. There is a tragic mismatch in America between national and academic interests in the sea: oceanography from space requires investment in the boundary layer, which is not fashionable.

The European Union has abundant money, but has not yet learned how to spend it effectively. The European oceanography budget for marine science and technology (MAST) is growing fast, but gets spent with high entropy, that is on lots of small projects rather than on a few large ones, against the spirit of subsidiarity. Without a European navy and a powerful scientific academy, there can be no European leadership in oceanography; there will be no force large enough to battle the EC Treasury. The new European Committee on Ocean and Polar Science (the European JOI, or joint oceanographic institutions) will help establish the EU alongside the USA as a strong player in the first, but not the premier division; it is not in the *Challenger* league. ECOPS has identified four grand challenges, any one of which would put Europe in the premier division, and is now developing a strategy for Framework 5, which will steer MAST 4 funds towards them.

Perhaps the future leadership will lie in Asia. Three countries stand out: Japan, India and China. Japan has the money and the political machinery to reach and implement grand decisions, but Japan no longer has a blue seas navy; its leadership in oceanography will be commercially driven, echoing *Discovery* investigations (see Chapter 4) rather than the *Challenger* Expedition. Its big investments – satellites, drill ships and computers – remain copycat products. India has the political desire to make the Indian Ocean a modern *mare nostrum*. It has great scientific talents, with a power to innovate; and India is growing fast economically as it sheds its socialist legacy of 1946. But the country to watch is China. As its economic strength blossoms, its political tradition of centralism, combined with the genius of its population, make China a potent force in any branch of science that it chooses to lead. The desire of China's leaders to play a more prominent role on the world stage and to have more influence on the Chinese diaspora throughout Southeast Asia, calls for the maritime card. In the fifteenth century, China promoted maritime exploration, as Britain did in the eighteenth. I expect they will do so again in the twenty-first, starting in the next twenty years. That will be spearheaded by a white navy of oceanographic research vessels, based on Qingdao, with its massive conjuncture of government, research council and university oceanographers, the model for the new Southampton Oceanography Centre. I expect to see Chinese research vessels visiting Southampton's Empress Dock with increasing frequency during the next twenty years.

What is being done in the UK?

Scientific exploration of the great oceans will be led by the new great powers with unlimited resources: in the Pacific by America, Japan and China, in the Indian Ocean by India, in the Atlantic and Arctic by Europe (with Russia), and in the Southern Ocean by all of them. What will be the UK's role in the next twenty years? As an island nation it stands to lose more than almost any other country through ignorance of the sea. Marine industry generates 5 per cent of the UK GDP. So it has permanent national interests that require it to stay in the first division of oceanography. That has led to the investment in the

Southampton Oceanography Centre and a modern fleet of first-rate research vessels. It will continue to support an oceanographic community that is about 2 per cent of the nation's scientists. The next twenty years will see the completion of the restructuring started in the 1980s. There will be three great national centres: Plymouth, Southampton and Bidston–Bracknell, linked to the academic world through the Marine Biological Association, the University of Southampton, and the universities of Liverpool and Reading. The leading academic centres in Wales and Scotland will be at Bangor and Saint Andrews. Government fisheries and marine environmental research will be concentrated in Lowestoft and Aberdeen.

The UK can no longer go it alone in mounting an oceanographic venture in the *Challenger* league. But its first-rate scientists may participate prominently in leading programmes led by others, as they do today in the ODP. That will help it stay in the first division. The aim should be to be welcome partners in such leading-edge ventures, whether sponsored by America, Europe, India or China. That policy cannot be sustained merely by having first-rate scientists. They must have unique skills, tools and knowledge. That means a carefully crafted strategy for investing the limited funds available for oceanography. The strategy must have three elements: (1) focus on a small number of long-term programmes of national importance with scope for international pre-eminence; (2) develop unique tools, such as Autosub; and (3) train the next generation of oceanographers to the highest standards.

The House of Lords Select Committee on Science and Technology established the fact that the UK has a permanent commitment to the subject because of well understood national interests. The Co-ordinating Committee on Marine Science and Technology developed the theme and the Inter-Agency Committee on Marine Science and Technology maintains a forum for debating new issues before the government allocates them to a particular department. The Office for Science and Technology has belatedly launched the Foresight Panel in Marine Science and Technology. The Natural Environment Research Council remains the principal government agency responsible for achieving excellence and relevance in UK oceanographic research. NERC continues to support the Community Research Programmes, but the portfolio needs to be reinvigorated as the projects started in the 1980s are concluded.

Support for new technology continues to be half-hearted in the UK; the total cost of Autosub is higher than necessary because it is being drip-fed rather than properly budgeted as a priority project. Because of similar extended rather than concentrated investment, Britain is also slipping back behind our competitors in two other areas that have high national priority: computer simulation (OCCAM) and molecular biology (PRIME).

As for teaching, the opening of the Southampton Oceanography Centre, with the great increase in resident expertise provided for teaching, makes it timely to review our teaching of oceanographers. Many of the leading oceanographers of today were trained as undergraduates in the subject, often at Kiel and Scripps. The tradition in Britain has been to rely on entrainment from other subjects that overproduce (mathematics, physics, biology), and to assume that these graduates will pick up the subject on the job. The leaders in oceanography around the world have enjoyed professional training from the start. The UK needs to reconsider its policy about training undergraduates in the subject.

Conclusion

Much has been achieved in the hundred years since the *Challenger* Expedition. Some of the major projects such as WOCE will lead to the working out of new paradigms in the next twenty years, as plate tectonics has in the past twenty years. But some of the basics have still not been addressed adequately, including global bathymetry, biodiversity and the distribution of life in the sea. New challenges on the agenda for the next twenty years include the changing climate of the ocean and global operational oceanography. The development of new technology based on remote sensing and robotic vehicles will transform our ability to address these challenges. The pace of progress will be limited not so much by imagination or technology as by funding. However, funding is declining in North America and the UK but rising in Europe and Asia. It may be that the leaders in the twenty-first century will come from Asia, perhaps with China in the vanguard.

What will be 'the grand challenge', comparable to the *Challenger* Expedition, which might leave a historical mark on the next twenty years? This has been considered by the European Committee on Ocean and Polar Sciences, which has come up with four candidates: (1) the deep ocean floor, (2) the Arctic Ocean (the only ocean with a lid on), (3) the distribution of life in the sea, and (4) ocean forecasting (EuroGOOS). We shall see what emerges.

The next twenty years will be an exciting time in the history of oceanography, with many interesting developments. It remains to be seen whether there will be any 'great leaps forward', and who will lead them.

Note

1 The Mohorivičic discontinuity, named after its discoverer, represents the seismic division between oceanic crust and the earth's mantle.

References

Deacon, M. (1997) *Scientists and the sea, 1650–1900: a study of marine science*, 2nd edn. Aldershot: Ashgate.

IOC (Intergovernmental Oceanographic Commission) (1995) *Workshop an ocean colour database requirements and utilisation*, IOC Workshop Report 119. Paris: UNESCO.

Jones, M. T., Tabor, A. R. and Weatherall, P. (1994) *GEBCO digital atlas*. Bidston: British Oceanographic Data Centre.

Rennell, J. (1832) *An investigation of the currents of the Atlantic Ocean*. London: Rivington.

Postscript
The *Challenger* Expedition on postage stamps

A. L. Rice

One of the most curious 'legacies' of the *Challenger* Expedition is its depiction on postage stamps. Since the first stamp with a ship design appeared in 1852 (issued by British Guiana only twelve years after the arrival of Rowland Hill's famous 'Penny Black'), ships and boats of a bewildering variety have appeared on an enormous number of stamps from all over the world. Indeed, 'ships' is one of the largest stamp collecting themes and the Stanley Gibbons ships thematic catalogue published in 1989 (Bolton 1989) listed over 11,000 stamps depicting more than 2400 different named ships. As you might expect, some particularly famous vessels appear many times in the catalogue. For example, Darwin's *Beagle* (14 times), Nelson's *Victory* (22 times), Cook's *Discovery* (21 times) and his *Endeavour* (36 times), Columbus's *Pinta* (27 times) and *Nina* (31 times) and the Royal Yacht *Britannia* (57 times). But vying for top of the list are Columbus's *Santa Maria* with 90 appearances, just pipped by Cook's *Resolution*, which has appeared on no fewer than 92 stamps, including 7 from Norfolk Island alone. Compared with these all-time favourites, the *Challenger*'s six appearances, including one since the 1989 catalogue was published, are not very impressive. Even more distressing for British oceanographer-philatelists, despite the importance of the expedition to marine science throughout the world, the ship has never appeared on a British stamp, nor even among the many research and exploratory vessels that have been depicted on stamps issued by the Falkland Islands Dependencies or its philatelic successor, the British Antarctic Territories.

The *Challenger*'s first philatelic appearance was in 1965 as one of a series of sixteen stamps issued by Tristan da Cunha and including a range of ships with significance in the island's history. The fourpenny stamp in this series (Fig. PS.1) is based rather faithfully on a watercolour by W. F. Mitchell (1845–1914), showing the *Challenger* heeling to port under reduced canvas, with only the foresail, fore and main topsails, and the inner jib set. The painting was originally published as a lithograph in 1880 by Griffin & Co., booksellers near the main

Figure PS.1 Tristan da Cunha 1965 4d stamp showing the *Challenger* under reduced sail and based on a watercolour by W. F. Mitchell. (Reproduced with the kind permission of the Crown Agents.)

gate of the Royal Naval Dockyard at Portsmouth, who took orders for ship paintings on Mitchell's behalf, mainly from naval officers (Archibald 1980). Mitchell was presumably, therefore, normally very accurate, but in the *Challenger* painting, and in the Tristan 1965 stamp, the ship's prominent funnel, between the fore and mainmasts, is totally missing, as is the dredging platform forward of the mainmast.

The next *Challenger* stamps to appear were also issued by Tristan da Cunha, this time in 1973 to mark the centenary of the ship's visit to the remote island on 15–17 October 1873. Tristan had a population at the time of only eighty-four, but the *Challenger* personnel had little enough time ashore to meet them, for most of the three days spent in the area was devoted to surveying around the nearby Nightingale Island, depicted on the 7½p stamp with the ship's pinnace, and Inaccessible Island some twenty miles (30km) to the south. The four stamps in the series were issued separately, of course, but were also available in a 'miniature sheet' (Fig. PS.2) with a decorative surround showing the *Challenger* approaching the island under full sail. The ship itself appears on the 5p stamp, again with the island in the background, but this time with all sails furled and presumably at anchor. The 12½p stamp shows the ship's track through the Atlantic during both the outward journey in 1873 and the return journey in 1876. Finally, the 4p stamp is based on a drawing by J. J. Wild, the expedition's artist, depicting the ship's tiny 10 feet by 5 feet (3m by 1.5m) chemical laboratory built into one of her gun bays and crammed with benches, cupboards and racking. Curiously, the stamps also each carry views of deep-sea sounders developed and used on HMS *Bulldog* in 1860 but with no connection with either the *Challenger* or Tristan da Cunha!

The same basic design as that on the first Tristan *Challenger* stamp was also used in the next one, issued by Bermuda in 1976 (Fig. PS.3). This time Mitchell's original painting was modified in several ways, including the addition of the missing funnel and now with the mainsail set. The Bermuda postal authorities had missed the opportunity to mark the centenary of the *Challenger*'s two quite long visits to the island on 3–21 April 1873 and also from 28 May to 12 June, during which she surveyed and collected biological material in and around the island and also refitted and took on stores. Instead, the 1976 *Challenger* stamp was

Figure PS.2 Tristan da Cunha 1973 miniature sheet with four stamps marking the centenary of the ship's visit. (Reproduced with the kind permission of the Crown Agents.)

Figure PS.3 Bermuda 1976 stamp modified from the same basic design used for the 1965 Tristan issue, one of four celebrating the fiftieth anniversary of the founding of the Bermuda Biological Station. (Reproduced with the kind permission of the Crown Agents.)

one of a set of four issued to mark the fiftieth anniversary of the establishment of the Bermuda Biological Station, renowned for the work its many visiting scientists have been able to carry out on deep-ocean animals so easily accessible from the island's steeply shelving coastline.

The most handsome *Challenger* stamp is undoubtedly a large black and white issue by the French Southern and Antarctic Territories in 1979 (Fig. PS.4). The *Challenger* visited the French islands in the southern Indian Ocean in early 1874, but the stamp shows the ship's visit to St Paul's Rocks in the mid-Atlantic on 28 and 29 August 1873. It is based on a woodcut that appeared in Thomson's (1877) account of the ship's work in the Atlantic and in the *Illustrated London News* and was in turn apparently based on a photograph. It shows the ship's boats carrying a mooring line to the rocks, battling against the current, which Moseley (1892: 59) described as 'rushing past the rocks like a mill-race'.

Also in 1979, Australian Antarctic Territory issued a series of sixteen stamps depicting important Antarctic surveying vessels, including one showing the *Challenger* steaming between icebergs (Fig. PS.5). Although the ship was among icebergs more or less continuously for several weeks in February and March 1874, there was no intention for the expedition to reach a particularly high latitude and no attempt was therefore made to enter the pack ice. Nevertheless, she experienced the most dangerous few hours of the whole voyage during this period when, on 24 February, she collided with one iceberg during a gale and was almost run down by a second.

Figure PS.4 The 1979 French Southern and Antarctic Territories stamp showing the *Challenger* being moored to St Paul's Rocks in August 1873. Although Thomson (1877) says that this view is based on a photograph, the various collections of extant *Challenger* photographs do not contain one even vaguely similar (see Brunton 1994). Instead, two photographs of a drawing (a watercolour?) by J. J. Wild occur in the collections, one labelled correctly 'HMS *Challenger* secured by a Hawser to St Paul's Rocks' (Brunton 1994: no. 207) and the other (Brunton 1994: no. 327) erroneously said to be the ship in ice. Brunton lists several photographs taken at St Paul's Rocks for which no images now remain; it is possible that Wild based his drawing on one of these. (Reproduced with the kind permission of Terres Australes et Antarctiques Françaises.)

Figure PS.5 Australian Antarctic Territories 1979 stamp showing the *Challenger* steaming among icebergs. (National Philatelic Collection, Australia Post. Copyright Australia Post.)

The most recent *Challenger* stamps (Fig. PS.6) are a pair issued by Christmas Island in the Indian Ocean in 1989 to mark the seventy-fifth anniversary of the death, in a road accident in 1914, of Sir John Murray. Murray had been a junior naturalist during the *Challenger* voyage, but after the death in 1882 of the expedition's chief scientist, Sir Charles Wyville Thomson, he assumed responsibility for overseeing the publication of the official reports and ultimately became the most famous marine scientist of his day. But he, and the *Challenger*, are celebrated on these Christmas Island stamps and, in the case of Murray, also on a 1977 issue (Fig. PS.7), for much more mundane reasons: the exploitation of the island's phosphate deposits by the Christmas Island Phosphate Company (see Burstyn 1975, Rice 1987).

During the expedition Murray's tasks had included overall responsibility for the collection of sediment samples; subsequently he co-authored the scientific report on the deep-sea deposits that has formed the basis for all later submarine geology (Murray and Renard 1891). In the process he became interested in the origins of coral reefs and he developed a theory that was at variance with the ideas put forward by Charles Darwin. Whereas Darwin's explanation involved the submergence of wide areas of the earth's surface, Murray sought to explain the origin of reefs without the necessity of such submergence.

After the voyage, when Murray started to look for support for his ideas in the *Challenger* collections, he found that he was very short of samples from mid-oceanic islands. Accordingly, he turned to his contacts in the Admiralty Hydrographic Department for help in making good this deficiency. As a result, in January 1887 the naval surveying vessel *Flying Fish* was ordered to visit Christmas Island on her way home after six years' service in the Far East.

Figure PS.6 Christmas Island 1989 issue to mark the seventy-fifth anniversary of the death of Sir John Murray. The depiction of Murray is based on a portrait by Sir George Reid, painted in 1913, the year before Murray's death. (See Linklater 1972.) (National Philatelic Collection, Australia Post. Copyright Australia Post.)

Figure PS.7 Christmas Island 1977 stamps from a series of sixteen depicting some of the island's 'Famous Visitors'. Sir John Murray is shown on the 4-cent stamp, based on the same Reid portrait used for the 1989 issue. Murray's *Challenger* shipmates, Pelham Aldrich and J. F. L. P. Maclear, who each collected phosphate samples for him, are on the 3-cent and 5-cent stamps. Captain (later Admiral Sir) William May, who took possession of the island in June 1888, is on the 8-cent stamp. (National Philatelic Collection, Australia Post. Copyright Australia Post.)

By chance, the ship's captain was J. F. L. P. Maclear (Fig. PS.7) who had been second-in-command during the *Challenger* Expedition.

Among the beach-rock samples that Maclear brought back, Murray found a small pebble of pure calcium phosphate, previously unknown from coral islands. Intrigued, Murray asked the Hydrographer to arrange for more specimens to be collected, including samples from the highest parts of the island. This time yet another of Murray's former *Challenger* shipmates, Captain Pelham Aldrich (Fig. PS.7), now in command of HMS *Egeria*, was ordered to visit Christmas Island during a sounding survey he was to undertake from Malaya to Mauritius. Aldrich's samples convinced Murray of the presence of rich beds of calcium phosphate on the island and changed his interest in it from a purely scientific to a definitely economic one.

The manufacture of superphosphate from phosphate rock for use in agriculture had begun in the 1840s. By the 1880s there was a great demand for it as British agriculture competed against the influx of cheap American farm products. Murray recognized the commercial possibilities of his discovery and promptly set about the task of obtaining a licence to exploit it. This was no small problem, for at the time Christmas Island officially belonged to no-one and the first objective was to ensure that Britain took possession of it. With Murray's personal contacts in the Admiralty and in the government, this turned out to be surprisingly easy and on 6 June 1888 the island was formally annexed in the name of Her Britannic Majesty Queen Victoria (Fig. PS.7). Getting a government lease was more difficult, mainly because there was a rival claimant, George Clunies-Ross, whose family occupied and controlled the nearby Cocos Islands. After a series of vicissitudes, the Colonial Office issued a joint lease to Murray and Clunies-Ross in 1891, though it was a further six years before the Christmas Island Phosphate Company was registered under Murray's chairmanship.

After a very slow beginning, in which only about ten tonnes of phosphate was exported in 1899, production increased rapidly, so that by 1911 one million tonnes had been extracted. In 1913 Murray could claim that the British Government had by that time received more revenue in the form of taxes from the Christmas Island venture than the £170,000 total cost of the *Challenger* Expedition and the publication of the results. At the same time, Murray had received very considerable funds himself, and was probably the first oceanographer to make his fortune directly as a result of his scientific endeavours. Even after his death in 1914, the Phosphate Company continued to produce a handsome income for Murray's family and contributed very significantly to the island's economy until recent years, supporting a resident population in excess of 3000 in the early 1980s. Not surprisingly, therefore, since Christmas Island began issuing its own stamps in 1958, several series have been devoted to the phosphate industry (see Rice 1987), to the personalities involved in its development, and particularly to its founder, Sir John Murray, and to the ship on which the story began. Surely it is time that the British postal authorities similarly recognized the importance of the *Challenger* expedition, not only to the United Kingdom but to the world at large.

References

Archibald, E. H. H. (ed.) (1980) *Dictionary of sea painters*. Woodbridge, Suffolk: Antique Collectors' Club.

Bolton, P. (1989) *Collect ships on stamps*. London: Stanley Gibbons.

Brunton, E. V. (1994) *The* Challenger *Expedition, 1872–1876: a visual index*. London: Natural History Museum.

Burstyn, H. L. (1975) 'Science pays off: Sir John Murray and the Christmas Island phosphate industry, 1886–1914', *Social Studies of Science* **5**: 5–34.

Linklater, E. (1972) *The voyage of the* Challenger. London: John Murray.

Moseley, H. N. (1892) *Notes by a naturalist. An account of observations made during the voyage of HMS* Challenger *round the world in the years 1872–1876*. London: John Murray.

Murray, J. and Renard, A. F. (1891) 'Report on deep sea deposits based on the specimens collected during the voyage of HMS *Challenger* in the years 1872 to 1876', *Report on the scientific results of the voyage of HMS* Challenger *during the years 1873–76*. London: HMSO.

Rice, A. L. (1987) 'Oceanography on stamps: the Christmas Island Phosphate Company', *Sea Frontiers* **33**: 444–53.

Thomson, C. W. (1877) *The voyage of HMS* Challenger: *the Atlantic*, London: Macmillan.

Index

Printed and bound by CPI Group (UK) Ltd, Croydon, CR0 4YY

23/10/2024

01777679-0004